MODERN OPTICAL METHODS IN GAS DYNAMIC RESEARCH

Contributors to This Volume

A. J. Alcock
S. J. Arnold
Daniel Bershader
R. O. Berthel
Terrill A. Cool
C. Forbes Dewey, Jr.
K. D. Foster
Edmund J. Gion
Martin C. E. Huber
T. V. Jacobson
Franz C. Jahoda
George H. Kimbell
David W. Koopman
Ralph H. Lovberg
Thomas J. McIlrath
Ralph W. Nicholls
Joseph H. Spurk
R. D. Suart
Gordon W. Wares
S. J. Wolnik
Walter H. Wurster

MODERN OPTICAL METHODS IN GAS DYNAMIC RESEARCH

Proceedings of an International Symposium held at Syracuse University, Syracuse, New York, May 25-26, 1970, supported by The New York State Science and Technology Foundation

Edited by Darshan S. Dosanjh
*Department of Mechanical
and Aerospace Engineering
Syracuse University*

PLENUM PRESS • NEW YORK–LONDON • 1971

Library of Congress Catalog Card Number 75-155352
SBN 306-30537-2

© 1971 Plenum Press, New York
A Division of Plenum Publishing Corporation
227 West 17th Street, New York, New York 10011

United Kingdom edition published by Plenum Press, London
A Division of Plenum Publishing Company, Ltd.
Davis House (4th Floor), 8 Scrubs Lane, Harlesden, NW10 6SE, England

All rights reserved
No part of this publication may be reproduced in any
form without written permission from the publisher
Printed in the United States of America

PREFACE

This volume is based on material prepared by the contributors to the symposium on "Progress in Gas Dynamic Research by Optical Methods", held on May 25-26, 1970 in the Department of Mechanical and Aerospace Engineering at Syracuse University. The contents focus on experimental and analytical aspects of contemporary optical methods as applied in modern research on high speed and/or high temperature gaseous flows. State of the art, recent research results and possible research applications of spectroscopy, spectral interferometry, pulse laser holographic interferometry, laser as a diagnostic and plasma generating tool and the analysis of plasma by light scattering constitute part of the subject matter of this volume. The emerging importance and impact of recent laser developments on optical diagnostics of gas dynamic and gas-physics phenomena is a recurring theme throughout the volume.

Diverse applications of the shock tube to process gases to high temperature equilibrium conditions and the study of important characteristics of these radiating gases by contemporary spectroscopic methods are discussed in papers by Nicholls, Wurster and Wares, et al. Refractivity index measurements have long been extensively used for investigating gas dynamic and aerodynamic flows. However, the recent availability of the laser as a light source has brought significant improvements in the more conventional optical methods such as schlieren photography and interferometry as reported here in Alcock's paper.

More recent laser developments have resulted in several completely new optical diagnostic methods. For example, the spectral behavior of the refractive index, coupled with a high degree of spectral purity of available laser light sources, has renewed interest in spectral interferometry as a possible diagnostic tool in experimental gas dynamics. Bershader, Huber and Spurk have discussed the analytical and experimental aspects of this sophisticated optical diagnostic technique. The use of pulse laser holographic interferometry to measure refractivity of gases is discussed here by Jahoda. The unprecedented brightness of laser sources has made practical for the first time the use of Thomson scattering of light as a plasma diagnostic tool. Lovberg introduces the basic analytical and experimental aspects of the scattering method. Koopman discusses the use of focused Q-switched pulse lasers for generating plasma flows.

The various laser excitation methods are also reviewed in this volume. Cool documents the different approaches of continuous-wave chemical laser excitation. Kimbell et al. have presented some recent studies of the continuous wave combustion laser. A comprehensive review by Dewey of resonant energy exchanges between gaseous media and externally applied radiation field, made possible by the recent development of wavelength tunable laser sources of high instantaneous power, is included. The papers have been arranged in a sequence intended to achieve as much continuity of the subject matter as possible.

We wish to express our appreciation to the New York State Science and Technology Foundation for supporting a program on Modern Developments in Aerospace Sciences at Syracuse University under which the symposium on optical methods was organized. Under the same grant Dr. Daniel Bershader of Stanford University was invited to Syracuse University as a Distinguished Visiting Professor for the fall term 1969. Dr. Bershader's help in the selection of the invited speakers is gratefully acknowledged. The cooperation of the authors in preparing their individual papers for publication is greatly appreciated. I would like to take this opportunity to thank Mrs. Jeannette Levey, whose cheerful, steady and efficient handling of the secretarial and related matters contributed greatly, first to the successful organization of the symposium, and then to the publication of its proceedings.

December 1970   Darshan S. Dosanjh

LIST OF CONTRIBUTORS

Numbers in parentheses indicate the pages on which the authors' contributions begin.

A. J. ALCOCK, Division of Pure Physics, National Research Council of Canada, Ottawa, Ontario (119)

DANIEL BERSHADER, Department of Aeronautics and Astronautics, Stanford University, Stanford, California (65)

TERRILL A. COOL, Department of Thermal Engineering, Upson Hall, Cornell University, Ithaca, New York (197)

C. FORBES DEWEY, JR., Department of Mechanical Engineering, Massachusetts Institute of Technology, Cambridge, Massachusetts (221)

EDMUND J. GION, Ballistic Research Laboratories, USAARDC Aberdeen Proving Ground, Maryland (61)

MARTIN C.E. HUBER, Harvard College Observatory, 60 Garden Street, Cambridge, Massachusetts (85)

FRANZ C. JAHODA, Los Alamos Scientific Laboratory, University of California, Los Alamos, New Mexico (137)

GEORGE H. KIMBELL, T.V. JACOBSON, R.D. SUART, S.J. ARNOLD AND K.D. FOSTER, Defence Research Board, Defence Research Establishment Valcartier, Courcelette, P.Q., Canada (279)

DAVID W. KOOPMAN, Institute for Fluid Dynamics and Applied Mathematics, University of Maryland, College Park, Maryland (177)

RALPH H. LOVBERG, Department of Physics, University of California - San Diego, La Jolla, California (155)

THOMAS J. McILRATH, Harvard College Observatory, 60 Garden Street, Cambridge, Massachusetts (271)

RALPH W. NICHOLLS, York University, 4700 Keele Street, Downsview, Ontario (1)

JOSEPH H. SPURK, Ballistic Research Laboratories, USAARDC Aberdeen Proving Ground, Maryland (113)

GORDON W. WARES, STANLEY J. WOLNIK, AND ROBERT O. BERTHEL, Air Force Cambridge Research Laboratories, Bedford, Massachusetts (49)

WALTER H. WURSTER, Cornell Aeronautical Laboratory, Inc., Buffalo, New York (33)

CONTENTS

PREFACE   v
    Darshan S. Dosanjh

LIST OF CONTRIBUTORS   vii

RECENT RESEARCHES IN SHOCK-TUBE SPECTROSCOPY
    Ralph W. Nicholls

    Abstract   1
    Introduction   1
    Shock Tubes and the Spectroscopic Method   2
    Shock-Tube Spectroscopy of Astrophysically
        Important Diatomic Molecules   14
    Diagnostic Shock-Tube Spectra and the Kinetic
        Study of Collision Processes   25
    References   27

RADIATIVE DIAGNOSTICS NONEQUILIBRIUM FLOWS
    Walter H. Wurster

    Introduction   33
    Nonequilibrium Emission Measurements   34
    Absorption Measurements in the Vacuum Ultraviolet   43
    Summary   47
    References   47

TEMPERATURE ERRORS IN PLASMA DIAGNOSTICS AND THEIR
POSSIBLE EFFECTS ON ABSOLUTE TRANSITION PROBABILITIES
    G.W. Wares, S.J. Wolnik and R.O. Berthel

    Introduction   49
    Determination of Plasma Temperature from
        Relative gf-Values   50
    The AFCRL Shock Tube Absolute gf-Value
        Program Using Ultrasonics   52
    Errors in the Corliss and Bozman NBA
        Free-Burning ARC gf-Values   53
    Conclusion   58
    References   60

IMPURITY MEASUREMENTS IN THE EXPANSION TUBE
  Edmund J. Gion   61

    References   64

SOME ASPECTS OF THE REFRACTIVE BEHAVIOR OF GASES
  Daniel Bershader

    Introduction   65
    Constitutive Properties and Wave Propagation
      in a Medium   67
    Lorentz Electron Theory of Dispersion   68
    The Dispersion Function   71
    Spectral Interferometry: The Hook Method   74
    Special Features   78
    References   82

INTERFEROMETRIC GAS DIAGNOSTICS BY THE HOOK METHOD
  Martin C. E. Huber

    Introduction   85
    The Experimental Setup   88
    Mach-Zehnder Interferometers   93
    Stigmatic Spectrographs, Detectors and
      Light Sources   95
    Theoretical Aspects of the Hook Method,
      Review of Experiments   97
    Summary   108
    References   109

HOOK INTERFEROMETRY USING A SINGLE PLATE INTERFEROMETER
  Joseph H. Spurk   113

    References   117

SOME OPTICAL DIAGNOSTIC TECHNIQUES INVOLVING HIGH
POWER LASERS
  A. J. Alcock

    Introduction   119
    Laser-Produced Plasma Studies in the Nanosecond Region 121
    Laser-Produced Plasma Studies with Subnanosecond
      Resolution   125
    Recent Developments   132
    References   134

CONTENTS

PULSE LASER HOLOGRAPHIC INTERFEROMETRY
Franz C. Jahoda

Principles of Holography	138
Holographic Interferometry	140
Cine-Interferometry	149
Bibliography	153

THE ANALYSIS OF PLASMA BY LIGHT SCATTERING
Ralph W. Lovberg

Introduction	155
Elements of the Theory	155
Experimental Procedures	160
A Scattering Experiment	169
References	176

LASER - GENERATION OF RAREFIED PLASMA FLOWS
David W. Koopman

Abstract	177
Introduction	178
Apparatus	179
Measurements on Laser-Produced Flows	181
Study of Counter-Streaming Ionized Flows	190
References	196

A SUMMARY OF RECENT RESEARCH ON CONTINUOUS-WAVE CHEMICAL LASERS
Terrill A. Cool

Introduction	197
Four Basic Processes Influence CW Chemical Laser Operation	197
Approach One: Laser Action Directly from Product Molecules in Fluid Mixing Chemical Lasers	199
A Second Approach: Laser Action by Selective Energy Transfer from Produce Molecules in Fluid Mixing Chemical Lasers	203
A Comparison of Approaches One and Two	209
Total Population Inversion	213
A Third Approach: A CW Explosion or Detonation Wave Chemical Laser	214
Conclusion	216
References	218

EXCITATION OF GASES USING WAVELENGTH-TUNABLE LASERS
C. Forbes Dewey, Jr.

    Abstract   221
    Introduction   221
    Organic Dye Lasers   222
    Semiconductor Lasers   238
    Nonlinear Optical Techniques   244
    Resonant Interactions Between Lasers and Gases   251
    References   258
    Bibliography of Dye Laser Papers   264

PRODUCTION OF NON-EQUILIBRIUM ATOMIC POPULATIONS USING TUNABLE LASER EXCITATION
Thomas J. McIlrath   271

    Notes   278

RECENT STUDIES OF THE CS_2/O_2 COMBUSTION LASER
G. H. Kimbell, T.V. Jacobson, R.D. Suart, S.J. Arnold and K.D. Foster

    Abstract   279
    Introduction   279
    Transversely Sparked Laser   280
    CW Combustion Laser   284
    Summary and Conclusions   287
    References   287

INDEX   289

RECENT RESEARCHES IN SHOCK-TUBE SPECTROSCOPY

R.W. NICHOLLS

CENTRE FOR RESEARCH IN EXPERIMENTAL SPACE SCIENCE

YORK UNIVERSITY, TORONTO, CANADA

ABSTRACT
The shock-tube has become an important tool in spectroscopic research both from a wavelength and also from an intensity standpoint. Following a review of recent atomic and molecular spectroscopic research in which shock-tube methods were used, the current work in the author's laboratory is described. It includes the application of computer-generated "synthetic spectra" to the interpretation of intensities of optically thick shock excited emission spectra of astrophysically important diatomic metal oxides.

INTRODUCTION
The shock-tube is primarily considered in this paper as an important tool of contemporary spectroscopic research. Recent diverse applications of the shock-tube to spectroscopy from a number of laboratories are reviewed. This is followed by a discussion of applications of shock-tube techniques to quantitative spectroscopic problems of astrophysical interest in the author's laboratory.

The shock-tube provides a simple means for the controlled production of high temperature ($\sim 10^4$ °K) gaseous environments. The light emitted by the shock-heated gas excited by energetic collision processes of many kinds, (Gaydon and Hurle 1963) has been studied spectroscopically in many laboratories. Spectroscopic methods have also been frequently used as diagnostic tools to follow the kinetics of the rate processes taking place in the hot gases. The shock-tube has thus become a very popular research tool in chemical physics, astrophysics, atomic collisions, gas dynamics and other fields.

Much experimental and theoretical work has been done over the past few years on the effects of individual collision processes by which energy exchange processes take place in hot real gases. An authoritative body of knowledge has thereby developed in this field. (Gaydon and Hurle 1963, Bradley 1962, Bond, Watson and Welch 1965, Vincenti and Kruger 1965, Zel'dovich and Raizer 1966, Glass 1970). Much of the spectroscopic work is done when the shock-heated gas had reached L.T.E. and for such work the shock-tube is a very useful and reproducible spectroscopic light source. Its major limitation is the transient nature (~1ms) of the light pulse. On the other hand many of the shock-tube kinetic studies have been performed in regimes before equilibrium had been completely attained and such results are much harder to interpret unambiguously.

SHOCK-TUBES AND THE SPECTROSCOPIC METHOD

One of the principal attractions of the shock-tube as a spectroscopic light source, is its ability to sustain, even if only for milliseconds, particularly if tailored conditions are used, predetermined high temperature L.T.E. conditions in the shocked gas. A $1000°$ - $20,000°K$ temperature range is most common for shock-tube spectroscopic work. In this temperature range the average kinetic energy per article is between 0.1 and 2.0 eV. Energetic atom-atom (or atom-molecule) collisions are thus the primary excitation mechanism (Bauer 1969, Gilmore, Bauer and McGowan 1969). Nevertheless the auxiliary effects of electrons, ions and radiation (Brabbs 1968), when produced by the primary collisions, cannot be overlooked. Appropriate choice of initial conditions allows one to preselect temperatures, and thus selectively to excite molecular, atomic or ionic spectra (or mixtures of them). Temperatures comparable to those in stellar envelopes are easily attained in shock-tubes which have thus become attractive research tools in laboratory astrophysical investigations.

The sequence of procedures (The Spectroscopic Method) which are common to all of spectroscopic research outlined in Table 1. are also followed in shock-tube spectroscopy.

The shock-tube has been used for some years as a convenient source for excitation of atomic, molecular and ionic spectra which are only readily produced in a high temperature environment (see lines 1 and 2 of Table 1). Many shock-excited-spectra are in emission. Flash absorption studies of the hot gases are also sometimes made. Much work has been done (and some is described later) on the shock excitation of metal oxide spectra. Such oxides are often refractory materials at room temperature. They are thus placed in the shock-tube in powdered form sometimes as a smoke. The spectroscopy of oxides is of great current interest and many problems of location and assignment of energy levels remain to be solved. Some aspects of the shock-tube spectroscopy of powdered solids has been reviewed by Nicholls, Parkinson and Reeves (1963).

Identification of spectral features is the next step (line 3 Figure 1) after the excitation and recording of the spectrum.

Table 1

The Spectroscopic Method

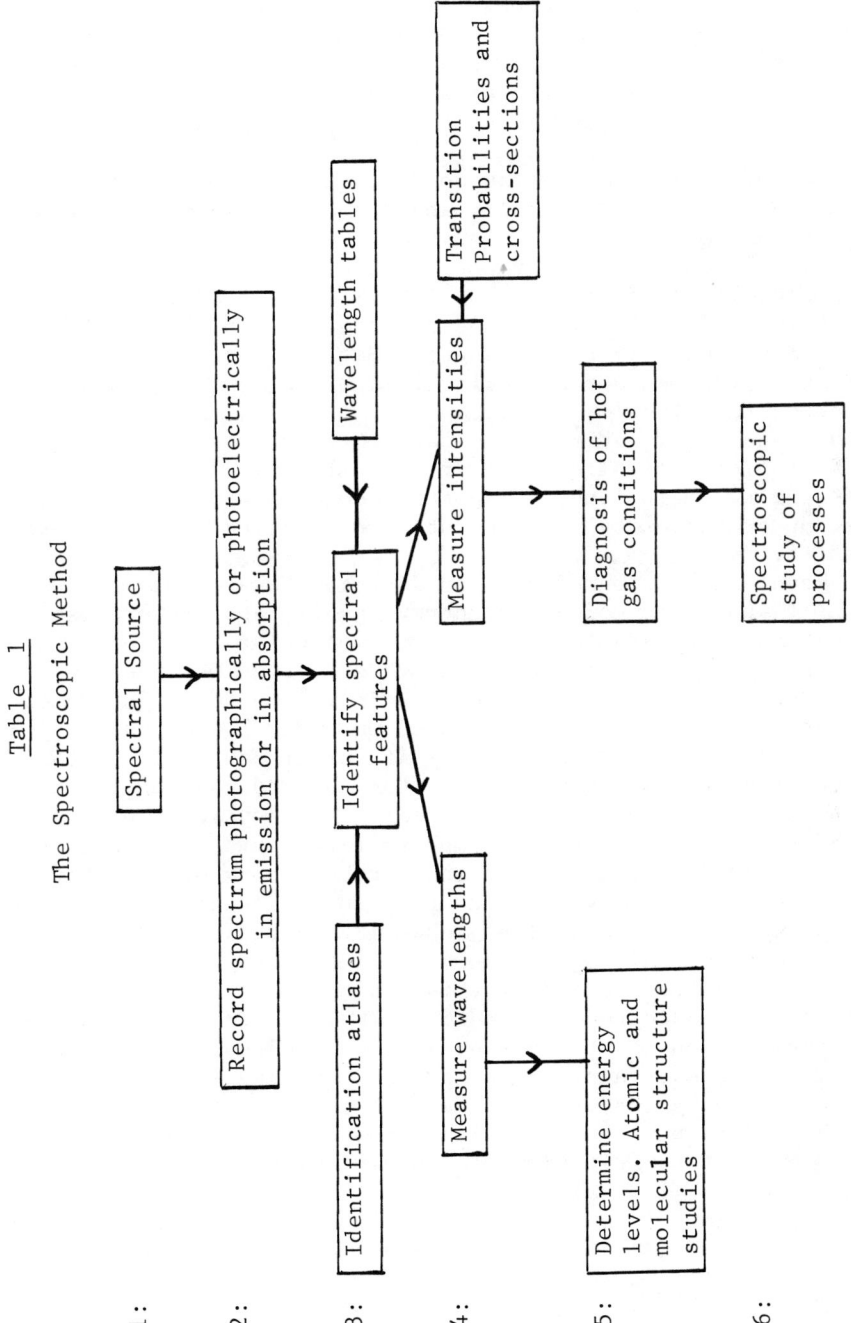

Identification procedures involve the use of wavelength lists and of comparison with atlases. A particularly fine example of such an atlas, and accompanying wavelength lists arising from shock-tube spectra is the important work of Weeks and Simpson (1967) on the vacuum ultraviolet spectrum of iron. This spectrum is astrophysically extremely important as it contributes strongly to the solar spectrum. Definitive identification, though simple in principle, is by no means trivial. It is the first major procedure in the interpretation of spectroscopic data. A set of identification atlases of astrophysically and aeronomically important molecular band systems (and associated compilations of data) is in progress in the author's laboratory. (Tyte and Nicholls 1964a, Tyte and Nicholls 1965a; Tyte and Nicholls 1965b; Hébert, Innanen and Nicholls 1967; Tyte, Innanen and Nicholls 1968; Degen, Hébert, Innanen and Nicholls 1969; Harrington, Seel, Hébert and Nicholls 1970; Hébert, Innanen and Nicholls 1971; Brocklehurst, Rosevear, Seel, Hébert and Nicholls 1971).

Lines 4 and 5 of Table 1, indicate that those spectroscopic procedures which are followed after identification depend upon which of the two main lines of spectroscopic endeavour (wavelength measurement or intensity measurement) are involved in the research in question.

The wavelength-oriented work is directed toward the precise location of atomic or molecular energy levels, and the precise determination of constants of atomic or molecular structure. Examples are recent work on Tl in the Harvard College Observatory Shock-Tube Laboratory, and work on TiO in the author's laboratory.

In contrast the intensity-oriented work can be motivated by a desire to provide transition probability or cross-section data for radiative transitions experienced by the shocked gas, or it can use such data to interpret diagnostically the physical conditions existing in the shocked gas. Examples of both types of investigation are found in a number of shock-tube laboratories. For many years programmes have been running in the author's laboratory for the provision of transition probability data for many molecular spectra of aeronomical and astronomical interest. Examples are discussed in a later section of this review.

Finally, as indicated in line 6 of Table 1, once diagnosis has been performed, conjectures can be made on the energy exchange mechanisms which maintain the gas in the measured state of energy partition. This is the area of chemical and excitation kinetics. Indeed the techniques of kinetic spectroscopy are often employed to unravel some of the excitation mechanisms which take place in hot luminous gases.

In what follows a review is made of spectroscopic work in some of these areas. Limitations of space prevent it from being exhaustive.

Shock Excitation Of Spectra, Their Character And Identification

The light emission associated with the propagation of shock waves in shock-tubes has been known for years, and originally was nearly always associated with impurities (e.g. Na, Ca, CN, C_2, OH, AlO, etc. in the tube). In fact the ease by which impurity spectra could be excited was exploited by Parkinson and the author in their first work on shock excitation of powdered solids (Nicholls and Parkinson 1957). Much of the early work on the absorption and emission spectroscopy of shock-excited gases has been summarized by Gaydon and Hurle (1963). See also Nicholls, Parkinson and Reeves (1963).

At relatively low temperature, the emission radiation from hot gases is dominated by discrete line or band emission features of large oscillator strength. They are easily excited by atomic collisions. At higher temperatures, where the degree or ionization becomes appreciable, plasma spectroscopic effects become evident, (Griem 1964, Marr 1968). These include the modification of the discrete features by various line broadening influences, the emission of Kramer's continua resulting from free-bound and free-free transitions of electrons in the fields of ions, (Allen, Textoris and Wilson 1965; McChesney and Al-Attar 1965; Hamberger and Johnson 1965), and the general extension of the wavelength range of interest into the vacuum ultraviolet (Marrone, Wurster and Shatton 1968). As the temperature rises, and the continuous spectra assume greater importance, the spectrum of the hot gas often approaches that of a black-body at the temperature in question (Taylor and Kane 1967).

Intense shock waves have been used for some years as sources of hot plasmas, and much astrophysically interesting work has been done with them. Particularly useful work has been done on the spectra of ions in various degrees of ionisation.

Hot shock-produced plasmas are often optically thick. This implies that the photon mean free path is less than a characteristic thickness of the slab of hot gas. For an optically thick slab the spectral emissivity $I_{em}(\lambda)$ at wavelength λ is determined by the solution of the equation of radiative transfer (Penner 1959). For a solid slab of thickness z this is:

$$I_{em}(\lambda) = B(\lambda, T)(1 - \exp{-k_\lambda' z}) \qquad (1)$$

$$\text{where } k_\lambda' = k_\lambda (1 - \exp{-hc/\lambda kt}) \qquad (2)$$

Here, k_λ is the spectral absorption coefficient or opacity and $B(\lambda, T)$ is the Planck function. There have been many experimental and theoretical studies, in recent years, of these aspects of shock-heated plasmas (Nelson and Goulard 1968). The emission spectrum of heated air has received particular attention (Landshoff and Magee 1969a, b; Armstrong and Nicholls 1967; Buttrey, McChesney and Hooker 1969).

Propagation of shock waves through dilute hydrocarbon mixtures has been used to excite such species as C_2 and CN (Fairbairn 1964).

Propagation of shock waves through dilute mixtures of various carbonyls (particularly of iron and of chromium) have been used to excite spectra of transition metals. Volatile compounds of other metals e.g. $VOCl_3$ have also been similarly employed in the excitation of specific molecular spectra.

The shock is a unique instrument for the excitation of lines of high rotation quantum number in molecular spectra, Gibson (1962) and Gibson and Buttrey (1962) have thus studied the excitation of N_2^+ bands to a much higher degree of rotational development than had previously been possible. Such work can lead to an increased knowledge of molecular lines (such as are needed for example for unambiguous identifications in stellar spectra), molecular constants, and perturbations which would not be uncovered in less extensive rotational developments associated with lower temperature sources.

Such highly developed spectra often have a different gross appearances from those which are common when they are less developed in, for example, discharge or flames. This can lead to difficulties in identification. One particularly interesting recent example of this is definitive identification by Goldberg, Parkinson and Reeves (1965) of bands of the CO Fourth Positive system in the solar spectrum between 1500 and 2000A. They compared the solar spectrum in this region photographed by a rocket-borne spectrograph (Tousey 1964) with a flash absorption spectrum of shock-heated CO at about $5000^{\circ}K$. They were thereby able to associate the sharp onset of opacity in the solar spectrum at 1500A with the very strong overlap in rotational structure of highly developed CO Fourth Positive bands in this region. This work is a fine example of the many important problems of vacuum ultraviolet spectroscopy in the solution of which the shock-tube is an excellent tool (Garton 1966).

Shock-Tube Spectroscopy And Atomic And Molecular Structure

<u>Atomic Spectra</u>. The one and two electron spectra of Gp I,II,III elements have been intensively studied in the past two decades particularly with respect to their extension into the vacuum ultraviolet (Garton 1966). Much of this work has been concerned with the definitive establishment of line series, energy levels and series limits by use of quantum defect plots (see e.g. Garton and Codling 1965, 1968), and also with the study of broad autoionisation features (called Beutler-Fano Resonances) which appear in the one electron ionisation continua as a result of mixing of two electron levels with the ionisation continua. (See e.g. Garton and Wilson 1966, Marr and Heppinstall 1966, Garton, Grasdalen, Parkinson and Reeves 1968, Newsom and Shore 1968). In such work it had often been usual to vapourise the absorbing metal in furnaces or in heat-pipes (Vidal and Cooper 1969). In recent years, however, the shock-tube and flash absorption techniques have been used at Harvard College Observatory to extend the work to materials and spectra which were not susceptible to the study by other techniques. Examples are the measurements of Garton, Parkinson and Reeves (1963, 1966) on Ba In, Tl and of Newsom (1966) on Ca.

Our understanding of the energy level schemes and of the structure of Gp I-III elements has thus been greatly increased in recent years by the application of far superior instrumentation and experimental techniques than was used in the early studies of these elements by traditional spectroscopic methods during the first three decades of this century. The transient nature of shock-tube experiments has called for the development of reliable reproducible high intensity microsecond flash sources (Garton 1959), microsecond operation pre-slit shutters (Camm, 1960, Clayton 1963, Grasdalen, Haber and Newsom 1968), and time resolving spectrographs for absorption and emission work in the visible (Buttrey 1967) and in the infrared (Camm, Taylor and Lynch 1967).

Molecular Spectra. The shock-tube has also been used in recent years for the excitation, identification and analysis of new band systems from which new molecular constants have been derived. Examples are found in the work of Tyte (1964) on AlO, Parkinson and Reeves (1963) on TiN, Nicholls and Tyte (1965) on WO, and Linton and Nicholls (1969) on TiO. An important molecular constant is the dissociation energy which has been measured for some molecules using shock-tube techniques. For example Seal and Gaydon (1966) made photographic intensity measurements on the NH absorption spectrum using absorption flashes through NH produced in reflected shocks through N_2-H_2-Kr mixtures. The strength of the absorption of the (0,0) band depends upon the number density of NH and upon the oscillator strength of the band. Oscillator strengths are inferred from independent lifetime measurements of Bennet and Dalby (1960) and Fink and Welge (1964). The number density, as determined from thermochemical considerations depends upon the dissociation energy in the exponent of a Boltzman expression. In this way they inferred a dissociation energy of 3.21 ± 0.16 ev.

Tyte (1967) used a direct method of interpretation of the flash absorption spectrum of AlO to place a lower limit of 4.60 ev on the dissociation energy of AlO. A continuum which extends from 2730 A to shorter wavelengths was interpreted by him as the photo-dissociation continuum. The position of its long wavelength edge located the dissociation limit.

Shock-Tube Spectroscopy And Transition Probability Measurements

Spectral intensities are inherently controlled by the product of population (or number density) of active species and transition probability parameters. The shock-tube is an excellent spectroscopic source for transition probability studies because of the established existence of L.T.E. between the various degrees of freedom accessible to all species (Parkinson and Reeves 1964, Garton, Parkinson and Reeves 1964, 1966). This implies that, in principle, number densities are known and thus transition probabilities can be at once inferred from spectral intensity measurements in emission or in absorption.

The application of L.T.E. to provide estimates of number

densities is sometimes hindered by ignorance of location, and sometimes of existence of and character of energy levels. Such information is necessary for calculation of the relevant partition functions. A comparable difficulty exists, particularly when powdered samples are used, if the relevant thermochemical constants are not available. In the optically thick case the above ideas are applied in such a way as to take account for radiation diffusion, often by using curve of growth techniques. Shock-tube spectroscopy has been much used in recent years for the provision of transition probability data of atomic and molecular transitions of importance to astrophysics, aeronomy, combustion, re-entry phenonmenology, etc.

In the discussion of spectral intensities, the fundamental transition probability parameter for atoms and molecules is the transition strength matrix element

$$S_{uL} = \left| \int \psi_u^* M \psi_L \, d\tau \right|^2 \quad (3)$$

where ψ_u, ψ_L are the wave functions of the upper and lower states involved. M is the dipole moment of the transition and $d\tau$ is the element of configuration space. It is well known (Nicholls 1969a) that commonly met transition probability parameters are

Emission intensity $\quad I_{UL} \propto S_{UL} N_U / \lambda_{UL}^4 \quad$ (4a)

Einstein A coefficient $\quad A_{UL} \propto S_{UL} / \lambda_{UL}^3 \quad$ (4b)

Einstein B coefficient $\quad B_{UL} \propto S_{UL} \quad$ (4c)

Oscillator strength $\quad f_{UL} \propto S_{UL} / \lambda_{UL} \quad$ (4d)

Absorption coefficients α_{UL} (atomic), k_{UL} (linear) K_{UL} (mass), τ_{UL} (optical depth), Λ_{UL} (photon mean free path) $\propto S_{UL} / \lambda_{UL} \quad$ (4e)

The absorption coefficients are related by the Beer Law for optically thin media

$$\ln \frac{I_o}{I} = k_{UL} x = K_{UL} \rho x = N_L \alpha_{UL} x = \frac{x}{\Lambda_{UL}} = \tau_{UL} \quad (5)$$

where x is the optical path length of the absorber.

Many shock-tube spectroscopic determinations of transition probabilities involve the direct interpretation of emission or absorption intensity measurements of shock-heated gases whose thermodynamic state is known. Great care has to be taken in such work to ensure whether optically thin or thick conditions apply and to use the correct interpretation. Care has also to be taken to account for the change in line profiles by collision and other broadening influences.

Another spectroscopic technique for oscillator strength measurements which has been applied with success to shock-tube spectra at Harvard College Observatory (Huber 1966), is the interferometric "Hook" method. The experimental equipment includes the usual absorption spectroscopy arrangement crossed with a Mach-Zehnder interferometer. The wavelength difference between "hooked" fringes on either side of an absorption line is a measure of the oscillator strength of the line. Marlowe (1967) has reviewed the method which has been used for many years in static situations on atomic spectra in the Soviet Union, (Meroz 1962). It has been applied in static situations to molecular spectra by Anketell and Pery-Thorne (1967), Pery-Thorne and Banfield (1970). It is similarly being applied in the author's laboratory (Farmer, Hasson and Nicholls 1970).

Atomic Spectra. Many recent shock-tube studies of atomic transition probabilities have been astrophysically motivated. They have been concerned with such elements as Fe, Cr, O, etc. (and their ions).

Byard (1967) reports quantitative flash absorption measurements on 43 lines (4603-6180A) of FeI produced by shock treatment of a dilute mixture of argon and ion carbonyl. He used a curve of growth method to interpret the optically thick spectra and obtained fair agreement with the free burning arc results of Corliss and Bozmann (1962) and Corliss and Warner (1964). Similar and extensive studies were made by Huber and Tobey (1968) on 38 FeI, 5 CrI and 2 CrII lines (3150-3780A) and by Grasdalen, Huber and Parkinson (1969) on 34 lines of FeI and 12 lines of FeII (for which Van der Waals broadening coefficients were also determined) (3196-3341A). Very great care was taken to define the experimental conditions, and the results indicate that below 4000A the results of Corliss and Bozmann and Corliss and Warner are about an order of magnitude (one dex) too high. Above 4000A the shock-tube and arc results seem to agree.

CrI and II spectra have been shock excited using chromium carbonyl $(Cr(CO)_6)$, and oscillator strengths determined in a similar manner by Shackleford (1965), Byard (1968) and Wolnick, Berthel, Larson, Carnevale and Wares (1968). The lines appear in the (3000-5400A) region. There appears to be general agreement in the shock-tube measurements on chromium which are lower by up to an order of magnitude than those of Corliss and Bozmann (1962) and Corliss and Warner (1964).

Some comparisons emphasize the need, in atomic physics, to make measurements on basic data in different ways in different laboratories. Some confidence may be placed on the shock-tube results because of the realistic number density estimates which could be made.

TiI and TiII oscillator strengths were studied by Boni (1968a) using dilute $TiBr_4$ in his test gas. The date for TiI agreed well with earlier work using other techniques. Earlier measurements on TiII lines by Corliss and Bozmann and Corliss and Warner, however, appeared to be too large.

13 prominent SiII line oscillator strengths have been measured

by Miller and Bengston (1969) using dilute shock excited mixtures of SiH_4 and $SiHCl_3$ in Ne. Relative measures were normalised to a good theoretical estimate from using Coulomb wave functions.

Coates and Gaydon (1966) measured the transition probabilities of 13 argon lines (4159-7635A). The very high temperatures which can be set up in shock waves has been used to excite ions, Berg, Eckerle, Burris and Wiese (1964) for example used an electrically driven T-tube to produce shock waves in $He-O_2$ mixtures and thereby to excite 18 OII lines (4649-3271A) and 18 OIII lines (3760-3722A). Huber (1966) has used the Hook method on shock excited FeII lines to measure oscillator strengths.

Extensive recent measurements have been made at Maryland by Bengston (1968) and Miller (1968) on lines of AIII, CI, CII, FI, NeI, OI, BrI, BrII, SiI, SiII, PI, PII, SI, SII.

All of the above work refers to oscillator strength measurements on discrete atomic lines. The shock-produced plasma is well suited for study of intensities, widths (and thus ionic lifetimes) of broad autoionisation features which occur as a result of level mixing, (Marr 1967). Garton, Parkinson and Reeves (1964) using flash absorption techniques on the hot gas produced shock-excitation of mixtures of calcium and aluminium sulphides measured widths of broad autoionisation features of Al from which lifetimes were inferred. They studied in particular the astrophysically important AII doublet at 1932 and 1936A, whose large width corresponds to an autoionisation lifetime of 1.1×10^{-13} sec. Newsom (1968) has extended this work to a study of the autoionised Ca lines near 6360A.

An example of the application of shock-tube spectroscopy to the study of photoionisation continua (bound-free transitions) and cross-sections is found in the work of Rich (1967). He shock-excited dilute trichlorsilane ($SiHCl_3$) and measured by flash absorption techniques the photoionisation continua arising from the ground (^3P) and the first excited (^1D) states. The continua correspond to cross-sections of 37 and 33 megabrans respectively at the ionisation limit.

<u>Molecular Spectra.</u> During the past decade there has been an increasing demand for firm transition probability data for diatomic molecular spectra which have applications in astrophysics, aeronomy, and the physics of re-entry into planetary atmospheres, (Nicholls 1964, 1969b). Many experimental and theoretical methods have been applied to provide these data. For the same reasons as outlined in the preceding section, the shock-tube method has been very successfully applied to provide some of the data. Flash absorption and emission studies have been made.

A molecular gas is in general much richer in lines than an atomic one, and hence the effects of opacity and optical thickness are often met particularly if the lowest state of an absorption transition is the ground state. Synthetic spectra methods which take into account lineshapes and the detailed spectral profile have been developed in a number of laboratories (including the author's) to account for this. In such work the specific form of equations

SHOCK-TUBE SPECTROSCOPY

(1) and (2) appropriate to the spectra in question are used. It is noticed in these equations that the spectral emissivity is entirely controlled by the absorption coefficient. That is why the spectral absorption coefficients are such important transition probability data for the discussion of radiative transport in hot molecular gases, particularly in cases where the gas dynamic flow is effected by coupling with radiation (as in the case of planetary re-entry processes and in the dynamics of the envelopes of variable stars). Nicholls (1969a) has summarized the basic concepts of molecular spectroscopy assumed in the review which follows. In this section a brief general review is made of a number of molecules which have been studied. In the final section a specific review is made of some work on oxides which has been made in the author's laboratory.

AlO. AlO is used in aeronomical studies where it is applied as a "thermometric" molecule to measure upper atmosphere temperatures (Harang 1967). The synthetic spectrum method (Drake, Tyte and Nicholls 1967) has been applied to the laboratory study of intensity distributions in the blue-green system of AlO by Linton and Nicholls (1969a) in which self-absorption effects are important. Relative band strengths for 28 emission bands have been measured. They were shock-excited in powdered Al in a dilute $A-O_2$ mixture. This work corrects a number of earlier arc and exploding wire studies in which the effects of self-absorption have been neglected. Details are given in the next section.

BeO. The BeO blue-green system has aeronomical, re-entry and astrophysical applications. It is strongly self-absorbed in shock-excitation. Drake, Tyte and Nicholls (1967) developed a synthetic spectrum method by which the detailed spectral emissivity is evaluated at a number of points per band-line, for comparison with experimentally excited spectra. The procedure of comparison between experimental and theoretical spectra allows for the determination of band strength and the delineation of the variation of electronic transition moment with internuclear separation. 19 BeO bands were studied in this work. Details are given in the next section.

CO. The CO fourth positive system is a very important contributor to the solar spectrum (Goldberg, Parkinson and Reeves (1965), and to the atmospheric spectra of Mars and Venus. Rich (1968), by a flash absorption study of shock-excited dilute mixtures of CO and $Cr(CO)_6$ inferred oscillator strengths for 12 strong bands of this system. His results confirm the suggestion of Nicholls (1960) that the electronic transition moment for this system is relatively insensitive to internuclear separation, and that Franck-Condon factors well represent the intensity distribution across the system.

C_2. The C_2 Swan bands contribute to the opacity of some stellar envelopes and to the radiatively coupled flow phenomena of high speed flight in some planetary atmospheres which contain CO_2. They are also important in many combustion applications. A number of workers have thus supplemented other intensity measurements on this system with shock-tube studies. Sviridov, Sobolev and Sutovski (1965), by interpretation of measurements on five band widths within the Δv = 2, -1,0,+1,+3 sequences of the shock-excited emission spectrum of C_2 from A-CO mixtures in terms of an optically thin model conclude that the average oscillator strength for the band system is 0.022 $\pm$ 0.008.

Harrington, Modica and Libby (1966) in a similar experiment in which A-CF_4,C_2F_4,C_2ClF_3 mixtures were shock used the optically thin smeared rotational structure band model of Keck, Camm, Kivel and Wentink (1959) to assign an oscillator strength at the (0,0) band of 0.028$\pm$ 0.009 to the Swan system and 0.0051$\pm$0.0095 to the CF system which they also excited. Modica (1968) later measured the oscillator strengths of the CF_2 system in a similar experiment. Faribairn (1966) in a broad waveband study of the system shock-excited in mixtures of C_2H_2, C_2N_2, CO, assigned an "average" oscillator strength to the whole system of 0.003$\pm$0.012

Arnold (1968) using a reliable detailed synthetic spectrum method (Whiting, Arnold and Lyle 1969) to interpret intensity measurements on C_2 Swan bands shock-excited in CO_2-A mixtures was able to determine the variation of the electronic transition moment, and assigned an oscillator strength of 0.035$\pm$0.005 to the (0,0) band of the system and an average of 0.043$\pm$0.006 over the Δv = +1 sequence. He presents a very useful comparison table of other measurements using shock-tube and other techniques.

In many of the above studies attention was centred, photoelectrically, on specific band pases of the Δv = 1, 0, +1 sequences of bands.

<u>CN.</u> The CN red and violet band systems, like the C_2 Swan system, are very important astrophysically (stars and comets) and in hypersonic re-entry spectra through CO_2 containing atmospheres. Gippius, Kudryavtsev, Pechenov, Sobolev and Fokeev (1964) have made flash absorption measurements on the CN red bands (6330-6550A) resulting from shock-excitation of CO-N_2 mixtures. They derived an average electronic transition moment of 0.19$\pm$0.09 atomic units. In a more detailed similar study which used image tubes, Dronov, Sobolev, Faizullov and Boiko (1966) showed that the electronic transition moment had a magnitude of 0.11 atomic units.

In a quantitative synthetic spectrum study of the CN radiation from the shock layer of a hypervelocity projectile and using a time-of-flight scanning spectrometer, Reis (1965) measured average oscillator strengths for the Δv = +1 and -1 progressions of Violet system as 0.09$\pm$0.004 and 0.02$\pm$0.004 respectively.

<u>N_2 and N_2^+</u>. Band systems of N_2 and N_2^+ make extremely important contributions to many aspects of atmospheric physics and aeronomy, Nicholls 1964, 1969b). Reis (1964) using a similar method to that used in his work on CN, studied N_2 second positive and N_2^+ first negative band emissions for the shock layer around a hypervelocity projectile using a synthetic spectrum method was able to assign oscillator strengths to 0.057$\pm$0.01 and 0.53$\pm$0.01 and to the Δv = 0 sequences of the second positive and first negative bands - respectively. Wray and Connolly (1965) studied a 66A band pass at 4272A of

the emission from shock heated N_2 from it inferred an oscillator strength of 0.034 ± 14 to N_2^+ in that region.

NH. Harrington, Modica and Libby (1966) made quantitative intensity studies on the (0,0), (1,1) and (2,2) bands of NH(A-X) system shock-excited in a dilute A-NH$_3$ mixture and assign oscillator strengths of $(4.17\pm0.98) \times 10^{-3}$, $(4.95\pm1.2) \times 10^{-3}$ and $(7.04\pm2.0) \times 10^{-3}$ to the bands.

NO. There have been many shock-tube studies of NO. Recently Cann and Dickerman (1968) have re-examined the shock-excited A-NO spectrum at 32 wavelengths between 2611-5263A and have redetermined the variation of electronic transition moment with internuclear separation, for the NO β system. Wray (1969) in a recent study using constricted arc jet in air has shown that much of the excess near-infrared (9000-12,000A) radiation in shock-tube studies of N_2 which could not be ascribed to N_2 First positive emission. He claims that this emission may well be due to transitions between highly excited Rydberg states of NO. He assigned average oscillator strengths to 9 such transitions.

OH. Hydroxyl has been studied by shock-tube spectroscopy for many years. Watson (1964) reports a quantitative absorption study on a 100A pass band near the 306A (0,0) band from shock excited A-H$_2$O mixtures. He assigned an average oscillator strength of $3.9\pm0.9 \times 10^{-3}$. Watson and Ferguson (1965) investigated the possible effect of the vibration-rotation in the OH molecule, and found a maximum effect of no more than 5%.

SiO. Main, Marswell and Hooker (1968) as part of a continuing study of oxide spectra report on flash absorption studies of shock-excited aerosols of phenolic refrasil and air. Relatively low dispersion spectra were obtained photographically and their intensity distribution was fitted to a synthetic profile. From this work an average oscillator strength for the system of 0.023 ± 0.013 was found.

TiO. Linton and Nicholls (1969b,c) using the synthetic spectrum method employed in the author's laboratory on AlO and BeO have obtained relative band strengths for the α and γ systems of TiO which are important features of the absorption spectra of the enveloped of M-type stars. Details are given in the next section.

VO. Harrington and Nicholls (1969a,b) have recently made intensity measurements on the red and yellow-green band systems of VO using the synthetic spectrum method to interpret the emission spectrum of VO excited in shock-treated A-VOCl$_3$ mixtures. VO is also an important contributor to the spectrum of cool stars. Band strengths, were determined. The electronic transition moment variation has been studied and chemical kinetics have been followed. Details of this work are given in the next section.

Spectral Absorption Coefficients Of Air And Other Gases.

The work reviewed immediately above relates to the determination of oscillator strengths of important diatomic molecular band systems.

There has in addition been significant experimental and theoretical effort to study the absorption coefficients or opacities of gases or gas mixtures with special application to planetary atmospheres, particularly air. The effect of line shapes are important in such work. Whiting, Arnold and Lyle (1969) describe a synthetic spectrum method for calculation of realistic spectral profiles which takes line shapes and band structure into account and which does not make an over-simplified assumption of smearing of band structure.

The SACHA (Spectral Absorption Coefficient of Hot Air) computer code has recently been developed at Lockheed Research Laboratories (Churchill, Hagstrom and Landshoff, 1964, Churchill and Meyerott, 1965, Churchill, Armstrong, Johnston and Muller, 1966). It has been discussed and tabulated in full by Landshoff and Magee (1969a, b). The physical basis of the calculations have been discussed at length by Armstrong and Nicholls (1967).

The SACHA code has been powerfully applied by Buttrey, McChesney and Hooker (1968) to the determination of Stark-broadened molecular line profiles of shock-excited $N_2^\pm$ lines in the $9000°$ - $14,000°K$ temperature region.

Total radiation from shock-heated air has been studied by Gruzezynski and Warren (1967) and by Wood, Hoshizaki, Andrews and Wilson (1969). Near infrared contributions to the radiation from air and N_2 have been measured by Wray and Connolly (1965) and for NO by Wurster and Marrone (1967).

Free-free and free-bound continua in the emission spectra of shock-excited N_2, O_2 and air in the temperature regime $8000-14,000°K$ have been studied by Allen, Textoris and Wilson (1965).

Infrared absorption coefficients of shock-excited CO_2 have been measured by Sulzmann (1964) and similar work on H_2O has been done by Patch (1965). Absorption coefficients of shock-excited Cl_2 have been made by Jacobs and Giedt. Voigt line shape studies on individual shock-excited OH band lines near 3090A have been made by Bird and Schott (1965). Intensity measurments on spectra of shock-heated CO_2N_2 mixtures have been made in a number of laboratories to simulate entry conditions of vehicles into Martian and Venusian atmospheres. Examples are the work of Fairbairn (1964), Thomas and Menard (1966) and Menees and McKenzie (1968). In the first of these estimates of the CN oscillator strengths are also made.

SHOCK-TUBE SPECTROSCOPY OF ASTROPHYSICALLY IMPORTANT DIATOMIC MOLECULES

Reference has been made in the previous sections to work in progress in the author's laboratory in which the techniques of quantitative shock-tube spectroscopy are used to determine the transition probability parameters band strengths. The species chosen are astrophysically important diatomics which exist in stellar atmospheres. Over the past few years, AlO, BeO, CN, TiO and VO have been studied. The method and recent results are described below.

The principle of the method is, through the use of a detailed version of equation (1), to compare a realistic computer-produced synthetic emission spectrum of the optically thick shock-excited gas with an experimentally measured spectrum. The scaling procedure employed to produce agreement between the synthetic spectrum and the measured spectrum gives information to the variation of the electronic transition moment Re(r) with internuclear separation.

A number of programmes exist for the line-by-line computation of synthetic spectra. Those used in our laboratories are a development of that written by Drake (Drake, Tyte and Nicholls 1967), together with the similar powerful programme developed by Whiting, Arnold and Lyle (1969).

For the (J',J'') line of the (v',v'') band of a diatomic molecular band system, it is well known (Nicholls 1969a) that the transition strength $S_{v'J''}^{v'J'}$ of line is the product of electronic, vibrational and rotational strength factors. That is

$$S_{v'J''}^{v'J'} = R_e^2(\bar{r}_{v'v''}) \, q_{v'v''} \, S(J'J'') \qquad (6)$$

where $\bar{r}_{v'v''}$ is a characteristic internuclear separation, the r-centroid, associated with the band, $q_{v'v''}$ is the Franck-Condon factor and $S(J'J'')$ is the Hönl-London factor or rotational line strength factor.

It will be recalled from equation (1) that the absorption coefficient k_λ controls the spectral emissivity of the gas. It is not often in shock-tube work on molecules that the radiation density is sufficiently high that the corrected form k'_z (of equation 2) which takes account of stimulated emission of radiation has to be used. In the case of a molecular spectrum, lines of which also exhibit environmental and instrumental broadening, the exponent or optical depth of equation (1) can be written (see Drake, Tyte and Nicholls 1967, Linton and Nicholls 1969, 1970).

$$k_\lambda z = \sum_{v'v''} \sum_{J'J''} \Omega(\bar{r}_{v'v''}) \cdot q_{v'v''} S(J',J'') \frac{1}{\lambda_{v'J'v''J''}} \cdot b(|\lambda - \lambda^o_{v'J'v''J''}|) \cdot x$$

$$\frac{\exp(-\Delta E_{v''J''}/kT)}{Q_{vib} \, Q_{rot} \, Q_{el}} \qquad (7)$$

where

$\lambda^o_{v'J'v''J''}$ = wavelength of the $v'J' \to v''J''$ rotational line at its peak intensity

$b(|\lambda - \lambda^o_{v'J'v''J''}|)$ = normalized distribution function of the rotational line shape

$\Delta E_{v''J''}$ = energy of level $v''J''$ above lowest level

and Q_{vib}, Q_{rot}, Q_{el} are the vibrational, rotational and electronic partition functions

In equation (7) $\Omega(\bar{r}_{v'v''})$ is a dimensionless optical depth parameter defined by

$$\Omega(\bar{r}_{v'v''}) = \frac{8\pi^3}{3hc} N_0 |R_e(\bar{r}_{v'v''})|^2 \cdot z \tag{8}$$

where

N_0 = total gas phase concentration of the active species $\Omega(\bar{r}_{v'v''})$ thus contains the electronic part of the transition strength.

The rotational line shape can be represented by the classical Voigt profile (Hummer 1965)

$$b(|\lambda-\lambda^\circ|) = 1/b_D \cdot (\ln(2/\pi))^{1/2} \cdot H(a,x) \tag{9}$$

where

$$b_D = \text{Doppler half width (cm}^{-1}) = \frac{(2kT\ln 2)^{1/2}}{mc^2} \frac{1}{\lambda^\circ} \tag{10}$$

and

$$H(a,x) = \frac{a}{\pi} \int_{-\infty}^{\infty} \frac{\exp(-y^2) dy}{a^2 + (x-y)^2} \tag{11}$$

Here $a = (\ln 2)^{1/2} \cdot b_c/b_D$, the line-shape parameter

$x = (1/\lambda - \lambda/\lambda^\circ)(\ln 2)^{1/2}/b_D$

b_c = collision half-width = $N_B \sigma (8\pi kT(1/M_A + 1/M_B))^{1/2}/\pi c$

N_B = number of molecules of type B per cm^3

M_A, M_B = molecular weights of colliding species

σ = collisional cross-section for rotational line broadening

The experimentally observed intensity, $I_{obs}(\lambda,T)$, at the focal plane of the spectrograph is the convolution of $I_{em}(\lambda,T)$, calculated from equations (1) and (7) with the normalized spectrograph slit function $g(|\lambda-\lambda^\circ|)$. That is

$$I_{obs}(\lambda,T) = \int_0^\infty I_{em}(\lambda',T) \cdot g(|\lambda-\lambda'|) d\lambda'. \tag{12}$$

A triangular slit function of base width 2δ is often a good representation

thus $g(|\lambda-\lambda^\circ|) = \begin{cases} (\delta-|\lambda-\lambda^\circ|)/\delta^2 & |\lambda-\lambda^\circ| \leq \delta \\ 0 & |\lambda-\lambda^\circ| \geq \delta \end{cases}$ (13)

A simpler representation of the slit function, particularly for relatively wide slits is a rectangular function of width 2δ

thus $g(|\lambda-\lambda^\circ|) = \begin{cases} \frac{1}{2\delta} & |\lambda-\lambda^\circ| \leq \delta \\ 0 & |\lambda-\lambda^\circ| \geq \delta \end{cases}$ (14)

The synthetic spectrum computer programme is easily adapted to any slit function.

In the application of the above method, $I_{obs}(\lambda,T)$ is generated by computer methods, line-by-line across a spectrum. This requires firstly a detailed knowledge of the band line wavelengths λ^o which implies that a definitive spectral analysis has been performed and the relevant vibrational and rotational constants which describe the spectrum are available. Sometimes the necessary wavelength spectroscopy has not been done to a sufficient degree of accuracy. Knowledge of location of energy levels is required to evaluate the partition functions.

Secondly the Franck-Condon factors $q_{v'v''}$ and r-centroids $\bar{r}_{v'v''}$ for each band and the Hönl-London factors $S(J',J'')$ for each line must be available. Theoretical studies have been made in a number of laboratories recently and a number of Franck-Condon factor and r-centroid arrays are thereby available. (See for example McCallum, Jarmain and Nicholls 1970). For well known band systems, Hönl-London factors are also available (Kovacs 1969, Tatum 1967). In some cases however, particularly when there is a change in Hund coupling case within a band, the situation is not always unambiguous.

In an actual application of the method one or more empirical "self-absorption indicators" are defined. Such an indicator is a characteristic of the spectrum, and is used as an index of comparison between the experimental and synthetic spectrum. A typical index is, for example, the ratio of intensity maxima in the P and R branches of a band. Others can be proposed depending upon the structure of the spectrum. The self-absorption indicators and integrated intensities may be used to evaluate $\Omega(\bar{r}_{v'v''})z$ or $N_o R_e^2(\bar{r}_{v'v''})z$ for each band as described below after integrated intensities have been measured for pre-selected wavelength intervals in each band. Often the kinetic temperature of the gas has been measured by self-reversal or other means. Two types of plot can then be made from the synthetic spectra:

a) Self-absorption indicator versus $\Omega(\bar{r}_\infty)$ for a number of temperatures spanning the measured temperature.

b) Emission intensity within the chosen wavelength interval of each (v',v'') band as a function of $\frac{1}{T}$ for each of a range of values of $\Omega(\bar{r}_{v'v''})$.

The curves (a) are often relatively insensitive to temperature and cross each other within a small temperature range. Alternative use of these curves with experimental data permit the fairly precise determination of temperature to a few percent. The set of curves (b) are computed for each (v',v'') band for a series of values of $\Omega(\bar{r}_{v'v''})$. Intersection between the known $\frac{1}{T}$ and measured intensity values occures on a curve of appropriate $\Omega(\bar{r}_{v'v''})$ value. Thus $\Omega(\bar{r}_{v'v''})$ or $N_o R_e^2(\bar{r}_{v'v''})z$ is determined for each band or band segment. $\Omega^{1/2}$ is then plotted against $\bar{r}_{v'v''}$ to obtain a smooth curve which delineates the $R_e(v)$ variation from which band strengths can be determined through

$$S_{v'v''} = R_e^2(\bar{r}_{v'v''}) q_{v'v''} \tag{15}$$

and the other transition probability parameters of each band can

Table 2

Relative Band Strengths For (AlO $A^2\Sigma - X^2\Sigma$) System

v' \ v''	0	1	2	3	4	5	6	7
0	1.00 1.00	0.332 0.331	- -	- -	- -	- -	- -	- -
1	0.394 0.384	0.489 0.474	0.477 0.488	- -	- -	- -	- -	- -
2	0.131 0.129	0.510 0.502	0.227 0.194	0.524 0.537	- -	- -	- -	- -
3	- -	0.265 0.266	0.499 0.476	- -	0.537 0.537	- -	- -	- -
4	- -	- -	0.363 0.360	0.447 0.395	- -	0.515 0.499	- -	- -
5	- -	- -	0.185 0.199	0.424 0.408	0.381 0.303	- -	0.491 0.450	- -
6	- -	- -	- -	0.266 0.283	0.455 0.421	0.317 0.222	- -	- -
7	- -	- -	- -	- -	0.336 0.353	0.468 0.409	0.267 0.154	- -
8	- -	- -	- -	- -	- -	0.394 0.407	0.465 0.376	0.221 0.101

Upper Entry Calculated Using Morse Potential
Lower Entry Calculated Using RKR Potential

Table 3

Band Strengths ($S_{v'v''} \times 10^2$ Atomic Units) For The
BeO ($B^1\Sigma - X^1\Sigma$) Band System

v'\v''	0	1	2	3	4	5
0	2.97	0.539	0.060	-	-	-
1	0.283	2.42	1.00	0.180	-	-
2	-	0.476	2.00	1.38	0.36	-
3	-	-	0.597	1.73	1.82	0.046
4	-	-	-	0.668	1.52	2.22
5	-	-	-	-	0.700	-

Table 4

Relative Band Strengths For The CN ($B^2\Sigma - X^2\Sigma$)

Violet System

v' \ v''	0	1	2	3	4	5	6
0	1.000	0.107	-	-	-	-	-
1	-	0.849	0.168	-	-	-	-
2	-	-	0.713	0.216	-	-	-
3	-	-	-	0.587	0.251	-	-
4	-	-	-	-	-	0.282	-
5	-	-	-	-	-	-	0.247

be inferred through application of equations (4). In cases where N_0 is known absolutely, through thermochemical or other arguments absolute transition probabilities can be inferred, otherwise relative values are derived.

The following sections review specific results obtained by the use of this method to specific band systems.

The AlO ($A^2\Sigma-X^2\Sigma$) Blue-Green System.

The AlO ($A^2\Sigma-X^2\Sigma$) blue-green system occurs in some stellar atmospheres, is used as a "thermometer" molecule in measuring upper atmosphere temperatures and is probably an important component of re-entry spectra. It is a relatively compact band system which thereby distributes its oscillator strength across a rather narrow wavelength range. Its shock-excited spectrum (using powdered Al in Ar-O_2 mixtures) has been studied by Linton and Nicholls (1969) using the method described above and the relative band strengths thereby derived from the system are displayed in Table 2.

Data analysis was made using Morse and RKR potential Franck-Condon factors and as seen in Table 2 there is very little difference in the band strengths. The strongly "diagonal" nature of the band system is clear from Table 2. This is another indication of the compactness of the band system in wavelength.

The BeO ($B^1\Sigma-X^1\Sigma$) Blue-Green System.

The method described above was developed during studies on shock-excited powdered BeO (Drake, Tyte and Nicholls 1967). The BeO ($B^1\Sigma-X^1\Sigma$) Blue-green system thereby excited is similar in appearance and characteristics to the AlO system. Both exhibit well marked sequences as seen in the spectra reproduced in the two respective papers cited. It has astrophysical and re-entry importance.

The relevant transition probability data are displayed in Table 3 in an absolute form.

Some simple thermochemical arguments were used to make an estimate of absolute number densities of BeO in this case. While they may be in error by perhaps a factor of 2, the relative band strengths are probably reliable.

The CN ($B^2\Sigma-X^2\Sigma$) Violet Band System.

The CN Violet ($B^2\Sigma-X^2\Sigma$) band system, similarly to the C_2 Swan System, plays important astrophysical roles in cool stellar atmospheres and in cometary emission spectra. It has a similar appearance to the AlO and BeO systems, exhibiting as it does a series of band sequences across a narrow wavelength range.

Myer (1968) shock-excited graphite containing N_2 and studied the intensities of the CN Violet bands thereby produced, by the methods described above. A set of the resulting relative band strengths is displayed in Table 4.

Table 5

Relative Band Strengths For The
TiO α ($C^3\Delta - X^3\Delta$) System

v' \ v''	0	1	2	3	4
0	1	0.978	0.443	0.123	0.0219
1	0.893	0.0820	0.623	0.664	-
2	0.440	0.448	0.208	-	-
3	0.161	0.470	-	-	-
4	0.049	-	-	-	-

Table 6

Relative Band Strengths For Bands Of The
TiO β ($c^1\Phi$-$a^1\Delta$) Band System

v'\v''	0	1	2	3	4	5	6
0	1.00	0.078	0.005	0.0002	-	-	-
1	0.087	0.839	0.148	0.014	0.001	-	-
2	0.002	0.158	0.684	0.209	0.028	0.002	-
3	-	0.008	0.228	0.540	0.257	0.046	0.004
4	-	-	0.017	0.288	0.409	0.292	0.683
5	-	-	-	0.031	0.337	0.294	0.314
6	-	-	-	0.001	0.050	0.372	0.197

Table 7

Band Strengths ($S_{v'v''}$, Atomic Units) For Bands

Of The VO ($A^4\Sigma^- - X^4\Sigma^-$) Yellow-Green System

v' \ v''	0	1	2	3	4	5	6
0	0.653	0.694	0.432	-	-	-	-
1	0.754	-	0.341	0.518	-	-	-
2	0.504	0.163	-	-	0.299	-	-
3	0.263	0.409	-	-	-	-	-
4	0.142	0.389	0.175	-	-	-	-
5	-	0.270	0.353	-	-	-	-
6	-	0.125	0.340	0.168	-	-	-
7	-	-	0.250	0.364	-	-	-

The TiO $(C^3\Delta-X^3\Delta)\alpha$ And $(c^2\Phi-a^2\Delta)\beta$ Systems.

The TiO Band Systems are very important contributors to the spectrum of M-type stars. The two strong band systems in the visible $\alpha(C^3\Delta-X^3\Delta)$ and $\beta(c^1\Phi-a^1\Delta)$ are the most important features. The recent spectroscopy of these systems have been discussed by Linton and Nicholls (1969). They also studied the intensities of the shock-excited spectra of TiO_2 in argon (Linton and Nicholls 1970). They concluded from an examination of the results of the data analysis procedure described above that for both of these systems the electronic transition moment is relatively constant and thus that the relative band strengths $S_{v'v''}$ are determined by Franck-Condon factors. Relative band strengths for these systems are thus displayed in Tables 5 and 6

The VO $(A^4\Sigma-X^4\Sigma)$ Yellow-Green System.

Similarly to TiO, VO is also an important contribution to the opacity of the envelopes of cool stars. Harrington and Nicholls thus shock-excited $VOCl_3$ vapour to produce bands of the yellow-green $(A^4\Sigma-X^4\Sigma)$ band system.

The intensities of these bands were subject to the analysis described above and relative band strengths derived. A parallel study of the chemical kinetics of dissociation of the shocked vapour was carried out to place relative measurements on an absolute scale. The results are displayed in Table 7.

An independent study of VO relative intensities excited in an electrical discharge confirmed the variation of electronic transition moment implied in Table 7.

At present studies on ZrO, ScO and YO band systems which are also astrophysically important, are in process in our laboratories.

DIAGNOSTICS SHOCK-TUBE SPECTRA AND THE KINETIC STUDY OF COLLISION PROCESSES

As indicated in the introduction, once a firm set of basic data on structure and transition probabilities has been established for atomic and molecular spectra, spectroscopic methods can then be employed diagnostically for the determination of physical conditions while exist in luminous plasmas. The most commonly sought-after diagnostic information is temperature. When it is defined, L.T.E. conditions exist, and it expresses compactly the populations distribution of species in various energy states available to them. Relative populations are the second most sought-after data, particularly in the discussion of kinetics. In L.T.E. conditions, populations are Boltzmann. Many interesting problems remain however in the discussion of non-equilibrium situations.

The most powerful and most commonly used method for determination of temperatures in shock-heated gases is the pyrometric line reversal method originally adapted to shock-tube studies by Clouston, Gaydon and Glass (1958), as reviewed by Gaydon and Hurle (1963). In such work the gas of interest is "seeded", often by Na-emitting material. The method has been extended by Parkinson and Reeves (1964) and Garton, Parkinson and Reeves (1965, 1966) to the thermometric use of lines of species already in the hot gas. The need for seeding is thereby eliminated. They have also established in this work that the physical conditions which occur in many spectroscopic shock-tube experiments are very closely L.T.E. for much of the experimental time. Thus the Boltzmann and Saha equations may be used for the specification of populations. Byard and Roll (1965) have also used this method. Blau and Pratt (1968) have suggested an extension of the line reversal method to band spectra. Coates and Gaydon's (1966) work on shock-excited organic gases should allow temperatures in the 6000-15,000A region to be measured by the two line method. A Kirchoff law method in which absorbance and radiance in the infrared of shock-heated CO_2 and H_2O has been used by Penzias, Dolin and Kruegle (1966) to measure temperatures in the $2000°$ range.

A vast literature exists on the study of atomic and molecular collision processes which take place within the hot gas of the shock wave (see e.g. Bauer 1969, Gordon, Klemperer and Steinfeld 1969). It is not the purpose of this section to review that work in depth but rather as an illustration of the final stage of the spectroscopic process (line 6 of Table 1), a few recent shock-tube studies of basic processes (dissociation, ionisation, etc.) which used spectroscopic methods of measurement will be mentioned.

Reference has been made earlier to the growing importance of the vacuum ultraviolet region of the spectrum. There is growing interest in this wavelength region in shock-tube spectroscopy. For example at the Cornell Aeronautical Laboratory, Marrone, Wurster and Stratton (1968) have developed the technology of a windowless vacuum ultraviolet spectrometer which can be used in a transient manner in conjunction with a shock-tube. They have used this in important studies on N^+ and O^+ combination radiation involved in study of the kinetics of recombinations of these ions. Dr. Wurster has discussed examples of this work at the symposium. Other examples include the measurement (Rothe 1968) of photoionisation cross sections for the 3p $^2P_{\frac{1}{2},\frac{3}{2}}$ state of Na, by monitoring the recombination radiation from the inverse process $e+Na^+=Na(^2P)$ which occurs in Na seeded shocks; also, the study by Boni (1968b) of the kinetics of ionisation of Ti by monitoring TiI and TiII lines emitted in optically thin conditions in shock-excited $TiBr_4$-A mixtures.

Dissociation and recombination are very important processes in shock waves, and many spectroscopic studies have been made on them. For example Appleton, Steinberg and Liquornik (1968) have studied the kinetics of nitrogen dissociation caused by heavy particle collisions in shock waves through nitrogen. They monitored the 1176A N_2 absorption Lyman-Birge-Hopefield band from v"=10 as a function of time to measure specific N_2 dissociation from that level. Carbetta and Palmer (1967a,b) have studied dissociation and recombination of chlorine in shock waves, by monitoring the recombination continuum in the visible spectrum. Similarly the kinetics of a number of possible recombination reactions of H and O were studied by Gutman, Hardwidge, Dougherty and Lutz, by monitoring the chemiluminescent O CO recombination in shock heated H_2-O_2-CO-A mixtures. A study of O-O recombination was made by Myers and Bartle (1968) by monitoring the recombination continuum as a function of time following the shock-treatment of O_3-A mixtures.

The kinetics of excitation processes in shock-heated nitrogen have been worked on in many laboratories. Examples of such studies is the work of Wray (1966) on the excitation of the $B^3\Pi$ state of N_2 and the $B^2\Sigma$ state of N_2^+ in shock-heated N-N_2 mixtures, and similar work by Smekhov and Losev (1968).

The growth of research activities on CO_2 lasers has stimulated much experimental and theoretical work on energy exchange processes in N_2-CO-CO_2 mixtures (Taylor and Bitterman 1969). It has also stimulated shock-tube studies of the kinetics of some of the processes. An example of this is the recent work of Russo and Watt (1968) on vibrational excitation and de-excitation of N_2 in the presence of CO.

In all studies of the kinetics of excitation or other processes in which the primary measurements are by photoelectric spectroscopy, extreme care has to be taken in identification of the spectral feature which is monitored. This is particularly difficult when various band passes of a continuum has to be studied and one only has theory to guide one as to the expected intensity distribution of the continuum. All spectroscopic measurements in hot gases are fraught with problems of optical thickness (which may well change rapidly with time and place in the plasma). Thus extreme care has to be exercised at all times to ensure that the conditions under which measurements made are understood.

REFERENCES

Allen, R.A., Textoris, A., and Wilson, J., JQSRT, 5, 95, 1965
Anketell, J., and Pery-Thorne, A., Proc.Roy.Soc. A 301, 343, 1967
Appleton, J.P., Steinberg, M., Liquornik, D.J., J. Chem. Phys. 48, 599-608, 1968
Armstrong, B.H., and Nicholls, R.W., Thermal Radiation Phenomena Vol. 2 The Equilibrium Radiative Properties of Air:Theory DASA 1917-2, Lockheed Missiles and Space Co. 1967
Arnold, J.O., JQSRT, 8, 1781, 1968
Bauer, E., JQSRT, 9, 499, 1969
Bengston, R.D.: The measurement of transition probabilities and stark widths for CI,FI,NeI,CII,CIII,BrI,BrII. TN:BN 599, Inst. for Fluid Dynamics and Applied Maths, Uni. Md. 1968.
Bennett, R.G. and Dalby, F.W., J. Chem.Phys. 32, 1716, 1970
Berg, H.F., Eckerle, K.L., Burris, R.W. and Wiese, W.L., Ap. J.139, 751, 1964
Bird, P.F. and Schott, G.L., JQSRT, 5, 783, 1965
Bond, J.W., Watson, K.M. and Welch, J.A.: Atomic Theory of Gas

Dynamics, Adison Wesley, 1965
Boni, A.A., JQSRT 8, 1385, 1968a
Boni, A.A., J. Chem. Phys. 49, 288, 1968b
Brabbs, T.A., J. Chem. Phys. 49, 1433, 1968
Bradley, J.N., Shock Waves in Chemistry and Physics, Methunen, 1962
Brocklehurst, B., Rosevear, A.H., Seel, R.M., Innanen, S.H., Hébert, G.R., and Nicholls, R.W., 1971, Identification Atlas of Molecular Spectra 9. The CN Violet and CN Red Band Systems (in preparation)
Buttrey, D.E., Applied Optics 5, 881, 1967
Buttrey, D.E. and Gibson, J., Unpublished work, 1962
Buttrey, D.E., McChesney, H.R. and Hooker, L.A., JQSRT 8, 717, 1968.
Byard, P.L. and Roll, R.E., JQSRT 5, 715, 1965
Byard, P.L., JQSRT 7, 559, 1967
Byard, P.L., JQSRT 8, 1543, 1968
Camm, J.C., Rev. Sci. Inst. 31, 278, 1960
Camm, J.C., Taylor, R.L. and Lynch, R., Applied Optics 6, 885, 1967.
Cann, M.W.P. and Dickermann, P.J.: Molecular f-numbers from high resolution spectra, AFWL-TR-67-76, Kirtland AFB, 1968
Carabetta, R.A. and Palmer, H.P., J. Chem. Phys. 46, 1325, 1967a
Carabetta, R.A. and Palmer, H.P., J. Chem. Phys. 46, 1333, 1967b
Churchill, D.R., Armstrong, B.H., Johnston, R.R. and Muller, K.G., JQSRT 6, 371, 1966
Churchill, D.R., Hagstrom, S.A., Landshoff, R.K.M., JQSRT 4, 291, 1964
Churchill, D.R. and Meyerott, R.E., JQSRT 5, 69, 1965
Clayton, J.O., Rev. Sci. Inst. 34, 1391, 1963
Clouston, J.G., Gaydon, A.G., Glass, I.I., Proc. Roy. Soc. A, 248, 429, 1958
Coates, P.B., and Gaydon, A.G., Proc. Roy. Soc. A 293, 452, 1966
Corliss, C.H. and Warner, B., Ap. L. Supp. 83, 395, 1964
Corliss, C.H. and Bozmann, W.R.: Transition Probabilities, NBS Monograph 53, 1962. (U.S. Govt. Printing Office)
Degen, V., Innanen, S.H., Hébert, G.R. and Nicholls, R.W., 1969 Identification Atlas of Molecular Spectra, 6. The O_2 Herzberg I system
Drake, G.W.F., Tyte, D.C. and Nicholls, R.W., JQSRT 7, 639, 1967
Dronov, A.P., Sobolev, N.N., Faizullov, F.S. and Boiko, V.A., Opt. Spect. 21, 397, 1966
Fairbairn, A.R., AIAA Journal 2, 1004, 1964
Fairbairn, A.R., 1966, JQSRT 6, 325
Farmer, A., Hasson, V. and Nicholls, R.W., (1970) Unpublished work
Fink, E. and Welge, K., Zeits Naturforsch. 19a, 1193, 1964
Garton, W.R.S., J. Sci. Inst. 36, 11, 1959
Garton, W.R.S., Spectroscopy in the vacuum ultraviolet, Advances in Atomic and Molecular Physics 2, 93-176, 1966. (Ed. D.R.Bates and I Estermann, Academic Press)
Garton, W.R.S., and Codling, K., Proc. Phys. Soc. 86, 1067, 1965
Garton, W.R.S., and Codling, K., J. Phys. B. Ser. 2 1, 106, 1968

Garton, W.R.S., Grasdalen, G.L., Parkinson, W.H. and Reeves, E.M., J. Phys. B. Ser. 2 $\underline{1}$, 114, 1968

Garton, W.R.S., Parkinson, W.H. and Reeves, E.M., Proc. Phys. Soc. $\underline{80}$, 860, 1963

Garton, W.R.S., Parkinson, W.H. and Reeves, E.M., Ap. J., $\underline{140}$, 1269, 1964

Garton, W.R.S., Parkinson, W.H. and Reeves, E.M., Proc. Phys. Soc. $\underline{88}$, 771, 1965

Garton, W.R.S., Parkinson, W.H. and Reeves, E.M., Can. J. Phys. $\underline{44}$, 1745, 1966a

Garton, W.R.S., Parkinson, W.H. and Reeves, E.M., Proc. Phys. Soc. $\underline{88}$, 771, 1966b

Garton, W.R.S. and Wilson, M., Proc. Phys. Soc. $\underline{87}$, 841-850, 1966

Gaydon, A.G. and Hurle, I.R., The Shock Tube in High Temperature Chemistry and Physics, Chapman and Hall, 1963

Gibson, J., MS Thesis, Stanford University, 1962

Gilmore, F.R., Bauer, E. and McGowan, J.W., JQSRT $\underline{9}$, 157, 1969

Gippius, E.F., Kudryavtsev, E.M., Pechenov, A.N., Sobolev, N.N. and Fokeev, V.P., High Temp. $\underline{2}$, 159-164, 1964

Glass, I.I. (Ed.) Shock Tubes, Proceedings of the 7th International Shock Tube Symposium (University of Toronto Press, 1970)

Goldberg, L., Parkinson, W.H. and Reeves, E.M., Ap. J. $\underline{141}$, 1293, 1965

Gordon, R.G., Klemperer, W. and Steinfeld, J.I.: Vibrational and Rotational Relaxation, Ann. Rev. Phys. Chem. $\underline{19}$, 215, 1968

Grasdalen, G.L., Huber, M. and Newsom, G.H., Rev. Sci. Inst. $\underline{39}$, 886, 1968

Grasdalen, G.L., Huber, M. and Parkinson, W.H., Ap. J. $\underline{156}$, June 1969 (in press)

Griem, H., Plasma Spectroscopy, McGraw Hill, 1964,

Gruszcznski, J. and Warren, R., AIAA Journal $\underline{5}$, 517, 1967

Gutman, D., Hardwidge, E.A., Dougherty, F.A. and Lutz, R.W., J.Chem. Phys. $\underline{47}$, 4400, 1967

Hamberger, S.M. and Johnson, A.W., JQSRT $\underline{5}$, 683, 1965

Harang, O.E., Physica Norvegica $\underline{2}$, 71, 1967

Harrington, J., Modica, A.P. and Libby, D.R., J. Chem. Phys. $\underline{44}$, 3380, 1966

Harrington, J., Modica, A.P. and Libby, D.R., JQSRT $\underline{6}$, 799, 1966

Harrington, J. and Nicholls, R.W., Proc. Nat. Com. for Canada of Int. Astron. Union meeting, UBC, May 2,3, 1969a

Harrington, J. and Nicholls, R.W. (1969b) Unpublished work

Harrington, J., Seel, R.M., Hébert, G.R. and Nicholls, R.W., 1970 Identification Atlas of Molecular Spectra 7: The VO Yellow-Green and Red Systems

Hébert, G.R., Innanen, S.H. and Nicholls, R.W., 1967 Identification Atlas of Molecular Spectra 4: The Schumann-Runge System of O_2

Hébert, G.R., Innanen, S.H. and Nicholls, R.W., 1971 Identification Atlas of Molecular Spectra 8: The CO Fourth Positive System (in preparation)

Huber, M., JOSA 56, 1428, 1966
Huber, M. and Tobey, F.L., Ap. J. 152, 609, 1968
Hummer, D.G. Mem. Roy. Astron. Soc. 70, 1, 1965
Jacobs, T.A., Cohen, N. and Giedt, R.R., J. Chem. Phys. 46, 1958, 1967
Jacobs, T.A. and Giedt, R.R., JQSRT 5, 457, 1965
Keck, J.C., Camm, J.C., Kivel, B. and Wentinck, T., Ann. Phys. 7, 1, 1959
Kovacs, I., Rotational Structure in the Spectra of Diatomic Molecules (Adam Hilger, London) 1969
Landshoff, R.K.M. and Magee, J.L., Thermal Radiation Phenomena Vol. 1, Radiative Properties of Air, Plenum Press 1969a
Landshoff, R.K.M. and Magee, J.L., Thermal Radiation Phenomena Vol. 2, Excitation and Non-Equilibrium Properties of Air, 1969b
Linton, C. and Nicholls, R.W., JQSRT 9, 1-11, 1969a
Linton, C. and Nicholls, R.W., J. Phys. B. Ser. 2, 2, 490, 1969b
Linton, C. and Nicholls, R.W., JQSRT 10, 311, 1970
Marlow, W.C., Applied Optics 6, 1715, 1967
Marr. G.V., Photoionisation processes in Gases, Academic Press, 1967
Marr. G.V., Plasma Spectroscopy, Elsevier, 1968
Marr, G.V. and Heppinstall, H., Proc. Phys. Soc. 87, 293, 547, 1966a
Marr. G.V. and Heppinstall, H., Proc. Phys. Soc. 87, 547, 1966b
Marrone, P.V., Wurster, W.H. and Stratton, J.E., 1968 Shock Tube Studies of N^+ and O^+ Recombination Radiation in the Vacuum Ultraviolet. CAL AG 1729-A-7 (Cornell Aeronautical Lab. Buffalo New York)
McCallum, J.C., Jarmain, R.W. and Nicholls, R.W., 1970 Spectroscopic Report No. 1 Franck-Condon Factors and Related Quantities for Diatomic Molecular Band Systems, CRESS, York University.
Nicholls, R.W., Nature 186, 958, 1960
Nicholls, R.W., Ann. de Geophys. 20, 144, 1964
Nicholls, R.W., 1969a Electronic Spectra of Diatomic Molecules, Chap. 6, Electronic Structure of Atoms and Molecules, Vol. III Physical Chemistry on advanced treatise (Ed. H. Eyring, D. Henderson and W. Jost) Academic Press
Nicholls, R.W., Can. J. Chem. 47, 1847, 1969b
Nicholls, R.W. and Parkinson, W.H., J. Chem. Phys. 26, 423, 1957
Nicholls, R.W., Parkinson, W. H. and Reeves, E.M., Applied Optics 2, 919, 1963
Nicholls, R.W. and Tyte, D.C., Proc. 5th International Shock Tube Symposium 111-121, 1965, US Naval Ordance Lab. Silver Spring, Md.
Parkinson, W.H., and Reeves, E.M., Can. J. Phys. 41, 702, 1963
Parkinson, W.H. and Reeves, E.M., Proc. Roy. Soc. A 282, 265, 1964
Patch, R.W., JQSRT 5, 137, 1965
Penner, S.S., Quantitative Molecular Spectroscopy and Gas Emissivities (Adison Wesley) 1959
Penzias, G.J., Dubin, S.A. and Kruegle, H.A., Applied Optics 5, 225, 1966

Pery-Thorne, A. and Banfield, F.P., 1970, J. Phys. B $\underline{3}$, 1011, 1970
Reis, V., JQSRT $\underline{4}$, 783, 1964
Reis, V., JQSRT $\underline{5}$, 585, 1965
Rich, J.C., Ap. J. $\underline{148}$, 275, 1967
Rothe, D.E., JQSRT $\underline{9}$, 49, 1969
Russo, A.L. and Watt, W.S., Infra Red Measurement of Vibrational Excitation and De-excitation Rates of N_2 using CO Additive Cal AF-2567-A-1, 1968. Cornell Aeronautical Laboratory, Buffalo, NY.
Seal, K.E. and Gaydon, A.G., Proc. Phys. Soc. $\underline{89}$, 459, 1966
Shackleford, W.L., JQSRT $\underline{5}$, 303, 1965
Smekhov, G.D. and Losev, S.A., High Temp. $\underline{6}$, 369, 1968
Sulzmann, K.G., JQSRT $\underline{4}$, 375, 1964
Sviridov, A.G., Sobolev, N.N. and Sulowski, E.M., JQSRT $\underline{5}$, 525, 1965
Tatum, J.B., Astrophys. J. Suppl. $\underline{14}$, 21, 1967
Taylor, R.L. and Bitterman, S., Rev. Mod. Phys. $\underline{41}$, 26-47, 1969
Taylor, W.H. and Kane, J.W., Applied Optics 6, 1493, 1967
Thomas, G.M. and Menard, W.A., AIAA Journal $\underline{4}$, 227, 1966
Tousey, R., Quart. J. Roy. Astron. Soc. $\underline{5}$, 123, 1964
Tyte, D.C., Nature $\underline{202}$, 383, 1964
Tyte, D.C., Proc. Phys. Soc. $\underline{92}$, 1134, 1967
Tyte, D.C. and Nicholls, R.W., Identification Atlas of Molecular Spectra 1. Blue green system of AlO, 1964a
Tyte, D.C. and Nicholls, R.W., 1964b, Identification Atlas of Molecular Spectra 2. The First Negative System of N_2^+
Tyte, D.C. and Nicholls, R.W., 1965, Identification Atlas of Molecular Spectra 3. The Second Positive System of N_2
Tyte, D.C., Innanen, S.H. and Nicholls, R.W., 1967, Identification Atlas of Molecular Spectra 5. The C_2 Swan System
Vidal, C.R. and Cooper, J., J. Appl. Phys. $\underline{40}$, 3370, 1969
Vincenti, W. and Kruger, C.H., Introduction to Physical Gas Dynamics, Wiley, 1965
Watson, R., JQSRT $\underline{4}$, 1, 1964
Watson, R. and Ferguson, W.R., JQSRT $\underline{5}$, 595, 1965
Weeks, Dorothy W. and Simpson, Edwina A., Absorption Spectrum of Iron in the Vacuum Ultraviolet 2950-1588A, Scientific Report No.19, Shock Tube Laboratory, Harvard College Observatory, 1967
Whiting, E.E., Arnold, J.O. and Lyle, G.C., JQSRT $\underline{9}$, 775, 1969
Wolnick, S.J., Berthel, R.O., Larson, G.S., Cornevale, E.H. and Wares, G.W., Phys. Fluids $\underline{11}$, 1002, 1968
Wood, A.D., Hoshizaki, H., Andrews, J.C., Wilson, K.H., AIAA Journal $\underline{7}$, 130, 1968
Wray, K.L., J. Chem. Phys. $\underline{44}$, 623, 1966
Wray, K.L., JQSRT $\underline{9}$, 255, 1969
Wray, K.L., and Connolly, T.J., JQSRT $\underline{5}$, 111, 1965
Wurster, W.H. and Marrone, P.V., JQSRT $\underline{7}$, 591, 1967
Zel'dovich, Y.B. and Raizer, Y.P., Physics of Shock Waves and High Temperature Hydrodynamic Phenomena Vols. 1 and 2, Academic Press, New York, 1966

RADIATIVE DIAGNOSTICS IN NONEQUILIBRIUM FLOWS

Walter H. Wurster

Cornell Aeronautical Laboratory, Inc.

Buffalo, New York

INTRODUCTION

Conventional shock tube spectroscopy generally proceeds along two lines. One of these has been stressed in the previous paper, in which the shock tube is used to provide a method for processing gases to high temperature equilibrium conditions, under which the radiation from the gases can be measured, and the spectroscopic parameters of the radiating species thereby deduced. The second aspect is that of using known radiating systems as diagnostics to deduce the kinetics of various gasdynamic or molecular processes. In this paper, examples of each will be discussed, with the emphasis placed on nonequilibrium measurements. These examples are taken in part from a current research program at this laboratory, wherein the objective of the work is to obtain the data necessary to calculate the vacuum ultraviolet (VUV) radiation flux behind strong shock waves in air. To do this, the problem involves two tasks: (a) to obtain the spectral distribution and the associated transition probabilities for the radiation, and (b) to measure the excitation rates that govern the populations of the relevant electronic states.

Because the spectral region of interest extends to 700A, which is well below the transmission limit of available window materials, these experiments employ several unique features which may be of interest at a Symposium dealing with optical methods in gasdynamic research. The discussion in this paper will be centered upon two specific techniques, namely the measurement of radiation profiles behind strong shock waves using narrow field of view radiometers, and the measurement of absorption spectra at wavelengths below 1000A. The data that are presented are current, and primarily in support of the validation of the experimental techniques; final measurements under the program have yet to be made.

NONEQUILIBRIUM EMISSION MEASUREMENTS

General Discussion and Instrumentation

At the temperatures of interest to the research problem, the most likely species capable of radiating in the VUV is molecular nitrogen, with transitions taking place from high lying electronic states. These states lie well above the dissociation energy of nitrogen, resulting in conditions that directly highlight the nonequilibrium nature of the problem.

Immediately behind a strong shock wave in air, the translational temperature is very high - approximately 18,000°K in the case of a 15 in. diameter sphere reentering at 22,000 ft/sec. This temperature very quickly begins to drop as the energy is fed into the dissociation of oxygen and nitrogen, and into the internal modes of excitation of the various gaseous species. The final equilibrium temperature is about 7000°K. The entire nonequilibrium flow behind the shock is therefore characterized by strong gradients in temperature, species concentrations and internal mode energy content, each of which affect both the spectral distribution and intensity of the emitted radiation. The more fundamental question arises also, of whether the electronic states of nitrogen at energies in excess of 12 eV will be populated and radiate before the molecules dissociate.

Some of the relevant energy levels of the nitrogen molecules are shown in Fig. 1. The dissociation energy of nitrogen is also indicated in the figure. As seen, states with energies in excess of 12 eV are those of interest. The excited states in nitrogen are either singlet or triplet in the characterization of the electronic spin, and it can be seen that the allowed transitions to the ground state (no spin change) involve the states labeled $b\,^1\Pi$ and $b'\,^1\Sigma$. These and several others (not shown) have been studied in absorption, and their spectroscopic constants are known (1,2). No intensity data are available, however, for the spectral distribution or strength for wavelengths below 1000A. Most radiation data in nitrogen have involved transitions among the triplet states, the most notable being those of the (1+) system shown in the figure. These bands are readily excited in the shock tube, and have been the subject of a number of studies (3-6). Although several of these studies have demonstrated the nonequilibrium excitation of the triplet states, i.e., $B\,^3\Pi$, the only quantitative measurements have been reported by Wray (5). In measurements designed to

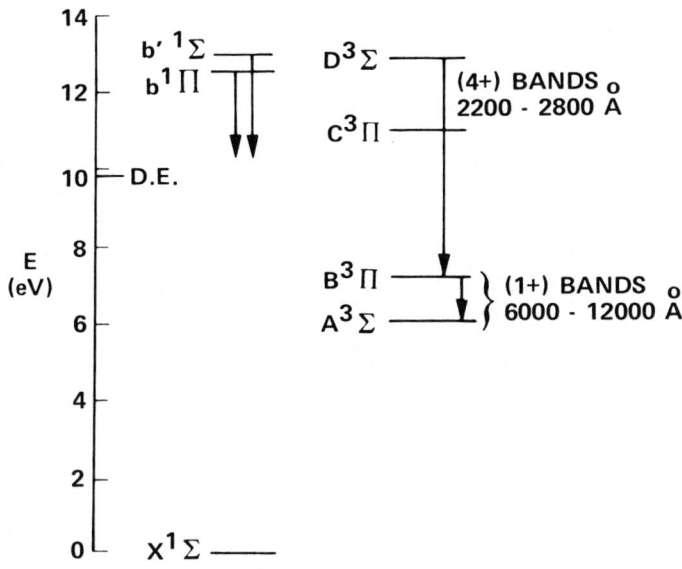

Fig. 1   Relevant Energy Levels in Nitrogen

elucidate the excitation mechanism for this state, he found that N atoms were some 100 times more effective than nitrogen molecules in collisionally populating these states. This higher efficiency is to be expected, since the molecule-atom collision can accommodate the required spin change.

The objective of the present studies is to determine the excitation mechanisms and the associated rates for the population of the singlet states. For this purpose it was planned to make emission measurements of the radiation from these states as they become populated by collisions taking place behind strong shock waves. Using the radiation as a diagnostic in such measurements relies upon the assumption that the radiation associated with a given transition is proportional to the instantaneous population of the upper energy level involved. This is generally the case, since at usual shock tube pressures the collision rates are sufficient to maintain local thermodynamic equilibrium throughout the shocked gas.

As an aid in analysis, and for direct comparison, it was decided also to measure simultaneously the populations of the $B^3\Pi$ and $D^3\Sigma$ states. The former is useful because of the capability of comparison with other available data from past studies. The $D^3\Sigma$ radiation had not heretofore been seen in shock tube studies, but was felt to be potentially useful because it represents a triplet state with an activation energy in excess of 12 eV, near that of the singlet states of interest.

The choice of wavelengths useful for the measurement of the populations of these states depends primarily on the knowledge of the associated spectra, and to a lesser extent on instrumentation parameters. The radiation from the (1+) system in shock-heated nitrogen is well documented (3) and posed no problems. The data of Ref. 7 were used as a guideline for the design of the radiometer to measure the (4+) band system. Several other considerations, however, were involved in the VUV measurements of the singlet states. A calculated "synthetic" spectrum for the (b' $^1\Sigma$ - X $^1\Sigma$) band system of nitrogen is shown in Fig. 2.

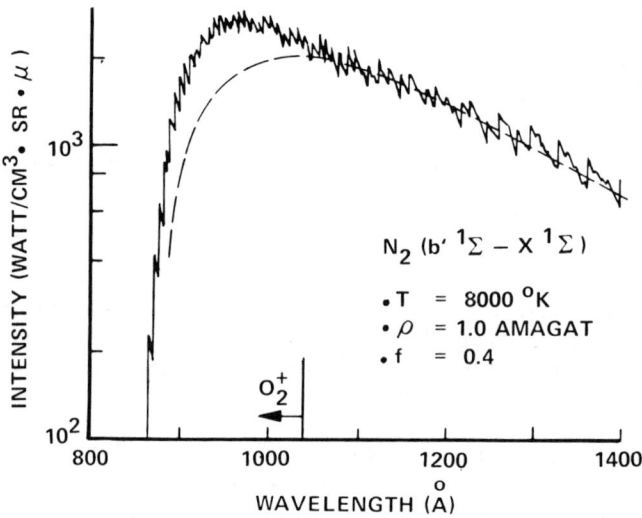

Fig. 2 Calculated Spectrum of the Nitrogen (b' $^1\Sigma$ - X $^1\Sigma$) Band System

This calculation was made using a computer program developed at this laboratory (8) and which used as inputs the recent spectroscopic parameters of Ref. 2.* Conditions were taken to be 8000°K, unit density and an arbitrary f-number of 0.4. An earlier calculation by Allen (9), based on older data, is shown as

---

*We are indebted to Prof. R. W. Nicholls of York University, Toronto, Canada for supplying the vibrational overlap integrals.

the dotted line. It can be seen that the difference is significant at wavelengths below 1000A, and that the determination of the radiation in this region requires quantitative spectral measurements. This question is addressed later in this paper.

For experimental convenience, it was decided to make the preliminary measurements of the VUV radiation from the b $^1\Pi$ and b' $^1\Sigma$ states at wavelengths above 1050A. This permitted the use of lithium fluoride windows in the shock tube and over the photomultiplier face. These measurements were meant primarily to see whether in fact these VUV states radiated at all. Final kinetic measurements will be made in the windowless spectral region below the 1050A LiF cutoff wavelength.

Three radiometers were designed and installed in a cruciform configuration at the same station in the shock tube. Data were thereby obtained from the same sample of gas that had been processed by the incident shock wave. Interference filters with 100A bandwidth, centered at 6600A and 2500A were used to define the wavelength intervals for the (1+) and (4+) band systems, respectively. The VUV radiometer bandpass was determined by the magnesium fluoride transmission limit at 1150A and the EMR 641G-09-18 phototube response cutoff between 1600 and 1800A.

The electronic time constant of the radiometers was kept below that determined by the spatial resolution of the optical system, which was about 1/4 μsec. The question of optimizing the signal-to-noise ratio in the system involves tradeoffs between spatial resolution and wavelength resolution in the face of low level and/or short duration light signals.

Fig. 3 demonstrates an interesting feature of these narrow field of view radiometers. The passing shock front is shown schematically, and the associated rising radiation front is the profile which constitutes the desired data. Thus, the spatial resolution defined by the field of view of the detecting system, should be narrow in comparison with the extent of the profile.

Two optical systems are shown in Fig. 3. Each utilizes two equal collimating slits, S, and each has the same light-gathering power (equal solid angles α and aperture slits, also S). The extreme rays in system A are shown as solid lines, and are seen to spread significantly in the shock tube. In system B we add a lens as shown, whose focal length is equal to the radius of the shock tube. It can be seen that this lens focuses the detector slit on the wall at the far side of the shock tube. The extreme rays in this case are thereby restricted as shown, and considerably enhance the spatial resolution of the system. Results of profile measurements with each of such optical systems will be shown later.

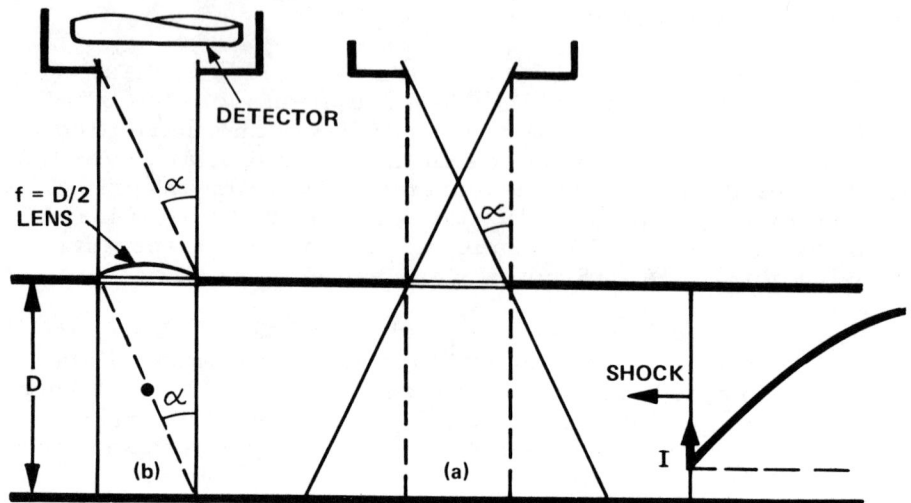

Fig. 3   Comparison of Two Radiometer Fields of View

It is of interest to consider the distribution of the radiation pattern in each of the systems. For this, one pictures each field of view as homogeneously filled with radiating particles. It can be seen that each particle along the near shock tube wall radiates uniformly into the detectors with a solid angle α over the entire extent of the aperture S. For system B, this is also true for the particles along the far wall of the tube. For system A, however, the radiators along the far wall and between the dotted lines radiate into a solid angle of $\alpha/4$, because the distance has doubled. Between the dotted lines and the extreme rays increased vignetting gradually reduces the effective solid angle to zero.

The largest nonuniformity in spatial response for system B takes place on the shock tube centerline. A particle in the center of the system has an effective solid angle of 4α, which also decreases to zero as the extreme rays of the system are approached.

Thus it can be seen that in no system are the particles equally weighted. The conclusion to be drawn is that once the spatial resolution has been specified, one is no longer permitted to ask for finer detail, because the system effectively does a nonuniform averaging over the particles contained in the overall field of view. In the cases to be discussed here, the most highly resolved data were taken with S = 1.0 mm, corresponding to an effective time constant of about 1/4 μsec.

A final note on the instrumentation concerns the use of slits near the windows in the shock tube. Conventional methods such as tape or razor blades were found to scatter light appreciably as the intensely radiating shock approached the observation station. This has the effect of causing a "foot" in the records at times prior to shock arrival, and distorts the leading edge of the radiation profile. The best solution found was to form the slits by aluminizing the windows themselves by vacuum deposition. This was not done to the LiF windows, and the effect of edge scattering will be seen in those records.

Measurements

Fig. 4 shows some typical results for the 6600A radiometer, monitoring the population of the $B\,^3\Pi$ state. Nominal shock tube operating conditions were an initial pressure p = 1.0 torr and a wave speed V = 15,000 ft/sec. Figure 4(a) is a record in pure nitrogen with a sweep speed of 10 μsec/cm. The complete profile is shown at two gains, showing the strong radiation overshoot, the decay due to the falling translational temperature, and the termination of test time as the driver gas interface passes the viewing station. Details of the leading edge profile are shown in Fig. 4(b), recorded at 2 μsec/cm. The good spatial resolution is evidenced by the clean break at shock wave arrival.

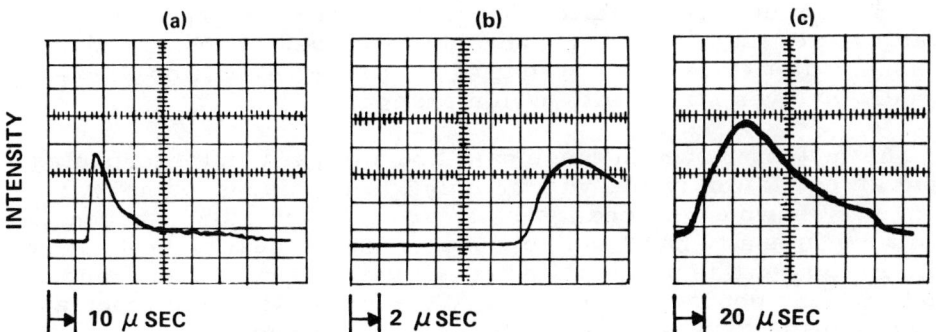

Fig. 4   Nonequilibrium Nitrogen (1+) Radiation Profiles.
(a), (b): Nitrogen, (c): 10% Nitrogen/90% Neon.
All Pressures: 1.0 Torr; Shock Speed: 15,000 Ft/Sec.

By way of comparison, Fig. 4(c) shows a similar record in a 10% nitrogen-90% neon mixture at 20 μsec/cm. It can be seen that the time to peak radiation has increased from about 3 to some 40 μsec. This lag is the result of a slower nitrogen dissociation rate in Ne, which inhibits N atom formation, and hence the rate of population buildup of the B $^3\pi$ state.

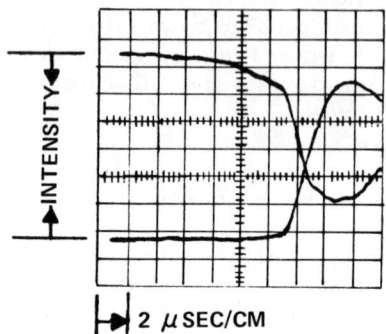

Fig. 5 Radiation Profiles in the Vacuum Ultraviolet. Upper Trace: 1300A; Lower Trace: 2500A. Initial Nitrogen Pressure : 1.0 Torr; Wave Speed: 15,000 Ft/Sec.

Records from the other two radiometers are shown in Fig. 5. The trace for the 1300A channel can be seen to have its leading edge severely distorted by slit-edge scattering of the radiation. Despite this resolution loss, it can still be seen that the 1300A radiation peaks ahead of that at 2500A. Each of these record radiation from electronic states at about the same energy, and so this faster excitation rate is reasonably explained by the fact that the 1300A radiation is an allowed singlet-singlet transition, while the 2500A involves states with a spin change.

The method by which these data are analyzed and interpreted to yield an excitation model will be only briefly discussed here. In Fig. 6 is shown a reduced intensity profile curve for nitrogen, wherein the intensity has been normalized by the peak overshoot value. Also shown is a calculated curve using an excitation model and rate for the population profile of the B $^3\pi$ state. The calculations were made using a nonequilibrium normal shock computer program (10), in which the excited states are treated as separate species and are assigned various rates for possible collision partners. For the case shown, both molecular and atomic nitrogen, as well as electrons were included as collision partners. Immediately behind the shock, the molecular nitrogen impact mechanism dominates. However, the N excitation soon

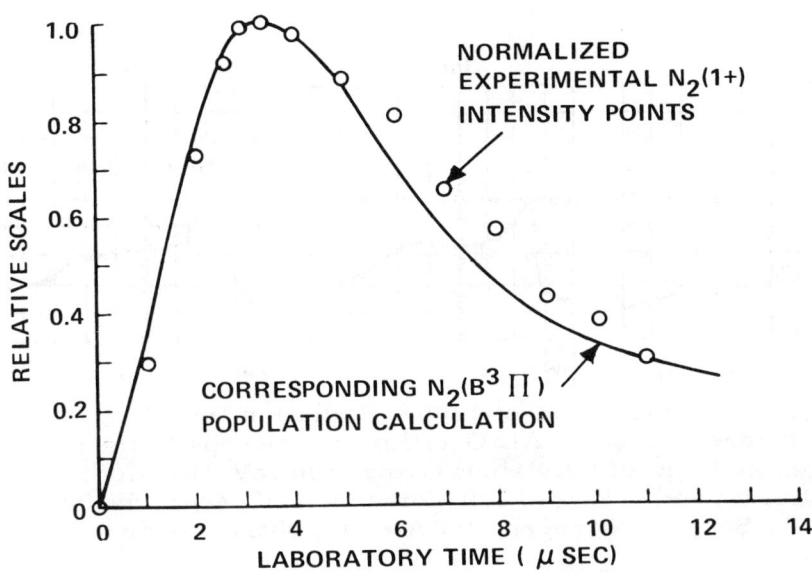

Fig. 6 Comparison of Excitation Model Calculation With Experimental Radiation Data

increases and becomes the dominant mechanism; its maximum effectiveness is approximately a factor of 3 greater than the electron impact mechanism for this wave speed. It is the matching of the computed curves for different mechanisms and rates with the experimental data that defines a consistent excitation model for the electronic states. As can be seen, the problem of the uniqueness of a derived model arises, since so many parameters are available for adjustment to the data.

For this reason, profile data are being taken for a number of different gas mixtures over a broad range of test conditions. Only by matching such an extended range of varied test data with the same model can a unique and consistent model be deduced. To date, for example, the 6600A profiles in pure nitrogen, and mixtures of nitrogen and neon or argon have been successfully matched by this technique. Further analysis is being pursued on tests in air and NO, whose radiation profiles in the VUV are shown in Fig. 7. Several interesting features are evident in the records. In going from (a) nitrogen to (b) air, it is clear that the radiation overshoot is much less. This effect could be anticipated because the translational temperature drops much more rapidly behind a shock wave in air, due to the faster oxygen dissociation rate. A feature yet unexplained however, is the steady plateau of radiation prior to the end of the test time. At these high

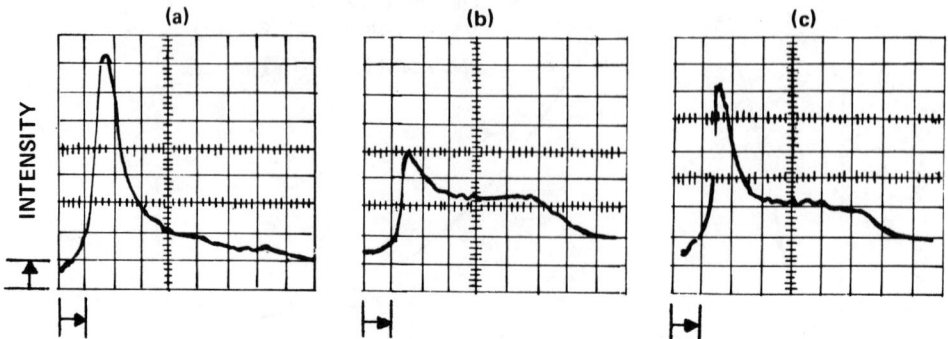

Fig. 7 Comparison of 1300A Radiation Profiles in Different Gases. All Oscilloscope Sweeps to the Right at 10 µSec/Division; Gains: 100 mV/Division; Initial Gas Pressures: 1.0 Torr; Shock Speed: 15,000 Ft/Sec; (a) Nitrogen, (b) Air, (c) Nitric Oxide

temperatures needed to excite radiation at these wavelengths, it would be expected that all the diatomic species would have dissociated appreciably. An identical shock tube test made using pure NO as the test gas, yields the record of 7(c). It can be seen that the overshoot is higher and steeper, which is reasonable on the grounds that in this case NO can be thermally excited prior to dissociation, while in air the NO must be chemically formed after oxygen dissociation. The radiation plateau, however, is still present, and its origin is the subject of current efforts. It seems definite, however, that there exists a radiating species in air and NO that is not present in the pure nitrogen tests.

By way of contrast, the record from a test in pure oxygen is shown in Fig. 8, for the 2500A radiometer. This wavelength region encompasses the bands of the Schumann-Runge band system. The striking feature here is that oxygen does not exhibit a radiation overshoot. This effect was already known; the upper electronic state crosses one with a repulsive potential. Thus, in this case the excited molecule has an additional fast path to dissociation which predominates over the radiative transition to the ground state. Finally, as equilibrium between molecules and atoms is approached, the radiation reaches its usual plateau value. The lack of an overshoot in oxygen also demonstrates the overall purity of the shock tube system. Impurities such as those caused by pump oil, poor vacuum levels or excessive outgassing generally result in "impurity spikes" at the shock front. These are readily recognizable and clearly could not be tolerated in cases where the actual radiation front is under study.

RADIATIVE DIAGNOSTICS

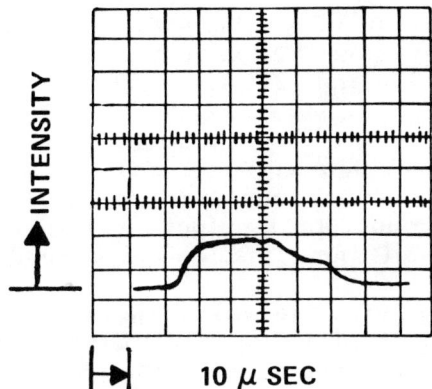

Fig. 8 2500A Radiation Profile in Oxygen. Initial Pressure: 1.0 Torr; Shock Speed: 15,000 Ft/Sec

Summarizing this section, the objective of these kinetic tests is to obtain an excitation model and the corresponding reaction rates for these high lying energy states in nitrogen. The necessary self-consistency and uniqueness for the model is achieved by testing the model against a number of cases involving different gases under varying conditions.

ABSORPTION MEASUREMENTS IN THE VACUUM ULTRAVIOLET

As was mentioned in the introduction, one of the tasks of the program is the determination of the spectral distribution and the associated transition probabilities for the radiation from nitrogen in the vacuum ultraviolet. Fig. 2 shows a calculated spectrum from one of the band systems and demonstrates the need for spectrally resolved data below 1000A which are vital to the determination of the radiative flux from shock waves. Further, the data must be comprehensive enough to permit the spectral intensity to be calculated for the entire range of temperatures encountered behind the shock wave, so that suitable integrations through the shocked gas can be performed.

It was decided to make these measurements in absorption, basically because these could be equilibrium measurements, with longer testing times than those afforded by the nonequilibrium emission profile experiments described in the previous section. The techniques chosen to obtain the data are somewhat unique and hence may be of general interest in gasdynamic-optical testing methods.

The general plan is to heat nitrogen to modest temperatures (4-8000°K), which would populate the vibrational levels of the ground state, and to measure the absorption spectra as a function of wavelength between 700 and 1200A.

One prime question involves the choice of a suitable light source, capable of delivering a strong continuum in this wavelength region. Normal VUV spark sources are inapplicable, and flash lamps, even when reproducible in output, lose their continuum nature and generally emit line radiation below 1000A.

To solve this problem, use was made of the results of a previous research program (11), conducted for the NASA, in which the radiation from nitrogen was measured at temperatures between 11-13,000°K. The source of radiation in this case is the radiative recombination of the nitrogen atomic ion with electrons. This radiation was shown to provide a strong continuum from 1100 to 700A. Thus, because both the light source and the test gas utilized nitrogen, the absorption experiment shown schematically in Fig. 9 was devised.

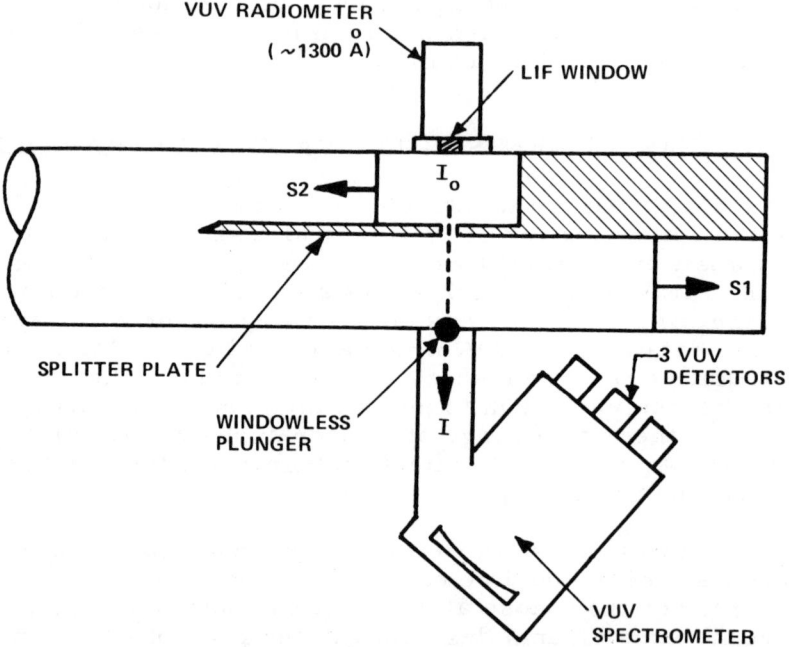

Fig. 9 The Vacuum Ultraviolet Absorption Experiment.
S1: Incident Shock Wave; S2: Reflected Shock Wave;
I_o: Reflected Shock Light Source at 11,300°K

A splitter plate was mounted in the end of the shock tube as shown. It divides the incident shock wave, which then proceeds down the two channels. One channel is obstructed, reflecting the shock, and produces a pocket of hot gas which serves as the light source. This light passes through an aperture in the plate and through the gas in the second channel, which has been heated by the incident shock only. Measurements of the spectrally resolved absorption spectra due to the heated gas are obtained using a multi-channel VUV spectrometer. The spectrometer is coupled to the shock tube using an explosively driven plunger, which acts as both valve and shutter, and exposes the spectrometer to the test gas for a period of about 120 μsec without interposed windows. The detectors are bare, windowless photomultiplier tubes operating in the vacuum of the spectrometer chamber. These tubes have no response at wavelengths longer than 1600A, and hence aid in spectral rejection of the strong near-UV and visible radiation from the shock tube. The system is thereby capable of making windowless measurements below 1000A. The spectrometer and plunger system are described in detail in Ref. 11.

In checking the operation of the splitter plate, the VUV radiometer depicted in Fig. 9 was deployed as shown, and also at the position shown occupied by the VUV spectrometer. In this way, measurements for system evaluation were made at 1300A. The same set of measurements were also made in the visible radiation at 5200A. These measurements confirmed the predictions that for the visible wavelengths, the radiation signal is the same, no matter which side of the splitter plate the detector was placed. This results from the fact that the shock-heated nitrogen in the test channel is completely transparent at 5200A. On the other hand, the 1300A radiation was found to be absorbed by about 50% when passed through the test gas.

Radiation records for the overall system are shown in Fig. 10. The light source output at 1300A is shown in the lower trace. Its features are the same in the visible wavelength region, and demonstrate the usefulness of the system as a light source during the test time of about 100 μsec.

Also shown, in the upper trace, is a record of the radiation recorded at 760A, which has passed through the test gas and been partially absorbed. By comparing this record with one taken without absorption (by rotating the splitter plate 180° in the shock tube), the net absorption can be obtained.

The overall system has just recently been tested and preliminary data obtained, of which these records are typical. To date, absorption measurements at 760, 1000 and 1088A have been made, with 20A resolution. Additional diagnostics, such as sidewall heat transfer measurements in the channel, have also been

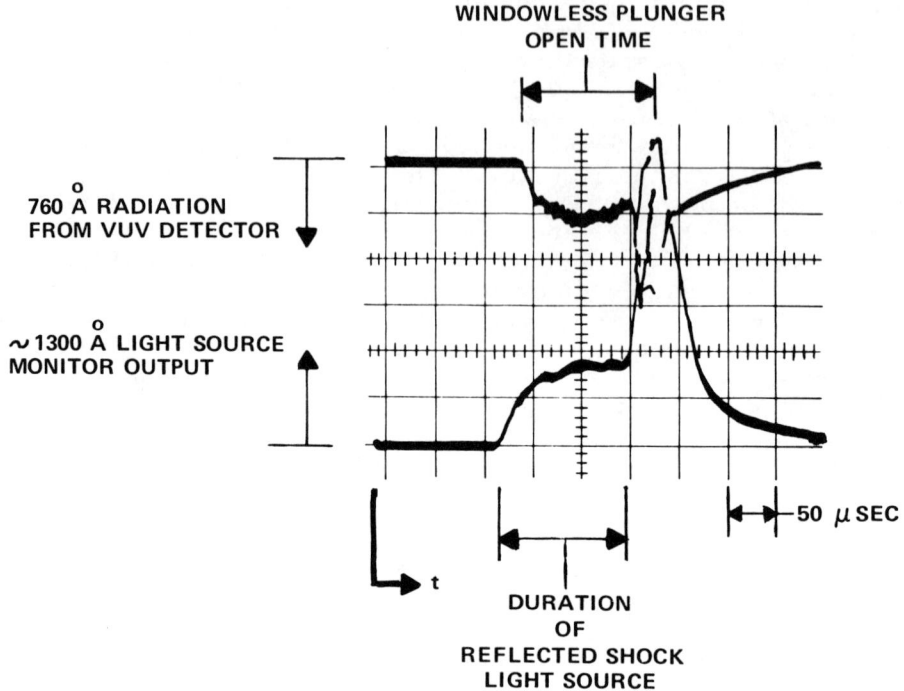

Fig. 10 Vacuum Ultraviolet Absorption Records

used. Together with the agreement of the measured absorption at 1088A, taken to compare with the only available measurements (12), they have shown the entire system to perform as planned. No adverse gasdynamic effects in the two channels or in the aperture have been found. This splitter plate technique thus appears to be a useful new tool in shock tube absorption spectroscopy.

In concluding this section, it should be remarked, that present plans include a modification to the system which will permit nonequilibrium emission spectra to be obtained with the spectrometer. For these studies the spectrometer will be viewing the advancing incident shock axially through an aperture in the end of the shock tube. Preliminary tests with the radiometers indicate the signal levels are sufficient to permit wavelength resolution of the emission radiation from the shock front. Indeed, the measurements to date show a steady radiation signal (once the detector field of view is filled), which corresponds to a pseudo-steady state measurement of a highly nonequilibrium gas sample. By this method one effectively has about 100 μsec of testing time to record the spectrum of the overshoot profile that passes a sidewall viewing port in about 10 μsec.

The VUV spectrometer measurements will thus provide a conclusive overall check on the entire program; they would be compared with calculations of the emitted intensity from the non-equilibrium gas, where the inputs to the calculation include both the excited state populations via the derived excitation model, as well as the measured absorption coefficients.

SUMMARY

The techniques developed in the course of current research on molecular excitation mechanisms and absorption spectroscopy in the VUV wavelength region have been used to demonstrate several features of optical diagnosis and measurements in various gasdynamic systems. In particular, a discussion of the use of narrow field of view radiometers and a method for the windowless viewing of shock-heated gaseous plasmas has been presented.

ACKNOWLEDGEMENTS

It is a pleasure to acknowledge the contributions of my friend and colleague Dr. Paul V. Marrone, who shares these research efforts with me.

REFERENCES

1. Carroll, P. K. and Collins, C. P., Can. Journ. Phys., Vol. 47, 563-589 (1969).
2. Dressler, K., Can. Journ. Phys., Vol. 47, 547-561 (1969).
3. Wurster, W. H., J. Chem. Phys., Vol. 36, 2111-2117 (1962).
4. Wray, K. L. and Connolly, T. J., J. Quant. Spect. Rad. Transfer, Vol. 5, 111 (1965).
5. Wray, K. L., J. Chem. Phys., Vol. 44, 623-632 (1966).
6. Smekhov, G. D. and Losev, S. A., High Temperature, (USSR) Vol. 6, 369-376 (1968).
7. Pearse, R. W. B and Gaydon, A. G., The Identification of Molecular Spectra, Chapman and Hall, Ldt., London, 3rd Edition, 1963.
8. Williams, M. J. and Treanor, C. E., Cornell Aeronautical Laboratory Rept. No. QM-1626-A-5, 1962.
9. Allen, R. A., Air Radiation Tables: NASA Contractor Report NASA-CR-557, 1966.
10. Marrone, P. V., Cornell Aeronautical Laboratory Rept. No. QM-1626-A-12(I), 1963.

11. Marrone, P. V. and Wurster, W. H., to be published in J. Quant. Spect. Rad. Transfer.

12. Appleton, J. P. and Steinberg, M., J. Chem. Phys., Vol. 46, 1521-1529 (1967).

TEMPERATURE ERRORS IN PLASMA DIAGNOSTICS AND THEIR POSSIBLE EFFECTS ON ABSOLUTE TRANSITION PROBABILITIES

Gordon W. Wares, Stanley J. Wolnik, and Robert O. Berthel

Air Force Cambridge Research Laboratories

Bedford, Massachusetts   01730

INTRODUCTION

A number of errors may occur in plasma diagnostics. Some of these errors are insidious, interact with one another, and tend to be self-perpetuating, their effects even being amplified in certain later applications. Self-absorption, lack of local thermodynamic equilibrium, systematic error of intensity measurement with respect to wavelength, incorrect absolute transition probabilities, and plasma temperature errors are among the commonest offenders. These have become inextricably intertwined with one another and frozen into the largest and most used compendium of what is probably the most crucial plasma diagnostic parameter, the absolute transition probability or f-value. We refer, in particular, to the Corliss and Bozman N.B.S. Monograph 53 of 1962, which gives some 25,000 experimental transition probabilities for 112 neutral and singly ionized species of 70 elements; also, to subsequent lists by Corliss and collaborators.

These tables have been very widely used for species concentrations, temperatures, etc., in physics and engineering and for stellar and solar abundances in astrophysics. Accordingly, the whole literature is suspect. The situation before 1962 was no better since relatively very few experimental transition probabilities were available and the errors in these, particularly in the King-type absorption furnace were a major contributor to the errors in the N.B.S. tables. Only a few previous measurements made by radiative lifetimes, anomolous dispersion, and atomic beam avoided these errors. Theoretical values were not numerous owing to lack of computer facilities and even they were often normalized by experimental values and became somewhat tainted thereby, in addition to their own well

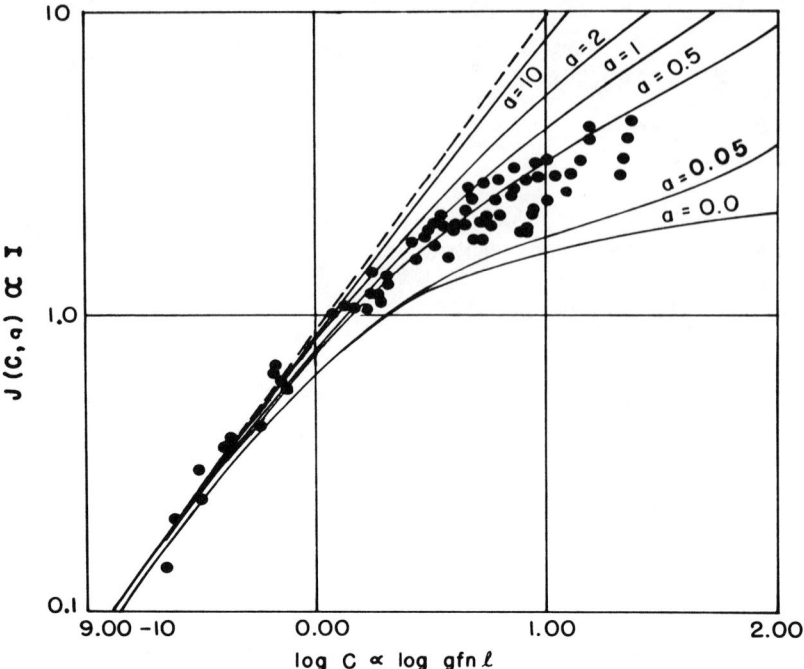

Fig. 1. Experimental curve of growth for Cr I: $n\ell$= concentration (cm^{-3}) X path length (cm); a = damping constant $\Delta v_L/\Delta v_D$. Reproduced by permission of The Astrophysical Journal, published by the University of Chicago Press.

known inherent errors. Recently, wall-stabilized arcs, shock tubes in which temperatures are measured, and more numerous radiative lifetime experiments as well as new ones such as beam-foil spectroscopy have begun to provide fairly reliable experimental absolute transition probabilities, even in the vacuum and extreme ultraviolet. These new values have recently led to a dramatic resolution of a paradox regarding the apparent 5- to 10-fold under-abundance of iron and chromium in the solar photosphere.

DETERMINATION OF PLASMA TEMPERATURE FROM RELATIVE gf-VALUES

Excitation temperatures of luminous plasmas in local thermodynamic equilibrium (LTE) can be determined from observed weak spectral lines, provided accurate relative transition probabilities (oscillator strengths or gf-values) are available. Knowledge of the relative line intensities (I) is required for a sufficient number of lines (of wavelength λ and statistical weight g of the initial state), which can normally be determined by standard methods of spectrophotometry. In a plot of log $I\lambda^3$/gf versus E_u (upper excitation potential), the slope of the least squares straight line fit is proportional to -5040/T and determines the temperature T. Obviously, a considerable range in excitation potential is desirable. Also, it is safest if this range in excitation potential is obtained with a sufficient number of lines over a short wavelength range in order to minimize the effects of any systematic errors in the intensity measurements with respect to wavelength.

Careful examination of these plots may point out strong lines that fall below the straight-line best fit because of self-absorption, although initially these lines were thought to be satisfactorily "optically thin". The safest way, however, is to do as we have recently done for Cr I in a shock tube plasma (Wolnik, Berthel, Carnevale, and Wares 1969): we actually derived a "curve of growth" (such as is illustrated in Figure 1 for Cr I) of intensity for the spectral lines of a given species as if we were going to use this derived curve to make corrections to partially self-absorbed lines. The latter fall farther and farther below the extension (dashed line in Fig. 1) of the linear portion as the effective number of oscillators $gfn\ell$ in the initial state increases and/or have smaller damping constants a. The parameter $a \equiv \Delta\nu_L/\Delta\nu_D$ is an inverse measure of how readily a line "saturates" due to the rapid-saturation tendency of the exponential line wings of the Doppler broadening component relative to the slow-saturation tendency of the quadratic wings of the Lorentzian broadening component. Having derived the experimental curve of growth for our plasma conditions we used it as a guide for selecting combinations of trace species concentration and degree of ionization thereof that would give us values of $gfn\ell$ corresponding to the linear portion of the curve of growth. The flexibility of shock tube work is such that we were able to vary n over a range of more than two hundred while keeping the product $gfn\ell$ in the linear region for both strong and weak lines, thereby demonstrating that our plasma was "optically thin" under the conditions actually used. Inspection of Figure 1 shows that our experimental damping constant (ignoring some scatter) lies in the range of a-value between 0.05 and 1.

An example of a possible systematic error in plasma diagnostics would be to calculate a theoretical a-value that would happen to be much larger than the experimental value would have been if it had actually been determined. The use of such an exaggerated value of a as the criterion of his linear region could lead the experimenter into serious self-absorption for his stronger lines. Since stronger and stronger lines are usually correlated with lower and lower excitation potential it will become apparent (Figs. 3 and 4) that such an error could simulate a temperature error in the resulting determination of gf-values.

Subsequent to such tests of optical thinness, it is necessary to check that LTE conditions prevail, i.e., that the excited states are populated in accordance with Boltzmann's law, and that they prevail throughout the light source actually used. This latter uniformity along the line of sight is one of the inherent advantages of the conventional shock tube relative to arcs, particularly free-burning arcs (Corliss and Bozman 1962, pp. XI-XIV; Takens 1970). Thus, although it is simple to plot log $I\lambda^3/gf$ versus E_u and determine a precise temperature, this technique must be used with care for meaningful temperatures to result, even if there is no question as to the accuracy of the relative gf-values. *(See note in proof.)

THE AFCRL SHOCK TUBE ABSOLUTE gf-VALUE PROGRAM USING ULTRASONICS

The first indication that serious systematic errors (presumably temperature errors) did indeed exist in generally accepted King furnace gf-value tabulations was provided by Wilkerson (1961), who used the Cr I relative gf-values of Hill and King (1951) to assign a temperature to his shock tube plasma. (In this paper "shock tube" always refers to a conventional pressure-driven shock tube in which a high-pressure driver gas compresses and heats a low-pressure driven gas that contains a trace amount of the experimental gas, the spectral lines of which are desired. Typically, the driver and driven gases are separated by a frangible diaphragm and remain separated until the experimental gas has been raised from a few mm of mercury pressure at room temperature to $\sim$ atmospheric pressure at temperature less than 15,000°K and has radiated in approximate LTE for many tens of microseconds behind the incident and/or reflected shock. Such conditions are favorable for providing the dominance of collisional interactions over radiative required for LTE.) Using these relative gf-values he obtained an excitation temperature of $\sim$5200°K, which contrasted with the temperature of $\sim$9300°K for the gas as derived from shock tube theory and the use of directly measured incident and reflected shock speeds and the assumption of LTE. Accordingly, when we began our long-term shock tube absolute gf-value program at AFCRL in the early 1960's we were determined to investigate LTE and the various sources of systematic error in the gf-value determinations, and in particular to make a concentrated effort to actually measure the temperature behind the reflected shock by optical line reversal (which was then a relatively new method for shock tubes) and/or by some independent method for every shock used for such determinations. We embarked on the development of a new method of ultrasonic temperature measurement, which was validated by simultaneous optical line reversal temperature measurements and found to be accurate and more convenient than line reversal for the conditions used in our gf-value research (Carnevale, Wolnik, Larson, Carey, and Wares 1967). In addition, it allowed us to make satisfying checks on the existence of LTE behind our reflected shocks. We are able to check that three "temperatures" measured simultaneously behind the same shock are the same: the gas kinetic temperature determined from the measured speed of ultrasonic pulses in a gas of known mean molecular weight and specific heat ratio, and the excitation temperature from line reversal. This latter excitation temperature was measured simultaneously for a high excitation line and for a low excitation line. The agreement of these two line reversal temperatures gave specific support to the LTE hypothesis that excited states are populated in accordance with Boltzmann's law. Slightly earlier, the Harvard College Observatory shock tube absorption group in studying autoionization in Ca I had given support to the hypothesis of LTE behind reflected shocks by demonstrating that the electron temperature agreed with a line reversal excitation temperature for the same shock (Garton, Parkinson, and Reeves 1966). The University of Maryland shock tube emission group under T. D. Wilkerson has also made a point of measuring more than one temperature behind the reflected shock, and their measurement of electron temperature has become an everyday tool.

ERRORS IN THE CORLISS AND BOZMAN NBS FREE-BURNING ARC gf-VALUES

In 1962 Corliss and Bozman (1962) published their comprehensive collection of some 25,000 absolute gf-values for 112 species of 70 chemical elements, which inspired numerous recalculations of laboratory plasma parameters and of astrophysical plasma parameters such as stellar abundances and solar photospheric abundances (Aller 1965; Grevesse 1969) without much change from the old calculations of Goldberg, Müller, and Aller (1960) who used gf-values mainly derived from absorption furnaces. Corliss and Bozman based their work on the Meggers, Corliss, and Scribner (1961) spectral intensity measurements in the N.B.S. low-current, free-burning copper arc. This arc was used as the source of excitation of a 1 atom per 1000 trace in the copper electrodes for each of the 70 elements in turn. The adopted N.B.S. arc temperature of $5100°K \pm 110°K$ was based on the average of some 31 free-burning arc temperature determinations (ranging from $3900°K$ to $6400°K$) in five laboratories in three countries. Both the adopted temperature and the absolute gf-values derive mainly from absolute gf-values determined in electric furnace absorption measurements for a number of elements.

That serious errors exist in this N.B.S. tabulation of absolute gf-values has been shown by the shock tube absorption work of Huber and Tobey (1968), who found discrepancies in the absolute scale of a factor of 10 or more for relatively high excitation lines of Fe I, Cr I and Cr II in the ultraviolet, and by our measurements for Cr I (Wolnik, et. al. 1968, 1969) in emission behind reflected shocks. Our data included a larger number of lines and large enough ranges in wavelength and in excitation to indicate a wavelength error (decreasing toward the red), which amounts to a factor of nearly 3 over our 3900Å to 5400Å range (Fig. 2, Wolnik, et. al. 1969), and an even larger excitation potential error amounting to a factor of ~5 over our range of upper excitation potential from 3.2 eV to 7.4 eV (Fig. 3, Wolnik, et. al. 1969), or, what amounts to the same thing, a

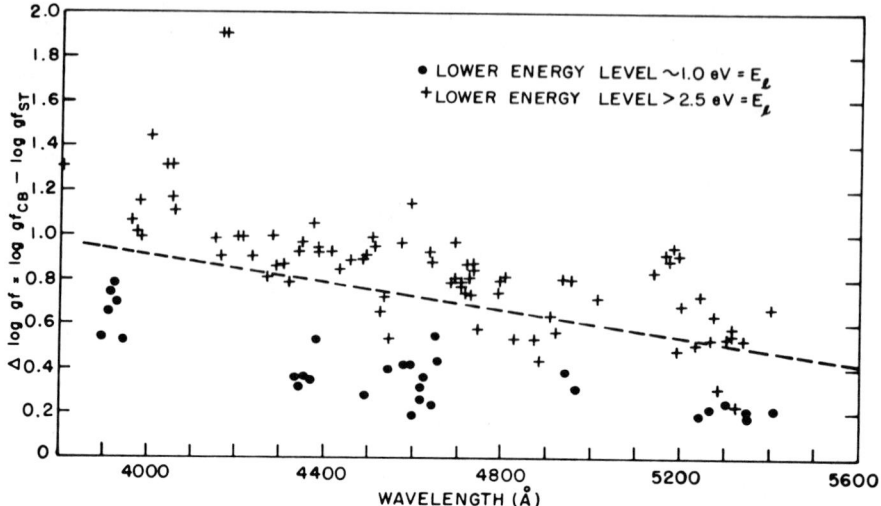

Fig. 2. A comparison of corrected Corliss and Bozman (CB) Cr I absolute gf-values with AFCRL shock tube (ST) measurements plotted against wavelength. Reproduced by permission of The Astrophysical Journal, published by the University of Chicago Press.

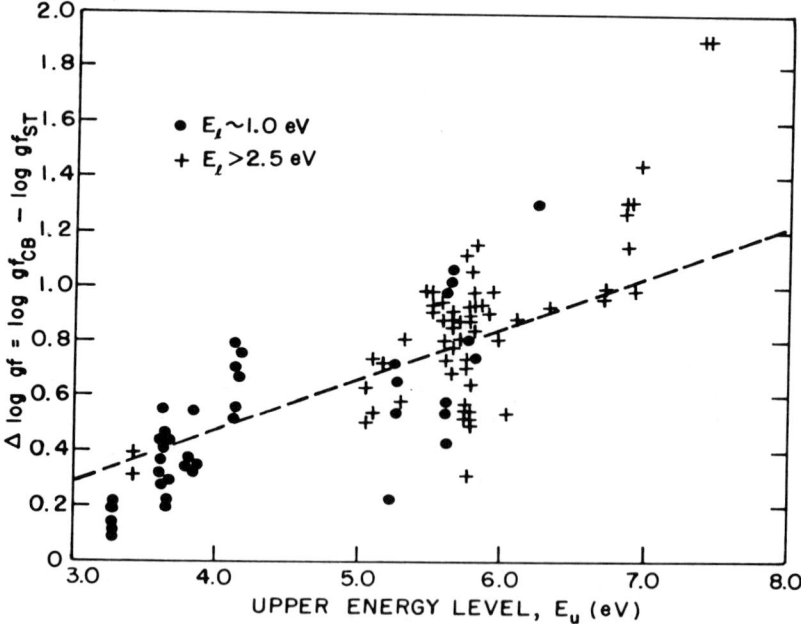

Fig. 3. A comparison of corrected Corliss and Bozman (CB) CR I absolute gf-values with AFCRL shock tube (ST) measurements plotted against upper energy level. Reproduced by permission of The Astrophysical Journal, published by the University of Chicago Press.

large temperature error in the N.B.S. arc. (We "correct" the Corliss and Bozman gf-values by removing their controversial "normalization function", as Corliss and Tech (1968) have done for Fe I. This function was an attempt to compensate for one of several departures from LTE. See Takens 1970). Using the N.B.S. spectral line intensities (Meggers, et. al. 1961) and our preliminary absolute gf-values for 30 Cr I lines in common, we derived an N.B.S. arc temperature of ~6900°K (Wolnik 1966), and from our published data (Wolnik, et. al. 1969) Corliss (1970) derived 6740°K. Our corresponding plot is shown in Fig. 4.

Our subsequent measurements for 118 Fe I lines (Wolnik, et. al. 1970) show from 89 lines in common with Meggers, et. al. (1961) a similar trend with excitation potential, indicating an arc temperature of ~6200°K, which is not appreciably changed by the addition of 24 faint unpublished lines of high excitation potential at longer wavelength. Figure 5 shows a plot of the combined data that gives 6209°K for the least squares determination. For Fe I, unlike the case for Cr I, there is only a slight systematic wavelength error indicated (in the same sense as for Cr I in Fig. 2) over the range from 3800Å to 5600Å (Wolnik, et. al. 1970). Here the separation of variables in the wavelength plot for Fe I is not as clear as in Fig. 2 for Cr I. Hopefully, iteration to partially correct the wavelength variation for excitation potential error and vice versa will move both values in the direction of reducing the temperature discrepancy with Cr I. *(See note added in proof.)

Our absolute gf-values for Fe I agree to within a factor of approximately 2 with gf-values reported by atomic-beam (Bell and Tubbs 1970), radiative lifetime (Whaling, Martinez-Garcia, Mickey, and Lawrence 1970) and wall-stabilized arc measurements (Garz and Kock 1969; Bridges and Wiese 1970). The wall-stabilized arc results include ~25 lines which can be directly compared. These comparisons show an excitation potential effect between the wall-stabilized arc and our measurements in the opposite sense to that observed when our values are compared with those of Corliss et. al., i.e., with increasing upper excitation potential, the wall-stabilized arc values decrease relative to our values. The magnitude of the discrepancy between our values and those of Garz and Kock is -0.09 dex/eV but only -0.04 dex/eV with the Bridges and Wiese measurements.

In Table 1 we have listed gf-values for a few lines recently measured in several investigations. Excluding the free-burning arc measurements, the values for the four high level lines agree to within a factor of 2, direct comparisons with our own values being within a factor 1.5. Our gf-value, 0.25, for the 3719.9Å resonance line is 0.17 dex lower than the revised atomic beam measurements, 0.37 (Bell and Tubbs 1970), which is larger than an earlier atomic beam value 0.29 (Bell, Davis, King, and Routley 1958). Bridges and Wiese converted their wall-stabilized arc relative gf-values to

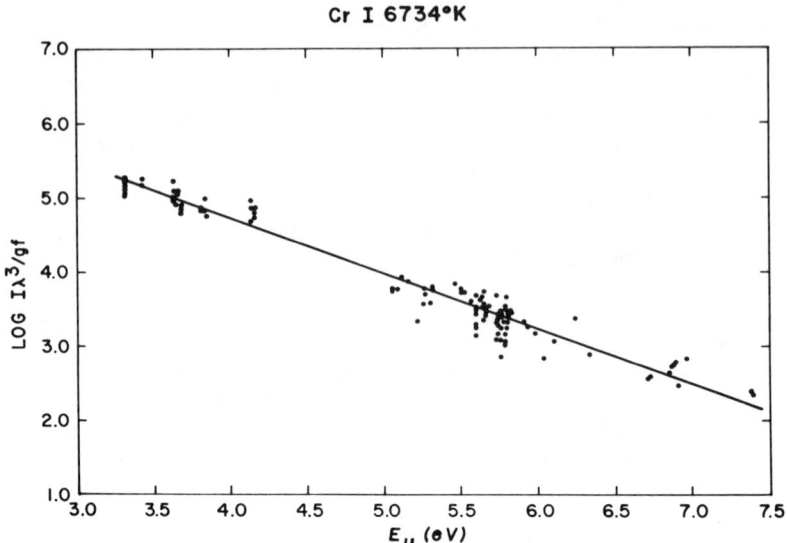

Fig. 4. Least squares determination of the N.B.S. free-burning arc temperature for Cr I from the slope (=-5040/T) of the plot of log $I\lambda^3/gf$ vs. E_u, using Meggers, et. al. N.B.S. Monograph 32 spectral line intensities and published AFCRL shock tube gf-values: $T \cong 6700°K$.

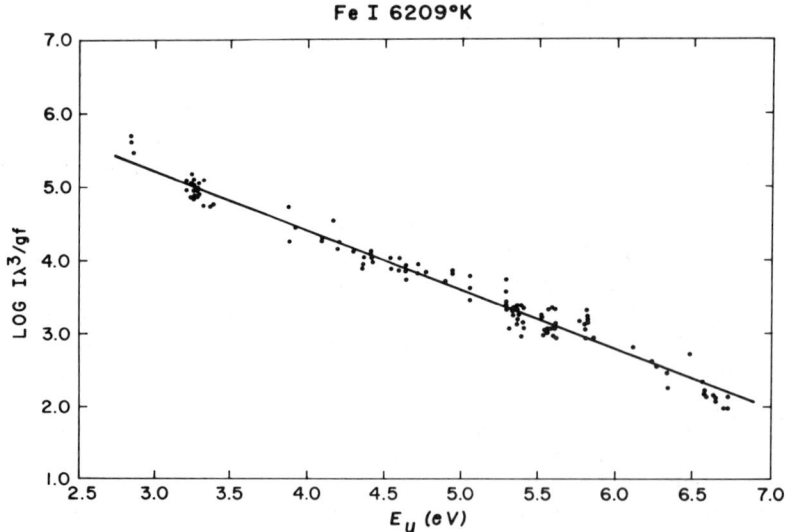

Fig. 5. Least squares determination of the N.B.S. free-burning arc temperature for Fe I from the slope (=-5040/T) of the plot of log $I\lambda^3/gf$ vs. E_u, using Meggers, et. al. N.B.S. Monograph 32 spectral line intensities and AFCRL shock tube gf-values: $T \cong 6200°K$.

Table 1

Comparison of Absolute Fe I gf-Value Scales

AFCRL Shock Tube Emission and Other Recent Measurements

Wavelength (Å)	RMT Mult. No.	Upper Exc. Pot. E_u (eV)	Shock Emis. AFCRL Wolnik, et. al. 1970	At. Beam (Revised) Mudd College Bell & Tubbs 1970	Lifetime Beam-Foil Cal. Tech. Whaling, et. al. 1970	Wall-Stabilized N.B.S. Arc Bridges & Wiese 1970	Shock Absorp. Grasdalen, et. al. 1969	Free-Burning N.B.S. Arc (Revised) Corliss & Tech 1968
3719.9	5	3.32	0.25	0.37	----	(0.37)	----	0.48
5242.5	843	5.97	0.15	----	0.16	0.15	----	0.64
4219.4	800	6.48	1.0	----	1.4	1.1	----	10.3
3233.1	620	7.04	----	----	1.4	1.2	0.76	26.7
3254.4	620	7.05	----	----	1.0	0.91	0.59	23.5

absolute values by normalizing them to the mean (f=0.041) of three recent absolute measurements for the resonance line 3719.9Å: the lifetime measurements by Wagner and Otten (1969, f=0.042) and by Klose (1970, f=0.039), and the Bell and Tubbs (1970) revised atomic beam value of f=0.041, which happens to be the same as the adopted mean. (The corresponding log gf is shown as "0.37" in Table 1.)

Under our experimental conditions 3719.9Å was stronger than convenient for measurement, thus we had to use extremely low concentrations of iron pentacarbonyl to avoid self-absorption effects. In addition, direct calibration with the standard carbon arc was difficult due to the shorter wavelength. Thus the 3719.9Å line, which was either experimentally highly favorable for the lifetime measurements or was of minimal error by definition of the normalizations chosen for the wall-stabilized arc determinations, is the least experimentally reliable of our measurements. This fact should be borne in mind in comparisons of absolute gf-values for Fe I such as in Table 1.

CONCLUSION

In summary, we may conclude that recent absolute gf-value determinations by shock tube, wall-stabilized arc, atomic beam, and radiative lifetime are coming into reasonable agreement. Whether the Corliss, et. al. free-burning arc data, which outweigh all else in quantity, can be "salvaged" is not certain, although Takens (1970) has made a big effort to do so. *(See note added in proof.)

One of the most exciting applications of recent chromium and iron absolute gf-values is in trying to resolve the paradox of the solar photosphere seeming to be 5- or 10-fold underabundant in iron and chromium relative to the solar chromosphere and corona and also relative to the appropriate type of meteorite. From our Cr I gf-values we determined an increased factor of approximately 4 (Wolnik, et. al., 1969), and from our Fe I gf-values Ross (1970) has determined a factor of 5. These values essentially resolve the paradox. The Kiel workers used their wall-stabilized arc Fe I gf-values (Garz and Kock, 1969) to determine an increased factor of 10 (Garz, Holweger, Kock, and Richter, 1969).

*Note added in proof: Corliss and Tech (1970) have announced that such attempts at correction of the Corliss and Bozman (1962) and Corliss and Tech (1968) N.B.S. absolute gf-values for systematic errors are hopeless - Takens'(1970) comprehensive effort in particular-because of several abrupt discontinuities in the gf-value scale for each of the 70 elements, which occur at different wavelengths for different elements. These discontinuities trace back to the Meggers, et. al., (1961) line intensity measurements for the 70 elements. In view of these systematic errors, combined with the large internal scatter and the recognized (Bridges and Wiese 1970) large departures from LTE in the N.B.S. low-current, free-burning arc, the scatter in our Figures 2 and 3 and the large difference in derived "temperature" between our Fig. 4 for Cr I and our Fig. 5 for Fe I are understandable. Figures 3 and 4 are instructive in that they illustrate a typical source of error in plasma temperature determination. In Fig. 4 the scatter seems moderate and the temperature statistically well determined. Actually, except for two points, the scatter is essentially the same in both cases but looks much smaller in Fig. 4 because of the approximately 3-fold larger range in the range of values of the ordinate log $I\lambda/gf$. Systematic errors of intensity measurements with respect to excitation potential do not show up at all as scatter in Fig. 4 and lead to a false impression of accuracy in the temperature determination.

REFERENCES

Aller, L. H. (1965), The Abundance of Elements in the Solar Atmosphere, in Kopal, Z., Ed., Advances in Astronomy and Astrophysics, Vol. 3, Academic Press, N. Y.

Bell, G. D., and Tubbs, E. F. (1970), Astrophys. J., 159, 1093.

Bridges, J. M., and Wiese, W. L., 1970, Astrophys. J. Letters, 161, 71.

Carnevale, E. H., Wolnik, S., Larson, G., Carey, C., and Wares, G. W. (1967), Physics of Fluids, 10, 1459 (July).

Corliss, C. H. (1970) Private communication.

Corliss, C. H., and Bozman, W. R. (1962), Experimental Transition Probabilities for Spectral Lines of Seventy Elements, N.B.S. Monograph 53, U. S. Government Printing Office, Washington, D.C.

Corliss, C. H., and Tech, J. L. (1968), N.B.S. Monograph 108, U. S. Government Printing Office, Washington 25, D.C.

⎯⎯⎯⎯⎯⎯ (1970), private communication (to Commission 14, Fundamental Spectroscopy, of the International Astronomical Union).

Garton, W. R. S., Parkinson, W. H., and Reeves, E. M. (1966), Proc. Phys. Soc., 88, 771.

Garz, T., and Kock, M. (1969), Astron. & Astrophys., 2, 274 (No. 3, July).

Garz, T., Holweger, H., Kock, M., and Richter, J. (1969) Astron. & Astrophys., 2, 446 (No. 4, Aug.)

Grevesse, N. (1969), Unpublished Ph.D. Thesis, University of Liège, Belgium.

Goldberg, L., Müller, E. A., and Aller, L. H. (1960), Astrophys. J. Suppl., 5, 1 (No. 45).

Hill, A. J., and King, R. B. (1951), J. Opt. Soc. Am., 41, 315.

Huber, M., and Tobey, F. L. (1968), Astrophys. J., 152, 609.

Klose, J. Z. 1970 (in preparation).

Meggers, W. F., Corliss, C. H., and Scribner, B. F. (1961), Tables of Spectral Line Intensities, N.B.S. Monograph 32, U. S. Government Printing Office, Washington 25, D.C.

Ross, J. E. (1970), Nature, 225, 610 (Feb. 14).

Takens, R. J. (1970), Astron. & Astrophys., 5, 244-63 (No. 2, Apr.).

Wagner, R., and Otten, E. W. (1969), Zs. f Physik, 220, 349.

Whaling, W., Martinez-Garcia, M., Mickey, D. L., and Lawrence, G. M., (1970), private communication (to be published in the Proceedings of the Second International Conference on Beam-Foil Spectroscopy, Lysekil, Sweden, June 7-12, 1970).

Wilkerson, T. (1961), Unpublished Ph.D. Thesis, University of Michigan, Ann Arbor, Michigan.

Wolnik, S. J. (1966), Unpublished.

Wolnik, S. J., Berthel, R. O., Larson, G. S., Carnevale, E. H., and Wares, G. W. (1968), Physics of Fluids, 11, 1002 (May).

Wolnik, S. J., Berthel, R. O., Carnevale, E. H. and Wares, G. W. (1969), Astrophys. J., 157, 983 (Aug).

Wolnik, S. J., Berthel, R. O., and Wares, G. W. (1970), Shock Tube Measurements of Absolute Fe I gf-Values (To be published in Astrophys. J., 162, 1970 (Dec).

IMPURITY MEASUREMENTS IN THE EXPANSION TUBE

Edmund J. Gion

Ballistic Research Laboratories, USAARDC

Aberdeen Proving Ground, Maryland 21005

An assessment is given on the importance of impurities in expansion tube flow, using the results of spectroscopic measurements. The expansion tube is a new type facility which produces high energy, high Mach number flows. It has been used at the Ballistic Research Laboratories for the study of nonequilibrium effects in high temperature gases. However, there has been a special problem regarding impurities and contamination of the expansion tube flow, which is related to the operation of the facility, shown schematically in Figure 1. A knowledge of the impurity concentration is desirable, since impurities are well known to be very prominent radiators, and even small concentrations could make invalid measurements of the total radiation, say, from shock layers about models.

Figure 1 shows what essentially looks like a double diaphragm shock tube. However, the operation cycle is considerably different

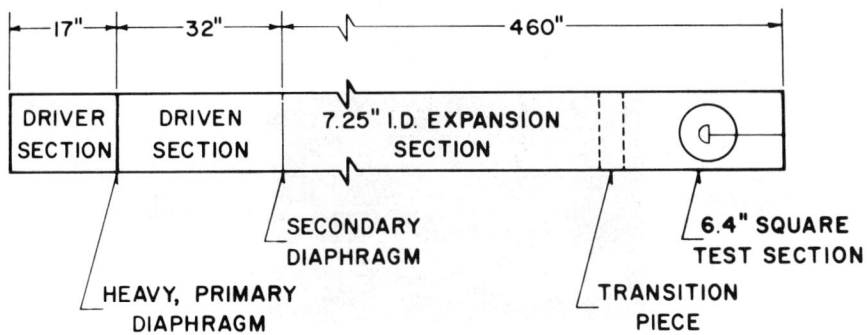

Figure 1. Schematic of the expansion tube.

from the shock tube, as will be seen; one is also testing in the region behind the contact front. The driven section contains the test gas, and a thin secondary diaphragm separates the test gas from the low pressure expansion section. The operation cycle proceeds: the primary shock from the driver processes the test gas, then goes on to rupture the secondary diaphragm. The heated and compressed test gas then expands unsteadily and runs downstream. The secondary diaphragm is usually broken into small bits and fragments, and, as the test gas accelerates and passes over these fragments, an obvious source of flow contamination is indicated.

Experimental measurements of the contamination levels were conducted in the following manner. The principal impurities were determined by time integrated spectra, and these were CN--from the mylar secondary diaphragm used, Fe, and Cr. The measurements were made on the shock layer about a 1-1/2 in spherical model. The shock layer was focused onto the entrance slit of a grating spectrograph. Four photomultipliers read the output at selected wavelengths. Sample photomultiplier traces of the impurity radiation are shown in Figure 2. A trace for the static pressure is shown to correlate the impurity radiation with the test flow. The test flow occurs roughly in the second hundred μsec from the trace beginning. Test time is ∿ 50 μsec. The Fe radiation was observed to be high before or during test time or low during test time and high afterwards. The CN radiation was always high during or before test time.

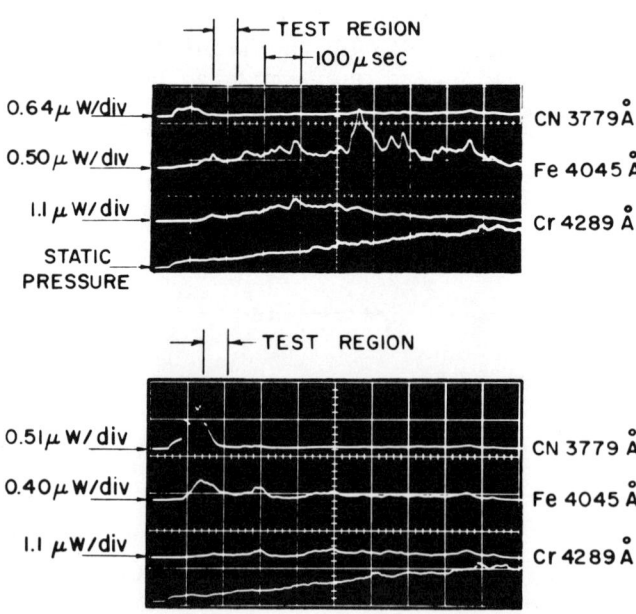

Figure 2. Typical radiation traces from impurities. $U_\infty \doteq 6500$ m/sec.

IMPURITY MEASUREMENTS

To calculate the number densities the Einstein-Boltzmann equation was used, relating the intensity of a spectral line to the number density of radiators and to the conditions of the experiment: (For the CN band there is a corresponding but more complicated relation.)

$$I = \frac{N\ell}{U} \frac{2\pi e^2 h}{m} \frac{gf}{\lambda^3} e^{-E/kT}$$

where ℓ is the depth of the shock layer, assumed equal to the model radius, U is the partition function, and the other symbols have their usual meaning. The absolute oscillator strengths for lines were taken from the NBS tables by Corliss and Bozmann.[1] The electronic f-value for CN is somewhat uncertain. A value of 0.086 was used from a review article by Soshnikov.[2] The temperature was determined from the total enthalpy assuming $T_\infty \simeq 1000°K$ and carrier gas in chemical equilibrium. This choice of T_∞ is based on previous measurements of p_∞, ρ_∞ made simultaneously under the same driving conditions. For the impurity measurements ρ_∞ was not measured simultaneously due to experimental restrictions. We estimate an error of $\sim 25\%$ in total number density (ion and atom) considering that the possible range of ρ_∞ may cause an error in T_∞ not greater than 500°K.

The assumptions made in using this equation are 1) the shock layer is homogeneous--for very strong shocks such as the case here, the layer is quite thin and this assumption is reasonable; 2) in the shock layer the carrier gas and impurities are in local thermodynamic equilibrium; and 3) the shock layer is optically thin--this assumption was checked using the curves of growth of Penner and Kavanagh[3] for the absorption in the shock layer. It was found that the experimental points lay on the linear portion of the curves of growth, i.e., the lines were optically thin. Absorption in the (cold) free stream as well as in the boundary layer was also considered, but the assumption of optical thinness could still be used since only a rough estimate of the concentrations was desired. For this reason no account was taken of the background radiation; likewise, no line wing corrections were applied.

The number densities for the Fe and Cr ions were not measured spectroscopically, since no provision had been made to measure these lines, practically all lying in the ultraviolet. Instead, the Saha equation was used to compute the ion number densities, assuming singly ionized species.

The results are shown in the Table. The total number density for the iron--neutral plus singly ionized--is $\sim 10^{14}$ particles/cm^3. The Cr and CN number densities are also of the same order. As the asterisks show, some runs were taken looking outside the boundary layer; some used a brass model in place of the stainless steel

U_S m/s	Fe 4045.8 Å NEUTRAL (1) ×10¹²	Fe 4404.8 Å NEUTRAL (1) ×10¹²	Fe 4404.8 Å NEUTRAL (2) ×10¹²	Fe 4404.8 Å TOTAL (1) ×10¹⁴	Cr 4289.7 Å NEUTRAL (1) ×10¹²	Cr 4289.7 Å TOTAL (1) ×10¹⁴	CN 3779 Å (1) ×10¹⁴	CN 3779 Å (2) ×10¹⁴
6120	.52	.96	6.9	1.7	-- --	-- --	.04	.64
6120	10.5	19.0	19.0	7.3	1.6	3.4	3.8	3.8
6750	1.5	2.7	2.7	2.8	1.1	2.8	.13	.76
6750	1.4	2.7	5.3	2.8	1.6	3.4	.58	3.0
6020 *	9.8	16.0	16.0	6.7	4.6	5.7	3.0	3.0
6330 *	.64	1.0	2.1	1.7	.64	2.1	.07	.76
6240 *	4.2	6.9	6.9	4.5	1.1	2.8	1.2	1.2
6400 *	3.2	4.8	5.6	3.8	1.6	3.4	1.7	2.5
6320 **	1.2	1.9	4.2	2.4	.51	1.9	.26	.84
6480 **	----	5.8	5.8	4.1	1.3	3.0	4.2	5.8
6910 *	1.7							
6160 *	2.2							
6520 *	1.5							
6520	.97							

(1) Test region   (2) Maximum

* Viewed outside of wall boundary layer
** Brass model; viewed outside of boundary layer

Table. Number densities of principal impurities.

model ordinarily used. For the most severe case of impurities the total number density is $\sim 15~(10^{14})$ particles/cm³. The carrier gas density is typically $\tilde{>} 3~(10^{18})$ particles/cm³. Thus the impurity concentration is ~ 500 ppm in the worst case. It is concluded that this impurity concentration is not important for chemical non-equilibrium studies. However, for radiation studies of an air-impurity mixture, say, such concentrations could be significant.

REFERENCES

1. C. H. Corliss and W. R. Bozmann, National Bureau of Standards Monograph 53, 1962.
2. V. N. Soshnikov, Soviet Phys.--Uspekhi 4, 425 (1961).
3. S. S. Penner and R. W. Kavanagh, J. Opt. Soc. Am. 43, 385 (1953).

SOME ASPECTS OF THE REFRACTIVE BEHAVIOR OF GASES

Daniel Bershader

Stanford University

I INTRODUCTION

When compressibility effects are present in fluid flow fields, the variation in density produces a corresponding change in optical refractivity. Two aspects of this behavior have been exploited in experimental gas dynamics, especially during the past 25 years. The first of these relates to the angular deflection of light in the presence of a transverse refractivity gradient. Starting from Fermat's Principle one can show in a straightforward manner (Ref. 1) that, given a medium in which the refractive index n is an arbitrary function of the coordinates, then the radius of curvature R of a light ray at any point (x,y,z) is given by

$$\frac{1}{R} = \operatorname{grad}\, \log\, n \cdot \vec{e} \qquad\qquad 1$$

where $\vec{e}$ is a unit vector perpendicular to the light path. This relation is the basis of the well-known schlieren and shadow techniques, now widely used in laboratory studies of fluid dynamic phenomena.

A second consequence of the optical interaction with a compressible flow is the variation of optical path through the medium. If light travels in the z-direction, and if we neglect deflections due to transverse gradients (usually a satisfactory assumption for this purpose), then the optical path Q(x,y) experienced by a ray entering the medium at the point (x,y,0) and leaving it at the point (x,y,L) is

$$Q(x,y) = \int_0^L n(x,y,z)dz \qquad 2$$

For plane flows in which z is an ignorable coordinate, the refractivity depends only on x and y, and we have

$$Q(x,y) = n(x,y)L \qquad 2a$$

To use the change of optical path associated with a change in refractivity for purposes of physical measurements, we measure the change of phase of the test beam with respect to a reference beam. That is done by the method of interferometry (Ref. 2). The phase change is then displayed as a shift of a pattern of interference fringes with respect to a reference pattern. If n_* is the reference index, then the fringe shift $S(x,y)$ corresponding to the optical path in equation 2 is

$$S(x,y) = \frac{1}{\lambda} \int_0^L n(x,y,z) - n_* \, dz \quad , \qquad 3$$

where λ is the wavelength of the light.

The bridge between the above considerations and the application to fluid dynamics is, of course, the relation between density and refractive index. In almost all of the work to date, it has proved satisfactory to use a very simple expression, known as the Dale-Gladstone Law which reads

$$n - 1 = K\rho \quad , \qquad 4$$

where ρ is the fluid density, and K is a constant for any given gas over a wide range pressures, with only a mild dependence on wavelength in the visible range. The same type of relation has been extended to gases at moderately high temperatures, say up to 20,000°K, with ionization present. In that case one may write

$$n - 1 = \sum_s K_s \rho_s + K'_e N_e \quad , \qquad 5$$

where the summation is over the different heavy particle species present, and where the last term describes the refractivity of free electrons of number density N_e. For air at the sodium D-line wavelength $\lambda = 5893 \text{Å}$, the value of K is $.226 \text{cm}^3/\text{gm}$., while the value of the electron specific refractivity K'_e is $-4.49 \times 10^{-14} \lambda^2 \text{cm}^3/\text{particle}$. Note the negative sign and the highly dispersive nature of K'_e.

The reason why equation 4 has generally served the purpose for analyzing refractive flows is that most of the experiments have been performed with air or other diatomic gases whose resonant wavelengths for electronic transitions are well into the ultraviolet, i.e., much shorter than the visible wavelengths used in the experiments. Further, thermal or other excitation of internal states has not been sufficient to produce appreciable percentage populations of excited states. However, the constant search to extend the range of the refractive methods to newer areas and the availability of the laser as a light source has aroused interest in examining more closely some fundamentals of the refractive behavior of gases.

In what follows, we will review basic features of the classical atomic dipole theory of optical dispersion with a view to indicating possible extensions of refractive studies to newer areas of gas-dynamic research.

II CONSTITUTIVE PROPERTIES AND WAVE PROPAGATION IN A MEDIUM

The electric field $\vec{E}$ of a light wave passing through a medium will distort the charge configurations of the atoms or molecules, thus inducing a dipole moment. The latter is characterized by a polarization vector $\vec{p}$ which is the induced dipole moment per unit volume. Generally, it is found that $\vec{p}$ is proportional to $\vec{E}$, but when the particle is surrounded by other molecular dipoles it sees a modified "effective field:

$$\vec{p} = N\alpha \vec{E}_{eff} = \frac{N\alpha}{1 - \frac{N\alpha}{3\epsilon_0}} \vec{E} \qquad 6$$

Here, N is the particle number density, ϵ_0 the permittivity of free space, and α is the induced electronic polarizability. For reference, we introduce three related and familiar quantities, namely χ_e, the electric susceptibility, ϵ, the permittivity; and ϵ_*, the relative dielectric constant. The pertinent relations among these quantities are (Ref. 3)

$$\chi_e = \frac{N\alpha/\epsilon_0}{1 - N\alpha/3\epsilon_0}, \qquad 7$$

$$\epsilon = \epsilon_0 (1 + \chi_e), \qquad 8$$

$$\epsilon_* = \epsilon/\epsilon_0 = \frac{1 + 2N\alpha/3\epsilon_0}{1 - N\alpha/3\epsilon_0}, \qquad 9$$

The particle density N and mass density ρ are related by

$$\rho = N \, m / \mathcal{L} \, , \qquad 10$$

where m is the molecular weight and $\mathcal{L}$ is Loschmidt's or Avogadro's number. We are then lead to the so-called Clausius-Mosotti relation between ϵ_* and ρ:

$$\frac{\epsilon_* - 1}{\epsilon_* + 2} = \frac{\mathcal{L}\alpha}{3\epsilon_0 m} \rho \qquad 11$$

This relation has been verified experimentally for many materials. Substitution of the expression relating ϵ_* and the refractive index n (discussed later) will led to the Dale-Gladstone Law, equation 4. The differential equation for $\vec{E}$ for the case of a non-conducting, non-magnetic material reads (Ref. 3)

$$\nabla \vec{E} - \mu_0 \epsilon \frac{\partial^2 \vec{E}}{\partial t^2} = 0 \, , \qquad 12$$

where μ_0 is the permeability of free space. In view of equation 9 and the relation for the velocity of light c, i.e.,

$$c = \frac{1}{\sqrt{\mu_0 \epsilon_0}} \qquad 13$$

it is evident that the index of refraction can be obtained as

$$n \equiv c/v = \sqrt{\epsilon_*} \qquad 14$$

Now, the induced polarization represents a response to the forcing function $\vec{E}$. That response is specified further below in the discussion of the Lorentz electron theory. The analysis will show that, in general, the refractive index must be represented as a complex, frequency-dependent quantity.

III LORENTZ ELECTRON THEORY OF DISPERSION

The Lorentz theory treats the molecular system as consisting of one or more damped simple electron oscillators; i.e., the system is characterized by inertia as well as dissipative and elastic restoring forces. Displacement of the electron from equilibrium by an external field generates a dipole moment. Each electron oscillator has a resonant frequency, and the damping produces the usual broadening of the amplitude response near resonance, as well as a phase difference between the applied and induced wave fields. The latter behavior results in the

REFRACTIVE BEHAVIOR OF GASES

constitutive parameters, whether expressed as polarizability, refractive index, etc., being represented as complex quantities. Solution of the usual equations yields the polarizability α as (Ref. 4)

$$\alpha_r(\omega) = \frac{e^2/m}{\omega_r^2 - \omega^2 + i\omega\gamma_r} , \qquad 15$$

where ω_r is the resonant frequency of the rth electron oscillator and γ_r is the damping force per unit mass per unit velocity. According to classical radiation theory for the so-called natural line width it is given approximately by

$$\gamma_r = \frac{e^2\omega_r^2}{6\pi\epsilon_0 mc^3} , \qquad 16$$

The complex refractive index, denoted by $\tilde{n}$, may be written

$$\tilde{n} = n - i\kappa , \qquad 17$$

where n is the usual phase refractive index, and κ is an extinction coefficient. That this is so becomes clear if we write the plane wave solution for $\vec{E}$ explicitly.

$$\vec{E} = \vec{E}_0 e^{-i(\tilde{n}\vec{k}_0 \cdot \vec{r} - \omega t)}$$
$$= \vec{E}_0 e^{-\kappa \vec{k}_0 \cdot \vec{r} - i(n\vec{k}_0 \cdot \vec{r} - \omega t)} , \qquad 18$$

where $|\vec{k}_0|$, the magnitude of the vacuum wave vector, $= \omega/c$.

The explicit expression for $\tilde{n}$ on the Lorentz model is

$$\frac{\tilde{n}^2(\omega) - 1}{\tilde{n}^2(\epsilon) + 2} = \frac{Ne^2/3\epsilon_0 m}{\omega_r^2 - \omega^2 + i\omega\gamma_r} \qquad 19$$

assuming only one electron contributes to the dispersion. More generally, there may be several electron oscillators per atom or molecule, with different resonant frequencies. Further, it was a result of experiment that the induced dipole moment is not the same for each oscillator. The expression for α_r in equation 15 has to be multiplied by an oscillator strength factor f_r which is a fraction between 0 and 1. The oscillator strengths are normalized such that

$$\sum_{r} f_{r'} = 1 \qquad 20$$

and the total polarizability α is

$$\alpha = \sum_{r'} \alpha_{r'} \qquad 21$$

The expression for the complex refractive index can now be written more fully as

$$\frac{\tilde{n}^2 - 1}{\tilde{n}^2 + 2} = \frac{Ne}{3\epsilon_0 m} \sum_{r'} \frac{f_{r'}}{\omega_{r'} - \omega + i\nu_{r'}\omega} \qquad 21$$

The classical theory of Lorentz gives remarkably good results. In the light of the correct solution for dispersion by bound electrons according to quantum mechanics, we should, however, note the following points:

(1) The concept of resonant oscillator frequencies is superseded by that of characteristic frequencies associated with transition between energy levels; $\omega_{nm} = (E_n - E_m)/\hbar$, where $\hbar$ = Planck's constant divided by 2π.

(2) The Lorentz theory cannot account for the refractive index of an excited state. Since the charge configuration varies with degree of electronic excitation, it is to be expected that the induced polarizability, and therefore the refractivity, would vary as well. The quantum mechanical solution for the refractive index contribution by the atoms populating the jth energy state, away from immediate resonance regions where absorption may be appreciable, is

$$n_j(\lambda) - 1 = \frac{N_j e^2}{8\pi \epsilon_0 mc^2} \left\{ \sum_{k>j} \frac{f_{jk}\lambda_{jk}^2 \lambda^2}{\lambda^2 - \lambda_{jk}^2} - \sum_{i<j} \frac{f_{ji}\lambda_{ji}^2 \lambda^2}{\lambda^2 - \lambda_{ji}^2} \right\} \qquad 22$$

where the first summation is over all states k higher than j, and the second summation is over all states i below j. Note that the contribution of the latter sum is negative. This so-called "negative dispersion" is purely a quantum phenomenon and has no analog in the classical theory. It was observed experimentally by Ladenburg and his collaborators (Ref. 5) about 40 years ago, and constituted an experimental verification of the quantum mechanics. See also some further remarks in Section VI below.

REFRACTIVE BEHAVIOR OF GASES

(3) The classical damping constant γ_r for Lorentz or natural broadening must be interpreted in terms of the Einstein coefficients for spontaneous emission. I.e., for the transition between states j and i

$$\nu_{ji} = \sum_{i'} A_{ii'} + \sum_{j'} A_{jj} \qquad 23$$

A_{nm} is the probability per particle per second of spontaneous transition from state n to state m. The quantity $\sum_m A_{nm}$ is the reciprocal of the lifetime of state n :

$$\sum_m A_{nm} = \frac{1}{\tau_n} \qquad 24$$

(4) The correct quantum mechanical analysis shows a basic relation between the Einstein A- coefficients and the oscillator strength or f-number. That there must be a relation stems from the fact that the f-number was originally introduced to explain the variation of intensity of spectral lines. The latter is also proportional to the A- coefficient. In quantum mechanics, the f_{nm} is a dipole transition moment, determined from a wave-function overlap integral. The relation between f_{nm} and A_{nm} is

$$f_{mn} = \frac{2\pi\epsilon_0 mc^3}{\omega_{nm}^3 e^2 \hbar} A_{nm} \qquad 25$$

IV THE DISPERSION FUNCTION

It is clear from equations 21 and 22 that the refractivity undergoes sharp changes around resonant frequencies. However, consider first what happens sufficiently far from resonance such that $\omega_r - \omega \gg \gamma_r \omega$; typically, this would be the case when the wavelength is several Angstron units off resonance. Then, the extinction coefficient $\kappa \ll n$, and with n very close to unity, equation 21 becomes

$$n - 1 = \frac{Ne^2}{2m\epsilon_0} \sum_{r'} \frac{f_{r'}}{\omega_{r'}^2 - \omega^2} \qquad 26$$

Converting frequencies to wavelengths, and using equation 10 as well, we obtain

$$n - 1 = \frac{\mathcal{L}}{m} \sum_{r'} \frac{a_{r'} \lambda^2}{\lambda^2 - \lambda_{r'}^2} \rho = K\rho \qquad 27$$

where

$$a_{r'} = \frac{e^2 f_{r'} \lambda_{r'}^2}{8\pi^2 \epsilon_0 mc^2} \qquad 28$$

Equation 27 gives the constituents of the Dale-Gladstone Law explicitly. For most gases of interest, the resonant wavelengths lie in the ultra-violet, i.e., $\lambda_{r'}^2 \ll \lambda^2$. Thus λ^2 in the numerator and denominator nearly cancel, and that is why K is only weakly dispersive.

Consider next the "near" region of the rth resonant line where $\omega_r^2 - \omega^2$ is still $\gg \gamma_r \omega$, but much smaller than $\omega_{r'}^2 - \omega^2$. I.e.,

$$\gamma_r \omega \ll \omega_r^2 - \omega^2 \ll \omega_{r'}^2 - \omega^2 \qquad 29$$

Further, we have again $\kappa \ll 1$, so that an expression of the form of equation 26 applies again. However, in this case we will separate the rth term from the sum. Replacing ρ by N again, and setting $\lambda = \lambda_r$ except in the difference term, we obtain

$$n - 1 = n_0 - 1 + \frac{e^2 \lambda_r^3 f_r N}{8\pi^2 \epsilon_0 mc^2 (\lambda - \lambda_r)} \qquad 30$$

where $n_0 - 1$ is the reduced refractivity due to the sum over $r' \neq r$. This expression is used in connection with the "Hook Method" of atomic analysis, described in a subsequent section. In particular, the slope of the dispersion curve is of interest, and we present it for reference:

$$\frac{dn}{d\lambda} = - \frac{r_0 \lambda_r^3 N f_r}{4\pi (\lambda - \lambda_r)^2} \qquad 31$$

where we have introduced r_0, the classical radius of the electron as

$$r_0 = \frac{e^2}{4\pi \epsilon_0 mc^2} \qquad 32$$

REFRACTIVE BEHAVIOR OF GASES

Finally, let us look at the resonant behavior. For simplicity it will be necessary to assume a weakly absorbing line, say $\kappa_{max} < 0.10$. Then, the real part of $\tilde{n}^2 - 1$, namely $n^2 - \kappa^2 - 1$, can be approximated again by $2(n - 1)$, while $\tilde{n}^2 + 2$ is set equal to 3. The imaginary part of $\tilde{n}^2 - 1$ is 2κ, since $n \approx 1$. With reference to equation 23, we obtain, upon equating real and imaginary parts, and setting $\omega = \omega_r$ except in the difference term

$$n - 1 = n_0 - 1 + \frac{e^2 N f_r (\omega_r - \omega)}{4 \omega_r m \epsilon_0 \left[(\omega_r - \omega)^2 + \gamma_r^2/4\right]} \qquad 33$$

and

$$\kappa = \frac{e^2 N f_r \gamma_r}{8 \omega_r m \epsilon_0 \left[(\omega_r - \omega)^2 + \gamma_r^2/4\right]} \qquad 34$$

The resonance curves for n and κ are plotted in Fig. 1. The refractive index is seen to increase with ω, except between the "half power" points around resonance. The latter behavior is termed anomalous dispersion. The extreme values of $n - n_0$ occur at frequencies $\omega_r \pm \gamma_r/2$, which are also the frequencies at which κ has half its maximum value:

$$(n - 1)_{extreme} = n_0 - 1 \pm \frac{e^2 N f_r}{4 m \epsilon_0 \omega_r \gamma_r} \qquad 35$$

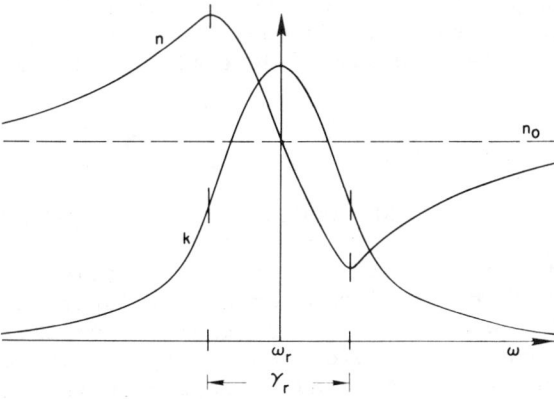

Fig. 1. Refractive index n and excitation coefficient κ around a resonant frequency ω_r for a Lorentzian oscillator. Half-width of the absorption curves is γ_r. The two curves are not scaled with respect to each other.

Suppose we examine some numerics in connection with the above formula. For some gases of principal interest $n_0 - 1$ is approximately 3×10^{-4} at atmospheric density, i.e., for $N = 2.7 \times 10^{25}/\text{met}^3$. Let us consider a dilute gas at 10^{-5} times atmospheric density, and let $Nf_r = 2 \times 10^{20}/\text{met}^3$. Further, consider a resonance at $\lambda_r = 4,000\text{Å}$, corresponding to $\omega_r = 4.71 \times 10^{15}$ radians/sec. The value of γ_r can be computed from equation 16 as 1.39×10^8 radians/sec. (lifetime $\sim 10^{-8}$ seconds), corresponding to a line width of $1.2 \times 10^{-4}\text{Å}$. The term $n_0 - 1$ is negligible by comparison to the resonant term and we obtain

$$(n - 1)_{\text{extreme}} = \pm .24 \qquad 35a$$

The result is actually too extreme because a correction for γ_r in accordance with quantum theory of radiation would give a larger value for that quantity. Further, the existence of collision broadening (pressure dependent) would increase γ_r still further except in extremely dilute gases.

Turning to the absorption profile, we note that the peak value of κ is

$$\kappa_{\text{max}} = \frac{Ne^2 f_r}{2m\epsilon_0 \omega_r \gamma_r} \qquad 36$$

Thus, the maximum value of κ is directly proporation to f_r and inversely proportional to γ_r. It turns out that the integral of the absorption profile over a line is independent of γ_r but is still proportional to f_r. Indeed, the integrated absorption measurement is a well-known way of determining f_r experimentally.

V SPECTRAL INTERFEROMETRY: THE HOOK METHOD

Assume light from a broad-band source is allowed to pass through a Mach-Zehnder Interferometer containing tubes of equal lengths L in the two legs. Suppose that one of the tubes is filled with a reference gas which has negligible dispersion in the spectral region to be investigated. Let the corresponding reference refractive index be n_*. The other contains test gas whose dispersion function is $n(\omega)$. We will not be concerned here with the absorption because the discussion that follows will deal with the "near" region of a spectral line, where $\kappa \ll 1$.

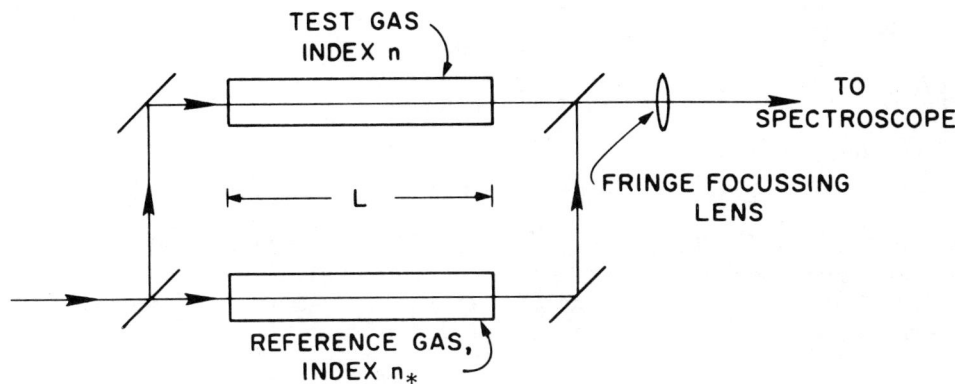

Fig. 2. Schematic representation of light path through interferometer and tubes with test and reference gases, respectively. For simplicity, we assume that the reference gas is non-dispersive.

Suppose that a lens is used to focus horizontal fringes on the vertical slit of a spectroscope. For this adjustment a narrow band filter may be inserted temporarily. The fringes which cross the slit produce a cosine-squared intensity variation over the length of the latter. If the source is sufficiently broad-band over the spectral region of interest, the images of the slit are spread continuously on the observation screen or film. Therefore, the fringes are reconstructed and appear superimposed on the spectrum. This arrangement permits some useful measurements as the following considerations will show.

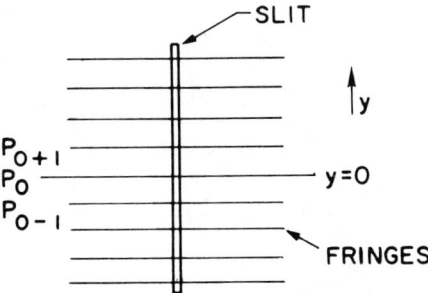

Fig. 3. Schematic representation of horizontal fringes focused on a vertical spectroscope slit.

Let the central pair of rays passing through the interferometer fall on the center of the slit. The optical path difference between the gas in the reference tube and that in the test tube may be assumed to correspond to p_0 fringes:

$$(n - n_*)L = p_0 \lambda \qquad 37$$

Now fringe number p lies at some distance y_p from the center of the slit where

$$y_p = (p - p_0)w, \qquad 38$$

where w is the fringe width or spacing. It is well known from the elementary theory of interference fringe formation that w is simply related to λ and to the angle $\emptyset$ between the interfering beams by

$$w = \lambda/\emptyset \qquad 39$$

Substitution of equations 37 and 39 into equation 38 yields

$$y_p = \frac{p}{\emptyset}\lambda - \frac{(n - n_*)L}{\emptyset} \qquad 40$$

If n were a constant independent of λ, i.e. no dispersion, then the fringes take the form of a system of divergent straight lines emanating from a point $(0, -(n - n_*)\frac{L}{\emptyset})$ in the λ-y plane. The difference in slope between successive fringes is $1/\emptyset$. Note that a change in p_0 by changing optical path difference in the interferometer slides the y - intercept on the $\lambda = 0$ axis up or down as the case may be: see Figure 4.

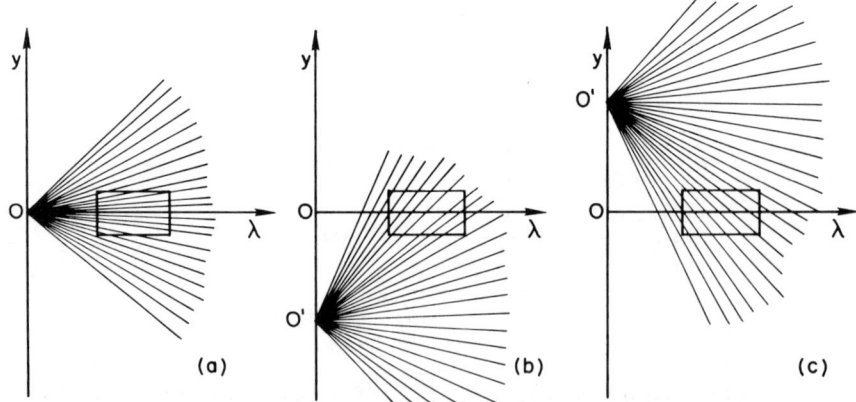

Fig. 4. Dispersion-free fringes seen as radial lines in the λ-y plane. The three cases correspond to variation of optical path difference in the interferometer. The absolute value of OO' is $(n-n_*)L/\emptyset$. The rectangles represent the observed region of the spectrum.

REFRACTIVE BEHAVIOR OF GASES

Case (a) as shown corresponds to equal optical paths in both legs of the interferometer, and the fringes are seen to have rather small slope. In case (b) the optical path in the test tube is greater than that for the reference tube; and vice versa for case (c). As the optical path difference increases, it is clear that the fringes viewed in the spectrum become more nearly vertical. Note, by the way, that the value of $\emptyset$, i.e., the fringe width, is a scaling factor in that replacement of y_p by $Y_p \equiv \emptyset y_p$ would give a relation independent of w.

Because of small dispersion away from resonances, the fringes in those regions do in fact appear as sloping lines. Now, however, let us examine what happens in the near region. For that purpose we substitute equation 30 in equation 40.

$$y_p = \frac{p\lambda}{\emptyset} - \frac{r_0 \lambda^3 f_r NL}{4\pi\emptyset(\lambda - \lambda_r)} + (n_* - n_0)\frac{L}{\emptyset} \qquad 41$$

where r_0 is given by equation 32. Clearly, the fringes are not straight any more. The slope of the pth fringe in the near region is

$$\frac{dy_p}{d\lambda} = \frac{p}{\emptyset} + \frac{r_0 \lambda^3 f_r NL}{4\pi\emptyset(\lambda - \lambda_r)^2} \qquad 42$$

Measurement of the fringe slope can be made to yield the product $N.f_r$ if other paramters are known. Indeed, this general approach was made by Puccianti around 1910. However, Roschdestvensky (Ref. 6) pointed out that the same information could be obtained more precisely by establishing rather sharp minima and maxima of fringe displacements in the near regions surrounding the spectral lines. In that connection note that the second term on the r.h.s. of equation 42 is always positive. For negative orders $p = -|p|$, that equation can be solved for two values of λ where the slope of the pth fringe vanishes. The configuration has been referred to as a "hook", and the corresponding values of λ, denoted by λ_H are

$$\lambda_H = \lambda_r \pm \sqrt{\frac{Lr_0 \lambda_r^3 N f_r}{4\pi|p|}} \qquad 43$$

It is customary to define a hook separation Δ by

$$\Delta \equiv \lambda_{H+} - \lambda_{H-} = \left\{ \frac{Lr_0 \lambda_r^3 N f_r}{\pi |p|} \right\}^{1/2} \qquad 44$$

Measurement of the hook separation has been used by several investigators (Ref. 7) to obtain values of f-numbers. The measured quantity is essentially the product Nf so that knowledge of N is required to determine f in an absolute manner. If N is not known, we can obtain relative f- numbers by looking at different transitions originating in the same state.

In recent years there has been a renewed interest in spectral interferometry as a tool in experimental gas dynamics. The reasons will be included among the remarks in the following section.

VI SPECIAL FEATURES

In this section we will treat briefly some further considerations which relate to current interest in the refractive behavior of gases.

1. Reciprocity between Refractive Index and Absorption Coefficient

In obtaining f-numbers by the hook method, one usually makes measurements at wavelength separations from the line center which are many times the line half-width, assuming an adequate value of N. (An exception would be the extensive Stark broadening associated with radiation from a dense plasma.) The procedure is in contrast to use of the integrated absorption profile for the same purpose (Ref. 7). Presumably, there is some advantage in using hooks over the absorption measurement, since less wavelength resolution is required.

The fact that refraction and absorption can be used to make the same measurement indicates that the real and imaginary parts of the complex refractive index are not independent of each other. The two are integral transforms of each other, as follows: (Ref. 4)

$$n(\omega) - 1 = \frac{1}{\pi} \wp \int_{-\infty}^{+\infty} \frac{\kappa(\omega')}{\omega' - \omega} d\omega'$$

$$\kappa(\omega) = \frac{1}{\pi} \wp \int_{-\infty}^{\infty} \frac{n(\omega') - 1}{\omega' - \omega} d\omega' \qquad 45$$

REFRACTIVE BEHAVIOR OF GASES

These expressions are called the Kramers-Kronig relations. The symbol ℘ denotes "principal parts", and means integration along the real axis, but excluding the pole at ω_r.

2. f-numbers of Excited State Transitions

Most of the early hook experiments dealt with ground-state resonance transitions. Typically, in sodium vapor, for example, the vapor pressure-temperature relation would be used to obtain vapor density. In such devices as discharge tubes carrying modest currents, the latter density would be equated with the ground state atom density. Thus, one could separate the latter factor from the product Nf in order to obtain f. Not only would the excited state population be small, but the probable lack of thermodynamic equilibrium made any determination of upper state populations difficult indeed. On the other hand, thermal devices such as the King Furnance have a limited temperature range.

In 1962, Dunaev et al (Ref. 8) produced equilibrium thermal excitation of mercury vapor in a shock tube, and were able to obtain f-numbers of several excited state transitions by the hook method. Shock-heating and arc-heating in some of the improved devices have extended the temperature range for such measurements and increased the capability for additional measurements of f-number. (See paper by Huber in this volume.) There is continuing interest, especially on the part of the astrophysicists, in f-number data.

3. Gas-Kinetic Studies by f-number Measurements

Tumakaev and Lazovskaya (Ref. 9) have shown that it is possible to time-resolve hook separation measurements behind a shock wave. In that case, one can obtain the rate of population of the excited state in question, since

$$\frac{d}{dt}(Nf) = f\frac{dN}{dt} \propto \frac{d}{dt}(\Delta^2) = 2\Delta\frac{d\Delta}{dt} \qquad 46$$

The quantity dN/dt can be expressed in terms of a rate constant for the process. The latter, in turn is given by a suitably averaged product of relative velocity and cross section of the colliding particles. Such gas-kinetic information for excited states would appear to be of considerable research interest.

4. Fringe Shift Near Resonance

Because the refractivity varies sharply over the profile of a spectral line, the index of refraction of sodium vapor, say, illuminated by a sodium lamp is quite difficult to compute. However, suppose one had a tuneable light source of width around 1Å for example. Examination of the dispersion relation in the near region of a line, say 5 Å from the center as compared with the expression far from any line yields the following approximate ratio:

$$\frac{(n-1)_{\text{near region}}}{(n-1)_{\text{far away}}} \approx \frac{\lambda}{\lambda - \lambda_r} \qquad 47$$

Thus at 5Å off resonance, $(n-1)$ would still be about 10^3 greater for visible light. One has the possibility, then, of obtaining greater sensitivity in fluid-dynamic experiments by the methods of shadow, schlieren or interferometry. Therefore, if one were able to use a suitable gas or gas mixture with a suitable contaminant, it would be possible to observe smaller flow perturbations such as turbulence or sound waves which are not otherwise visible. On the other hand, such increased sensitivity might permit one to check relations such as the vapor pressure-temperature curve by a direct density determination.

5. Negative Dispersion

This phenomenon was described in section III. By examination of hooks near the wavelengths λ_{ji} as contrasted with λ_{ik} in equation 22, one would observe a decreasing hook width as the population increased. The resultant index would be

$$n(\lambda \text{ near } \lambda_{ji}) - 1 = \frac{e^2 \lambda_{ji}^3 N_i f_{ij}}{16\pi\epsilon_0 mc^2 (\lambda - \lambda_{ji})} \left\{ \frac{N_j/g_j}{N_i'/g_j} \right\} \qquad 48$$

if we neglect other terms in the summation. Comparing this relation with equation 30, we see that except for the constant $n_0 - 1$, the negative dispersion is effectively represented by the bracketed factor above. Note that a population inversion would change the sign of $n - 1$. Therefore, it is evident that negative dispersion might be useful for determination of population inversions if the numerics are such as to give some sensitivity to the measurements.

6. Non-Linear Refractivity

The refractive behavior of gases as discussed in the present paper and in almost all other treatments to date is based on the assumption that the induced polarization is a linear process; i.e., the polarizability α or more strictly, the susceptibility χ_e

REFRACTIVE BEHAVIOR OF GASES

(see equations 6 and 7) is assumed to be independent of the field $\vec{E}$. This assumption must break down at field strengths of sufficient intensity, since we know that in the most extreme cases of focused high-powered lasers, it has been possible to ionize gases by the action of the field.

If a Taylor expansion is made for $P(E)$, we write for the case of molecules with no static moment

$$P(E) = E\left(\frac{\partial P}{\partial E}\right)_0 + \frac{1}{2} E^2 \left(\frac{\partial^2 P}{\partial E^2}\right)_0 + , \qquad 49$$

In accordance with equations 6 and 7, we may write

$$P(E) = \epsilon_0 \left(\chi_e^0 E + \frac{1}{2} \chi_e^1 E^2 + \ldots\right), \qquad 50$$

which may still be written in the old form

$$P(E) = \epsilon_0 \chi_e E, \qquad 51$$

if we define the electric susceptibility χ_e by

$$\chi_e = \left(\chi_e^0 + \frac{1}{2} \chi_e^1 E + \ldots\right) \qquad 52$$

For a plane wave whose time dependence is given by $E = E_0 e^{i\omega t}$, we have

$$P(E) = \epsilon_0 \left\{ \chi_e^0 E_0 e^{i\omega t} + \frac{1}{2} \chi_e^1 E_0^2 e^{2i\omega t} + \ldots \right\} \qquad 53$$

Thus, it is evident that $P(E)$ has some harmonic content, in that the oscillating dipole will radiate a frequency 2ω in addition to ω. Note that because of the E_0^2 term, the double frequency component is enhanced at higher intensity. This component is initially coherent with the ω wave, but there is an additional question with respect to differences in velocities of the ω wave and its harmonic as a result of dispersive properties of the gas. Only if the phase velicities of the two waves are equal, will they add constructively to produce effective harmonic generation. Note further, by the way, that if the incident light contains two frequencies, then the non-linear polarizability will result in the generation of sum and difference frequencies.

The non-linear refractive behavior of certain solids such as potassium dihydrogen phosphate is well known and is by now widely utilized for the generation of harmonic frequenices. Very few studies of this sort have been reported for gases (Ref. 10). As a first experiment, one might consider a very simple experiment with an interference pattern produced with test gas in one arm of the

Mach-Zehnder interferometer. The idea would be to look for a fringe shift with increasing intensity of light. One would use non-resonant frequencies at first in order to avoid appreciable absorption. The use of a short pulsed light source would be necessary in order to avoid effects due to internal excitation of the gas.

On the other hand, if frequencies at or near resonance are used, sufficient excitation of one or more upper states would produce a non-linear refractive behavior. This is so because the polarizability may be a sensitive function of level quantum number even at low field strengths, as in atomic hydrogen. Matters become more involved when high-intensity coherent pulses are used whose widths are not long compared to characteristic damping times for the resonances. Populations become anomalous and have a strong time dependence (the usual rate equations to describe such behavior are inadequate) with the result that the resonant dispersion becomes a sensitive function of the field strength. In such cases, the use of customary absorption and emission coefficients to describe pulsed response of resonant systems is not suitable. Any further detail on such effects is beyond the scope of the present discussion, but they have led to several exciting new physical phenomena which include photon echoes (Ref. 11), self-induced transparency (Ref. 12) and parametric amplification of signals (Ref. 10).

REFERENCES

1. Heath, A., <u>Geometrical Optics</u>, Cambridge, 1897.

2. Ladenburg, R. and Bershader, D., "Interferometry", Section A3 of <u>Physical Measurements in Gas Dynamics</u>, Princeton University Press, 1954.

3. Panofsky, W. and Phillips, M., <u>Classical Theory of Electricity and Magnetism</u>, Addison-Wesley, 1955.

4. Ditchburn, R., <u>Light</u>, vol. 2, Interscience, 1963

5. Ladenburg, R., "Dispersion in Electrically Excited Gases", Rev. Mod. Physics, $\underline{5}$, p. 243, October 1933.

6. Roschdestvensky, D. Ann. Physik $\underline{39}$, 307 (1912)

7. Foster, E., Reports of Progr. in Physics $\underline{27}$, 469 (1964).

8. Dunaev, Y. et al, Sov. Phys - Tech. Phys. **6**, 815 (1962).

9. Tumakaev, G. and Lazovskaya, V., Sov. Phys - Tech. Phys. **9**, 1449 (1965).

10. Gordon, P. et al, "Continuous Infrared Parametric Amplification in a Saturable Absorber", Appl. Phys. Lett. **14**, 235, April 1969.

11. Patel, C. and Slusher, R., "Photon Echoes in Gases," Phys. Rev. Lett. **20**, 1087, May 1968.

12. Patel, C. and Slusher, R., "Self-Induced Transparency in Gases," Phys. Rev. Lett. **19**, 1019, Oct. 1967.

INTERFEROMETRIC GAS DIAGNOSTICS BY THE HOOK METHOD

Martin C.E. Huber

Harvard College Observatory

Cambridge, Massachusetts

INTRODUCTION

Refractive index measurements have long been a well established tool for investigating aerodynamic flow patterns where number densities were determined from interferograms taken with a light source of moderate spectral bandwidth. However, the number density so derived gives no indication of the composition nor thermodynamic state of the gas under investigation; rather, its properties and composition have to be assumed.

Much more sophisticated diagnostics become possible by exploiting the *spectral behavior* of the refractive index, especially the anomalous dispersion near spectral lines. Measurements of anomalous dispersion are more useful in diagnostics than are conventional intensity measurements, mainly because the gas need not be optically thin in the center of spectral lines.

The hook method is the most convenient for anomalous dispersion measurements. It produces spectral interferograms that form hook-like patterns (Figure 1); the measured quantity is the separation between hooks.

We begin this paper with a short historical introduction and a brief review of the pertinent literature. This is followed by an explanation of the experimental aspects of the hook method, and a discussion of the two vital pieces of apparatus needed to perform hook method experiments — the Mach-Zehnder interferometer and stig-

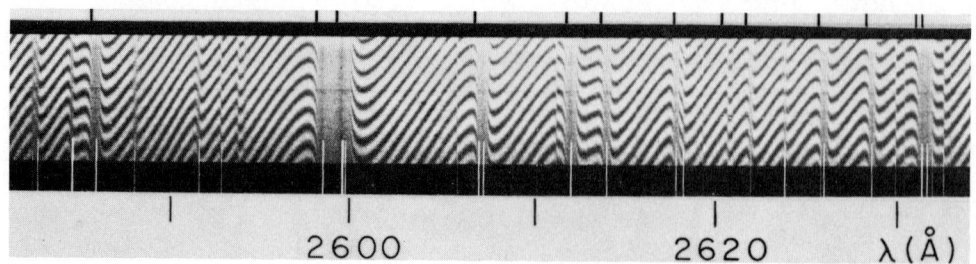

Fig. 1 Spectral interferogram taken through shock-heated iron at λλ 2580-2630. The transitions marked belong to UV multiplet number 1 of singly ionized iron (Fe II).

matic spectrographs. Finally, we describe recent theoretical and experimental work done with the hook method and its implications for gas diagnostics.

This paper, however, does not deal with refractivity due to free electrons. For a description of methods leading to the determination of electron density (including its spatial distribution), the reader is referred to Alpher and White (1965) or Ascoli-Bartoli (1970).

Historical Development of the Hook Method and Its Applications

The hook method has been used extensively for *f-value measurements*, i.e., to measure atomic properties using gases in a known state, yet this classical technique of quantitative spectroscopy (Rozhdestvenskii 1912) has rarely been employed in *gas diagnostics*. The first application of the hook method as a diagnostic tool was made by Ladenburg and co-workers in the late 1920's (Ladenburg and Kopfermann 1928; Kopfermann and Ladenburg 1928, 1930; Ladenburg and Levy 1930). These investigations demonstrated the phenomenon of negative dispersion, the refractive-index analogue to stimulated emission (Bershader 1971). Later on, the hook method was sometimes used to determine number densities or to investigate excitation rates (Tumakaev and Lazovskaya 1965; Cary and Nickels 1963).

At this point it seems suitable to stress the advantages of the hook method in determining properties of a gas, namely, the *number densities* and the *excitation*

conditions of its components. This is the inverse of an
f-value measurement by the hook method; instead of deducing oscillator strengths from anomalous dispersion in a
gas with known number densities, one determines number
densities of a gas with known oscillator strengths. This
may appear to be a rather trivial and obvious application
of the fact that the observable parameter in quantitative
spectroscopy is the integral

$$\int_\ell N(x) \, f \, dx \, ,$$

where $N(x)$ is the number density of atoms in the pertinent state at a given position x along the beam; f is
the oscillator strength; and ℓ is the total extent of
the gas layer observed. But we shall see that it is
feasible to use the hook method for both atomic and molecular species and for determining number densities of
ground as well as excited states. The two most important
features that make the hook method an ideal tool for
number density determinations of both atomic and molecular
species are: 1) its very large dynamic range and that it
can be used on saturated (i.e., optically thick) lines,
2) its relative insensitivity to line shape and thus to
collisions and Doppler broadening. Moreover, since it
does not require intensity measurements, the hook method
conveniently avoids the sundry difficulties of photometry.

Reliable oscillator strengths, however, are required
to determine number densities. Such values have become
available within the past years (Wiese et al. 1966, 1969).
It is now estimated that many f-values, especially those
for strong resonance lines accessible to the more accurate methods of f-value determination, are now known to
within better than 25 percent, in some cases even to
10 percent. Therefore, the hook method can be used to
sample number densities of single atomic states, by measuring the integral

$$\int_\ell N(x) \, f \, dx \, ,$$

where N is the number density of atoms in the lower state
of the given transition, even from an optically thick
line. Depending on the wavelength range used, this information can be obtained for a large number of transitions that, in general, will have a variety of lower
levels. Thus it is easy to sample *relative* number densities of different levels in one or more species under
different conditions. This makes it possible, for example,
to follow the course of excitation and of ionization with
time. Furthermore, if absolute oscillator strengths are

known, we can determine the integrated *absolute* number density of atomic as well as molecular states along the sampling optical beam.

Literature

Most of the early literature on the hook method was first published in Russian. A review paper by Penkin (1964) surveys this pioneer work in Russia, concerned mainly with the determination of oscillator strengths of gases kept in cuvettes, vacuum furnaces, or in some cases, discharge tubes. The Penkin review (Tech 1964) and many of the Russian papers referred to by Penkin had been translated into English earlier, thanks to the Israel Program for Scientific Translations (Meroz 1962, 1963). Another excellent review paper (Marlow 1967) stressed the theoretical aspects of the hook method, in particular, the influence of Lorentz as well as Doppler broadening on hook measurements. References to the literature published before 1965 are cited in the two review papers mentioned above.

More recent work on atomic spectra has been reported by Penkin and Shabanova (1965, 1967) as well as by Slavenas (1966). Anketell and Pery-Thorne (1967) and Pery-Thorne and Banfield (1970) applied the hook method to molecular spectra. The technique has also been used with shock tubes to determine oscillator strengths (Huber 1966) and to check on the number density of a given element in a shock-heated gas (Huber and Tobey 1968).

THE EXPERIMENTAL SETUP

In principle, the hook method provides a spectral display of refractive index from which to measure $Nf\ell$; (for the sake of simplicity, let us assume the case of a homogeneous layer of well-defined extent). The physical phenomenon observed is the anomalous dispersion at a wavelength near the spectral line in question, with fringes so arranged that they form hooks. One can show (e.g., Wiese 1968) that the square of the hook separation Δ (Figure 2) is proportional to the product $Nf\ell$:

$$Nf\ell = \pi K \Delta^2 / r_o \lambda_o^3$$

where r_o is the classical electron radius; λ_o is the wavelength of the resonance observed; and K is the so-

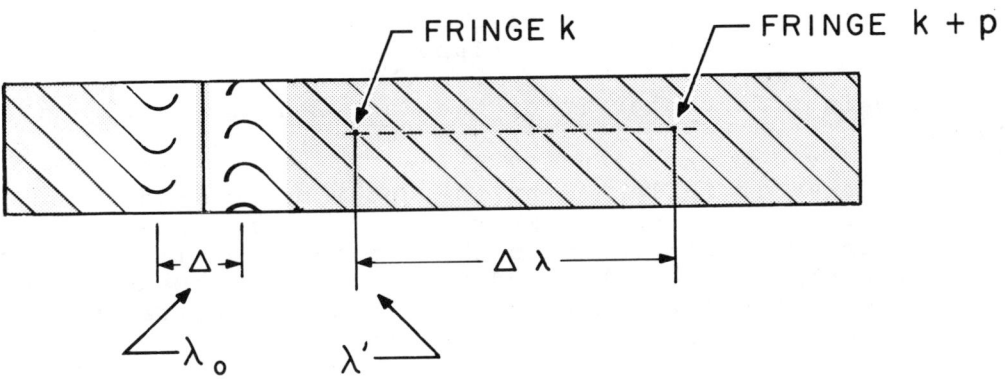

Fig. 2 Schematic hook spectrum illustrating parameters used in text.

called hook method constant that is related to the k of the fringes seen in the field. The hook method constant $K = p\lambda'/\Delta\lambda$ can be determined from the spectrum itself by counting the number p of fringes that occur within a line-free wavelength interval $\Delta\lambda$ at a wavelength λ' (Figure 2).

The optical setup for hook-method experiments shown in Figure 3 consists of the chamber (in this case a shock tube) containing the gas under investigation, a background light source producing a continuous spectrum; a Mach-Zehnder interferometer; and a stigmatic spectrograph (here mounted in a Littrow arrangement). In this example, the light source is imaged with an enlargement onto a diaphragm that can be considered the effective light source that illuminates interferometer and spectrograph. The light which is collimated before it enters the interferometer passes through an objective lens on the other side of the interferometer, forming an image of the diaphragm on the spectrograph entrance slit. Part of this image, namely the portion passing through the slit, reappears spectrally analyzed in the focal plane of the spectrograph. Since the interferometer produces fringes on the spectrograph slit, the slit is nonuniformly illuminated along its length, and the spectrograph must be stigmatic to image the fringes in the focal plane. In the example given, this is accomplished by the Littrow mounting. The light coming from the entrance slit and a totally reflecting 45° prism is collimated by a large lens — the so-called Littrow lens that serves simultaneously as a collimator and objective lens. The parallel light dispersed by a plane grating again passes through the Littrow lens and is focused onto a photographic plate or film mounted in the focal plane.

The interferometer is of the Mach-Zehnder type. It consists of two beamsplitters and two mirrors. The wavefront is divided in intensity at the first beamsplitter. The test beam is reflected by the upper mirror and then passes through the test chamber; the reference beam passes through a compensation chamber that equalizes the optical pathlength of the windows of the test chamber, is reflected by the lower mirror, and is then reunited with the test beam on the second beamsplitter.

If all interferometer plates are parallel to each other and the two beams have equal optical pathlengths, the field on the spectrograph slit will appear white due to constructive interference of the two beams. If, however, the optical paths of the two beams differ by half a wavelength, destructive interference occurs, and the field will appear dark.

To obtain fringes on the entrance slit of the spectrograph, a small angle must be introduced between pairs of parallel plates. Thus, fringes will actually be located at infinity and will be focused in the focal plane of the objective lens, i.e., in the slit plane. If the illumination is monochromatic, such fringes can be seen filling the entire field of view. By properly adjusting the interferometer plates, we can orient these fringes perpendicular to the slit. Bright fringes will appear at the following heights along the slit:

$$y = \frac{k\lambda - (n-1)\ell + (n'-1)\ell'}{a} \quad , \quad k = 0, \pm 1, \pm 2, \ldots$$

where n and ℓ are refractive index and length of the gas layer, respectively, and unprimed and primed symbols refer to test and compensation chamber, respectively; a is proportional to the angle between the two interfering wavefronts; k is the fringe order. The order of the central fringe in the field is given by the difference in optical pathlength between the two beams along the optical axis.

Since the position of a fringe (especially if it is of high order) may depend strongly on wavelength, a point in the field illuminated by light of one wavelength with a given fringe order may also be illuminated by light of another wavelength with a different fringe order. High order fringes, therefore, wash out and only two or three fringes on each side of zero order can be seen in undis-

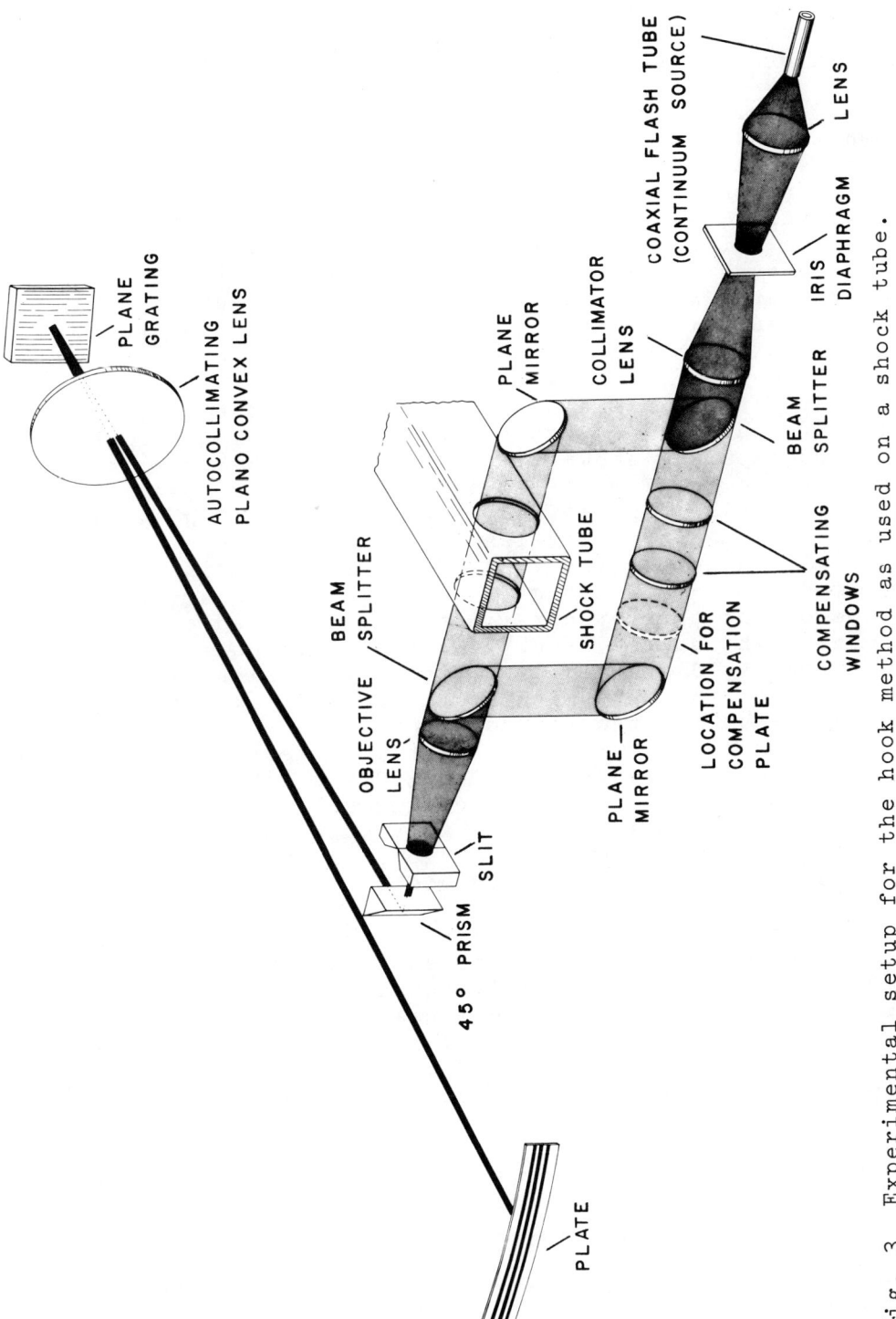

Fig. 3 Experimental setup for the hook method as used on a shock tube.

persed white light, which by its nature contains a wide continuous range of wavelengths.

To obtain fringes of high order, we must create a pathlength difference between reference and test beams. This is done either by changing the geometrical pathlength of one of the two interferometer beams or by inserting a plane-parallel plate made of glass or fused silica to increase optical pathlength. In the context of the hook method, such a plate is called a compensation plate, because at a certain wavelength (actually, where hooks are formed), it compensates the phase shift caused by anomalous dispersion.

Effects of anomalous dispersion and the compensation plate on the fringe pattern in the focal plane of the spectrograph are shown in Figure 4. In the uppermost spectrum we see fringes close to zero order, displaying constant height except for slightly increasing separations toward longer wavelengths. The next spectrogram, made with a gas in the test beam, shows the interference fringes bent at wavelengths close to absorbing lines. (Some of the dark spectral lines were caused by absorption in the light source itself and these, of course, display no anomalous dispersion.) If a compensation plate is inserted into the test beam, high-order fringes are obtained whose height is strongly dependent on wavelength. Two such cases are shown in the next two spectrograms, where the hook-method constant K is 2300 and 4600, respectively. The introduction of an absorbing gas into the test beam yields the final spectrogram in a superposition of the fringe patterns seen in the second and fourth spectra producing the characteristic hooks.

Note that the hooks form far from the absorbing core of the line; this makes it possible to use the hook method for diagnostics of optically thick lines. Since the absorption spectrum of such lines shows core saturation, absorption measurements require an accurate knowledge of the damping parameter, i.e., the ratio of Lorentz to Doppler line width. This problem does not arise with hook measurements where the influence of line broadening can usually be neglected as long as the hook separation is about ten times larger than the full width at half the maximum optical depth (FWHM) of the absorption line. The amount of hook separation can be controlled by varying the thickness of the compensation plate, and in cases where line broadening remains important, a correction can be applied (see section on Theoretical Aspects of the Hook Method, Review of Experiments).

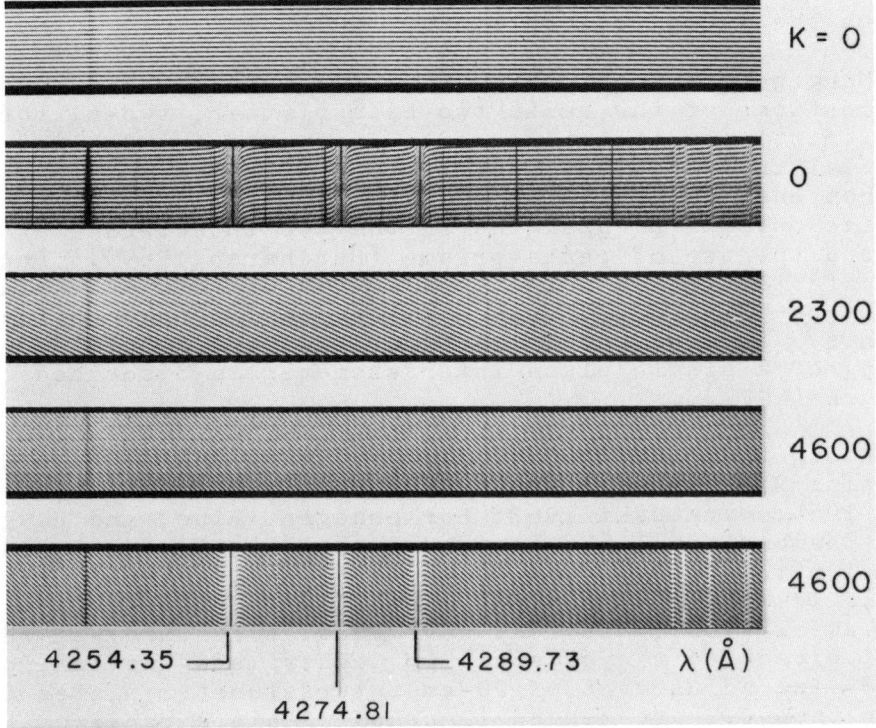

Fig. 4 Spectra taken with the arrangement shown in
Figure 3, using argon with a 0.02 percent admixture of
chromium hexacarbonyl [Cr (CO)$_6$] and an absorbing layer
5 cm thick. Reading from top to bottom, the spectra are
of: (a) zero order fringes; (b) zero order fringes with
absorbing medium in the test beam; (c) high order fringes
with 2 mm fused silica compensation plate in reference
beam; (d) same as (c) but with 4 mm compensation plate;
(e) same as (d) but with absorbing medium in the test beam.

Mach-Zehnder Interferometers

A remarkable amount of literature has been published
on the theory and design of Mach-Zehnder interferometers.
The two most recent summaries were published by Ladenburg
and Bershader (1955) and independently by Tanner (1968).
Both papers discuss theory, design, and optical specifications and adjustment. An additional comment on astigmatism in the Mach-Zehnder interferometer by Prowse
(1967a) pointed out that in order to avoid astigmatism
introduced by refraction at an oblique plane, the object
(and thus the test beam) should be positioned so that the

image is seen by reflection at rather than by transmission through the second beamsplitter.

Many papers and reports have also been published on modifications of the basic two-beamsplitter, two-mirror arrangement; new techniques of adjustment; and designs of actual instruments. Modifications include an adjustable compensator (Stein and Shultz 1965) to conveniently equalize optical pathlengths in the two interferometer arms and the use of penta-prisms (Hariharan 1969). Penta-prisms facilitate adjustment, allowing up to 1 cm variation in pathlength difference without additional optical elements. A novel arrangement for the hook method consisting of a single plate interferometer is described by Spurk (1971).

Among the many reports on designing, building, and adjusting Mach-Zehnder interferometers of various sizes (Hall 1954; Wygnanski 1963; Borkenhagen, Ribe, and Sawyer 1963; Chamberlain 1963; Lee and Park 1964; El-Wakil and Jaeck 1964; Johnstone and Smith 1965; Kaspar 1965; Gebhart and Knowles 1966; Rowe 1966; Prowse 1967b; Brooks, Probert and Maxwell 1968), the most novel design feature was the use of piezoelectric crystals by Gebhart and Knowles for remote fine adjustment of 20-cm interferometer plates to better than 0.2 arc second (1 μrad) within a range of 45 arc seconds (225 μrad).

It may also be useful to point out the novel design features employed by N.L. Hazen and G.U. Nystrom at Harvard College Observatory where two Mach-Zehnder interferometers were built: one for the visible and near ultraviolet wavelength regions and one for the vacuum ultraviolet. Made of clamped magnesium alloy tooling plates, both instruments are lightweight and have good thermal conductivity. All four corner-blocks, which contain the plate drives, are interchangeable. The same is true for the cells of such auxiliary optics as lenses and compensation plates, which can easily be inserted into rotational locking devices that position the elements on axis and perpendicular to the optical axis within 10 arc seconds. Eccentric shafts drive the four interferometer plates around two perpendicular axes that lie in the plane of the reflecting layer. Hinging is provided by flex-pivots which, being totally determinate, are perfect elements from a static point of view. In addition, such pivots allow rotation in one axis; since this motion is accomplished by pure bending of metal straps, the flex-pivots have no positional hysteresis in angle or axis.

In principle, interferometer plate drives do not pose too difficult a problem. They must provide very small motions (i.e., good resolution), but there are no special requirements on other drive parameters such as drive ratio, linearity, or repeatability, since the operator can rely on optical feedback rather than mechanical parameters to adjust the instrument. The total range of motion is also very small, in our case 30 arc minutes for the plates (or 0.75 mm for the pathlength adjustment); compared to the 0.2 arc second (or 0.075 μm) resolution required, this results in a "dynamic range" of roughly 10,000. It proved feasible to cover this dynamic range with a single mechanical system, allowing both coarse and fine adjustments at the same time. The piezoelectric drive (mentioned above) provides the same resolution for 20 cm plates as the mechanical system does for 5 cm plates and a useful dynamic range of only 225.

The critical areas of such a mechanical drive are a smooth finish of the eccentric shaft actually driving the plates, and a sufficiently precise bearing for the shaft carrying the eccentric. All remaining parts can have tolerances that require no extra care.

With the Harvard Mach-Zehnder interferometer for visible and near ultraviolet observations, pathlength adjustment is achieved by horizontal translation of a corner block; the vacuum interferometer utilizes a flex hinge consisting of two leaf-springs, a system that provides similar advantages for translation as the flex-pivot does for rotation.

Stigmatic Spectrographs, Detectors, and Light Sources

As mentioned earlier, spectrographs used in hook method experiments must be stigmatic. In addition, they should have good resolution in order to display fringes of high order, and high dispersion to allow a convenient measurement of hook separations. These conditions require a grating instrument. As a rule of thumb, the dispersion should be roughly 1 mm $Å^{-1}$, which can be accomplished using an 1800 line mm^{-1} grating in second order and a focal length of two meters. Since the optical speed of such an instrument will in general be large (in the class of $f/17$ or better), the grating width should be about 100 mm or more. A grating of this size, ruled with 1800 lines mm^{-1} and used in second order, will have a theoretical resolving power of 360,000. This number equals the highest fringe order that can be resolved theoretically and gives a safety factor of 50 over what may be needed in practice. The high dispersion, nevertheless, is required for the accurate measurement of hook separations.

The best stigmatic mountings for diffraction gratings are the Littrow and the Wadsworth mounts (Sawyer 1963). Other nearly astigmatic mountings are the Czerny-Turner and the very similar Ebert mounts. The Littrow instrument is the most compact, but requires transmission optics of rather large size (for example, the lens in our 2 m Littrow instrument is 18 cm in diameter). Thus a Littrow spectrograph with the dispersion required by the hook method is rather impractical for vacuum ultraviolet wavelengths. We therefore chose a commercially available Czerny-Turner spectrograph (McPherson Co.) for use with our vacuum Mach-Zehnder interferometer.

If stigmatic properties are required over a small wavelength region only, any spectrograph can be used as a stigmatic instrument as long as the fringes can be focused in the sagittal plane for the particular wavelength used.

Recent developments in holographic gratings (Labeyrie and Flamand 1969; Cordelle et al. 1970) will make it feasible to obtain stigmatic spectra over extended wavelength ranges using a holographic grating as the only optical element.

While photographic film or plates are conventionally employed for recording hook spectra, the introduction of image-tubes with beam deflection would make it possible to record hook spectra in sufficiently rapid succession to sample formation and excitation rates within a single event.

Finally, a few words about the requirements for a continuous background source in hook spectroscopy would be appropriate since they differ from those for absorption spectroscopy. When hot gases or plasmas are being investigated, absorption spectroscopy is possible only as long as the brightness temperature of the background continuum exceeds the thermodynamic temperature of the plasma. This does not apply rigorously to hook spectra, because fringes are recorded and measured only outside spectral lines, i.e., where the emissivity of a plasma is usually negligible. Furthermore, in absorption spectroscopy, background continua should be free of structure; in particular, no absorption or emission features at the wavelengths investigated can be tolerated. Here again, since requirements for hook spectra are not so stringent, quasi-continua can be used as they are emitted by dye-lasers (Bradley et al. 1968; Snavely 1969). Such continua can have durations as short as tens of picoseconds to milliseconds or, as reported recently (Peterson, Tuccio, and Snavely 1970), they can be emitted continuously.

THEORETICAL ASPECTS OF THE HOOK METHOD, REVIEW OF EXPERIMENTS

In this section, we consider theoretical aspects of the hook method and conditions for hook formation. Further, we show how line broadening affects hook positions and thus, experimental results. We also discuss hook formation in the case of crowded spectra, with emphasis on the situation in molecular bands. Each of these theoretical discussions is illustrated with examples of relevant experimental work.

We have seen that the vertical position y of a fringe on the spectrograph entrance slit is given by

$$y = \frac{k\lambda - (n-1)\ell + (n'-1)\ell'}{a}, \quad k = 0, \pm 1, \pm 2, \ldots$$

and noted that the difference $(n'-1)\ell' - (n-1)\ell$ describes the difference in optical pathlength between reference and test beams. In the hook method setup, $(n-1)\ell$ corresponds to the optical pathlength added to the test beam by the refractive index of the gas under investigation, and $(n'-1)\ell'$ stands for the optical pathlength added to the reference beam by the compensation plate.

A hook position is defined as the wavelength at which the vertical fringe position y goes through a maximum or minimum, i.e., where $dy/d\lambda = 0$. This yields

$$\ell \frac{d(n-1)}{d\lambda} = \ell' \frac{d(n'-1)}{d\lambda} + k = -K .$$

The right-hand side of the equation describes the experimental setup. Since this expression does not change rapidly with wavelength, it is commonly replaced by a single number K called the *hook-method constant*.

If we substitute the quantum-mechanical expression for refractivity $(n-1)$ of the gas under investigation (see, for example, Marlow 1967), the condition for hook formation becomes

$$\frac{d(n-1)}{d\lambda}\ell = \frac{d}{d\lambda}\left[\frac{r_o \ell}{4\pi}\sum_m \frac{N_m f_m \lambda_m^3}{\lambda-\lambda_m}\right] = -\frac{r_o \ell}{4\pi}\sum_m \frac{N_m f_m \lambda_m^3}{(\lambda-\lambda_m)^2} = -K$$

where λ_m, f_m, and N_m indicate respectively, the wavelength, absorption oscillator strength, and number density of the lower level for all transitions — in an atom or molecule. The following assumptions were made to obtain this expression: 1) that the refractivity at wavelengths within or close to the half-width (FWHM) is of no interest; 2) that no Doppler broadening is present; and 3) that negative dispersion can be neglected. The equation is further simplified if the spectral line under study is well separated from other lines, in which case, the squared wavelength difference in the denominator renders all but one of the summation terms negligible. The two hooks formed on each side of a single line indicate the two roots of this quadratic equation. Hence, if the hook-method constant K is known, and the wavelength of the line and one hook are measured, the product $Nf\ell$ can be determined.

In practice, one uses the fact that the two hook positions are symmetrical to the line center and measures their separation Δ_0 (see Figure 2). The formula for $Nf\ell$ then becomes $Nf\ell = \pi K \Delta_0^2 / (r_o \lambda_o^3)$.

As mentioned above, the constant

$$K = -\ell \cdot \frac{d(n'-1)}{d\lambda} - k = \frac{p\lambda'}{\Delta\lambda}$$

can be determined from the spectrum by counting fringes in a line-free wavelength interval. Because of the dispersion in the compensation plate, the hook-method constant is not simply the negative fringe order but can be smaller than the fringe order (taken as absolute number) by 5 to 10 percent or more.

Normally, the hook-method constant can be determined to an accuracy within one-half percent or better. The precision of hook position measurements depends on several factors, the most important ones being hook separation and grain size of the photographic emulsion. As mentioned earlier, the hook separation can be varied using a compensation plate of the appropriate thickness. In addition, by choosing the vertical fringe spacing (through interferometer adjustment), one can produce sharp or shallow maxima and minima of fringes (i.e., hooks), while hook separation remains fixed. Increasing separation between hooks (by using a thinner compensation plate) results in shallow extrema. These can be made sharper by enlarging the vertical fringe spacing, but this will reduce the number of fringes seen in the field of view. For reliable measurements, it is advisable to have about three to six fringes at a particular wavelength on hook spectra (see e.g., Forbrich 1967). Different methods of measuring the hook position have been described by Rozhdestvenskii (1951) and Shukhtin (1965) who use the information contained in the course of the fringes to determine more accurately the position of the fringe maxima.

For strong lines, the product $Nf\ell$ can usually be determined to within better than one percent. Two-percent measurements are quite routine. However, when the hook separation becomes smaller than 300 μm on the plate, measurements rapidly become very difficult and accordingly lose precision.

The fact that hook separation (with constant K) goes with the square root of the product $Nf\ell$ accounts for the large dynamic range of the method; the ratio of the largest and smallest $Nf\ell$ values that can be measured with the hook method is about 10^6. It is worthwhile comparing this range to that obtained with the absorption and emission methods, for which this ratio hardly exceeds a factor 10. Unfortunately, the two methods do not overlap; there is a gap of about a factor of 10 between the strongest transition that can be measured on or near the linear part of the curve of growth and the weakest line that can reliably be measured by the hook method.

Penkin (1964) describes a method for simultaneous absorption and hook-method measurements that can be applied to resonance lines. It is ideal for f-value measurements, because it does not require that number densities be known; but it does require an absence of collisions, i.e., very low pressures, and may therefore be of little interest in gas diagnostics. However, it is easy to observe hook and absorption spectra simultaneously; thus $Nf\ell$ for strong and weak lines can be determined with a single exposure. The arrangement needed is shown in Figure 5. The first beamsplitter in the interferometer is replaced by a plate that is half mirror and half beamsplitter. By imaging this "beamsplitter" onto the spectrograph slit, one also obtains a sharp dividing line in the focal plane of the spectrograph. Below this line, the image of the beamsplitter portion produces a

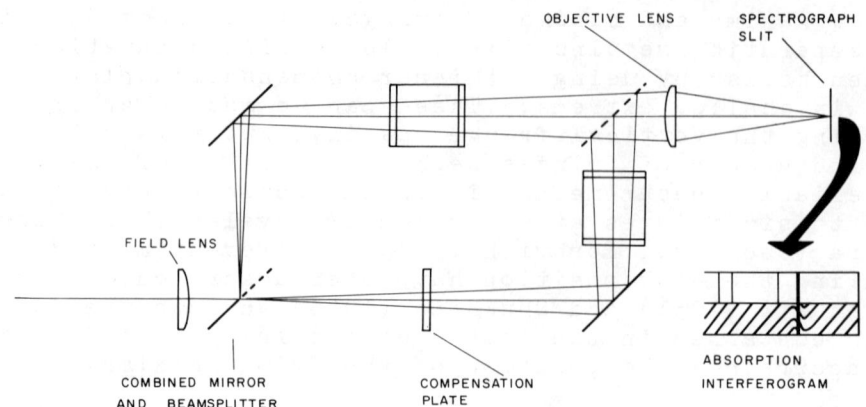

Fig. 5 Optical arrangement for simultaneous recording of hook and absorption spectra.

hook spectrum; above the line, there appears an absorption spectrum that shows no fringes, because the reference beam is blocked by the mirror portion. Such a combined spectrum, obtained at Harvard, is seen in Figure 6. The absorbing material in this case was shock-heated iron. Note how hooks form only near the strong lines and that almost no effect is seen on the fringe pattern near weak lines. This spectrum was recorded using a flash tube providing a continuous background in a 3 μs exposure (Wheaton 1964) and a fast spectrograph shutter (Grasdalen, Huber, and Newsom 1968) to avoid excessive line emission from the shock-heated gas.

The fact that fringes need not be focused at any particular location (in Figures 3 and 5 they were focused in infinity and on the first beamsplitter, respectively) could in principle be used to obtain spatial resolution within the absorbing region investigated. In this case, the fringes would have to be focused at the location of

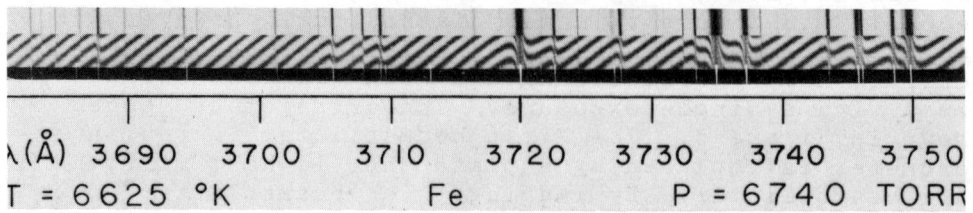

Fig. 6 Absorption and hook spectra, photographed simultaneously through a shock-heated gas.

the test chamber and onto the spectrograph slit. The
long dimension of the slit would then be aligned with an
axis in the source containing nonuniformities of interest.
To the best of our knowledge such use of the hook method
has not yet been reported; the evaluation of resulting
hook spectra is more complicated, since hook separations
will vary with height y on the spectrogram.

We have premised that another important feature of
the hook method (besides its large dynamic range) is its
low sensitivity to line broadening. The effect of line
broadening on hook separation is best shown schematically.
Figure 7 gives the ratio $\mu = \Delta/\Delta_0$ between the actual hook
separation Δ and the separation Δ_0 for the case of an
unbroadened line. The abscissa is the ratio of the Lor-
entzian half-intensity width (FWHM), $\Delta\lambda_L$ to Δ_0, and the
curve parameter is the corresponding ratio $\Delta\lambda_D/\Delta_0$ for the
Doppler width. One recognizes that the effect on hook
separation is below 10 percent as long as $\Delta\lambda_L/\Delta_0 < 0.2$,
i.e., as long as hook separation exceeds line width by a
factor of five or more. Since the correction needed is
small, an *estimate* of line-widths will often be sufficient
to obtain reliable results on Nfl. The diagram inserted
in the upper right-hand corner of Figure 7 shows the de-
pendence of μ on the ratio $\Delta\lambda_D/\Delta_0$ for the case where hook
separation is ten times larger than Lorentzian half-width.

The theoretical treatment of simultaneous Doppler
and Lorentz broadening was first given by Marlow (1967).
To introduce Doppler broadening, the number density N in
the quantum-mechanical expression for the refractivity
is replaced by an integral representing a Maxwellian velocity distribution. The damping giving rise to the
Lorentz broadening is entered into the same expression by
adding it to the damping constant γ, which represents
damping due to the natural lifetime alone. With these
changes, the assumption that observations are made only
far from the half-intensity point becomes unnecessary.
The course of a fringe followed through the resonance can
be seen in Figure 8. This diagram, which is similar to
Figure 10 of Marlow's (1967) review, also shows the
"inner hooks," and inner and outer hooks approaching each
other when the line is broadened.

The position of the inner hooks with a hydrogen discharge is currently being investigated by Forbrich (1970)
at Colorado Springs. Earlier observations of inner hooks
were reported by Prokof'ev (1924). To obtain fringes in
the region of absorption, the reference beam must be
attenuated with a filter, so that the two interfering

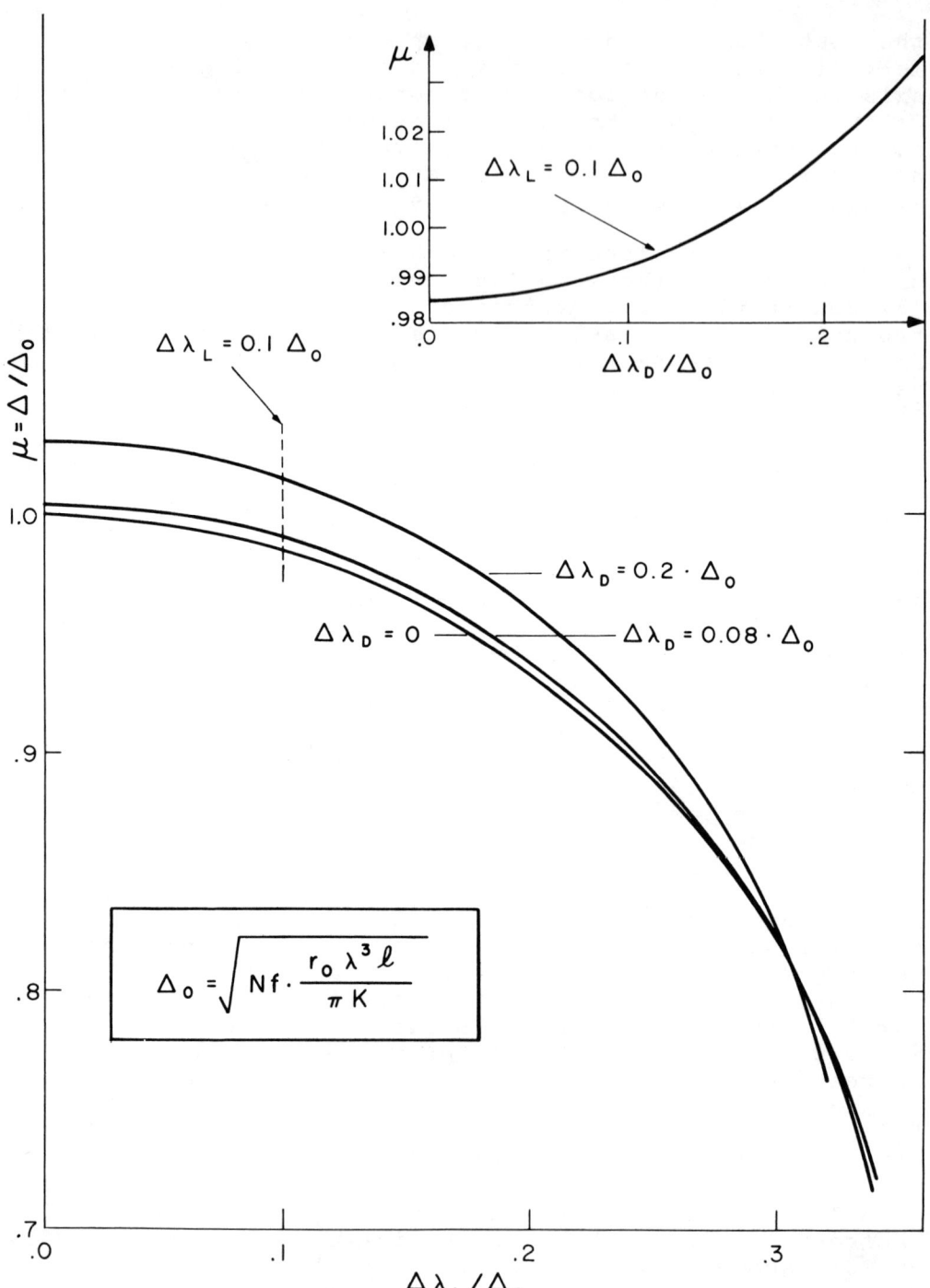

Fig. 7 Effect of line broadening on hook separation.

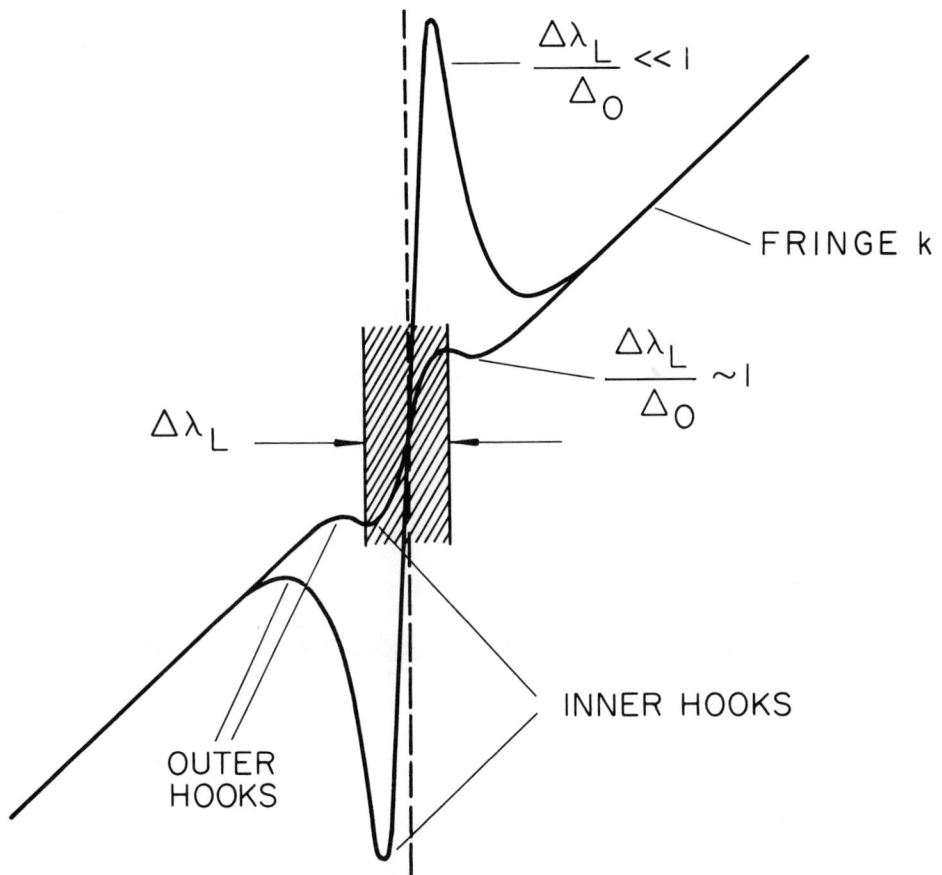

Fig. 8 Theoretical course of a fringe through an absorption feature for negligible broadening ($\Delta\lambda_L/\Delta_0 \ll 1$) and strong broadening ($\Delta\lambda_L/\Delta_0 \approx 1$).

beams are of similar intensity and yield good contrast (Figure 9).

A similar investigation on the 2537 Å line of mercury was made by Fork and Bradley (1964). To obtain the high resolution required near the line center, these authors used a mercury discharge as background source, and they varied the wavelength by using one Zeeman component and a variable magnetic field.

It should be mentioned that Day (1967) has computed the refractivity for the case of asymmetrically broadened

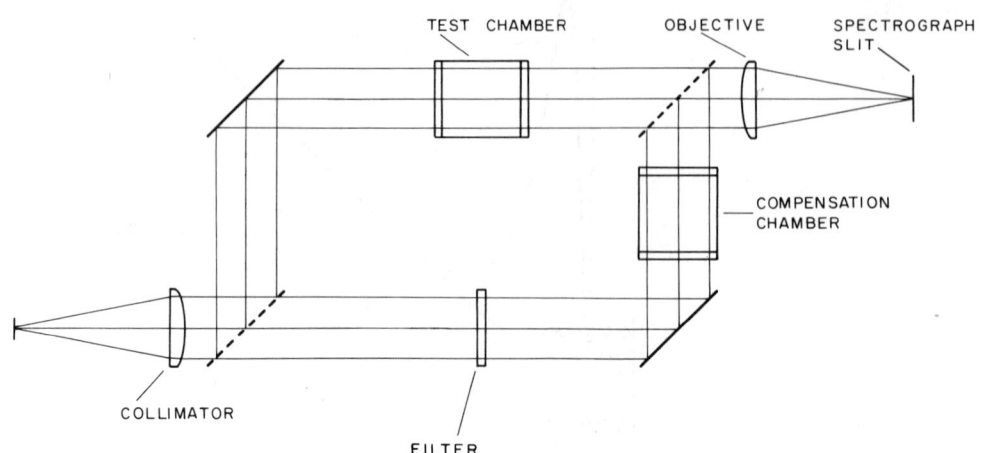

Fig. 9   Optical arrangement for the observation of inner hooks.

line profiles resulting from the quasistatic approximation for Stark broadening, and found that the hook separation is somewhat larger than for symmetrical profiles. Owing to the nature of the transform linking absorption and refractivity, the asymmetry most pronounced in the line wings of the absorption profile is most pronounced near the core of the refractivity profile and, therefore, does not lead to great differences from the case of hooks formed near a symmetrical Lorentz profile.

Finally, we take up the case of several closely spaced lines (Figure 10). We have shown above that hooks occur at wavelengths λ that are roots of the equation:

$$\psi(\lambda) = \frac{N_1 f_1 \lambda_1^3}{(\lambda-\lambda_1)^2} + \frac{N_2 f_2 \lambda_2^3}{(\lambda-\lambda_2)^2} + \frac{N_3 f_3 \lambda_3^3}{(\lambda-\lambda_3)^2} + \ldots = \frac{4\pi K}{\ell r_o}$$

that contains a sum $\psi(\lambda)$ of terms, each term representing one line. Hooks occur at those wavelengths where $\psi(\lambda)$ assumes the value $4\pi K/r_o \ell$, a constant that can be represented by a straight horizontal line. At those wavelengths where $\psi(\lambda)$ and $4\pi K/r_o \ell$ cross, there are real roots, and hooks are formed. If the lines are spaced too closely, the roots will be imaginary and no hooks will form. Here the influence of nearby lines becomes strong; i.e., the contribution of their summation terms is no longer negli-

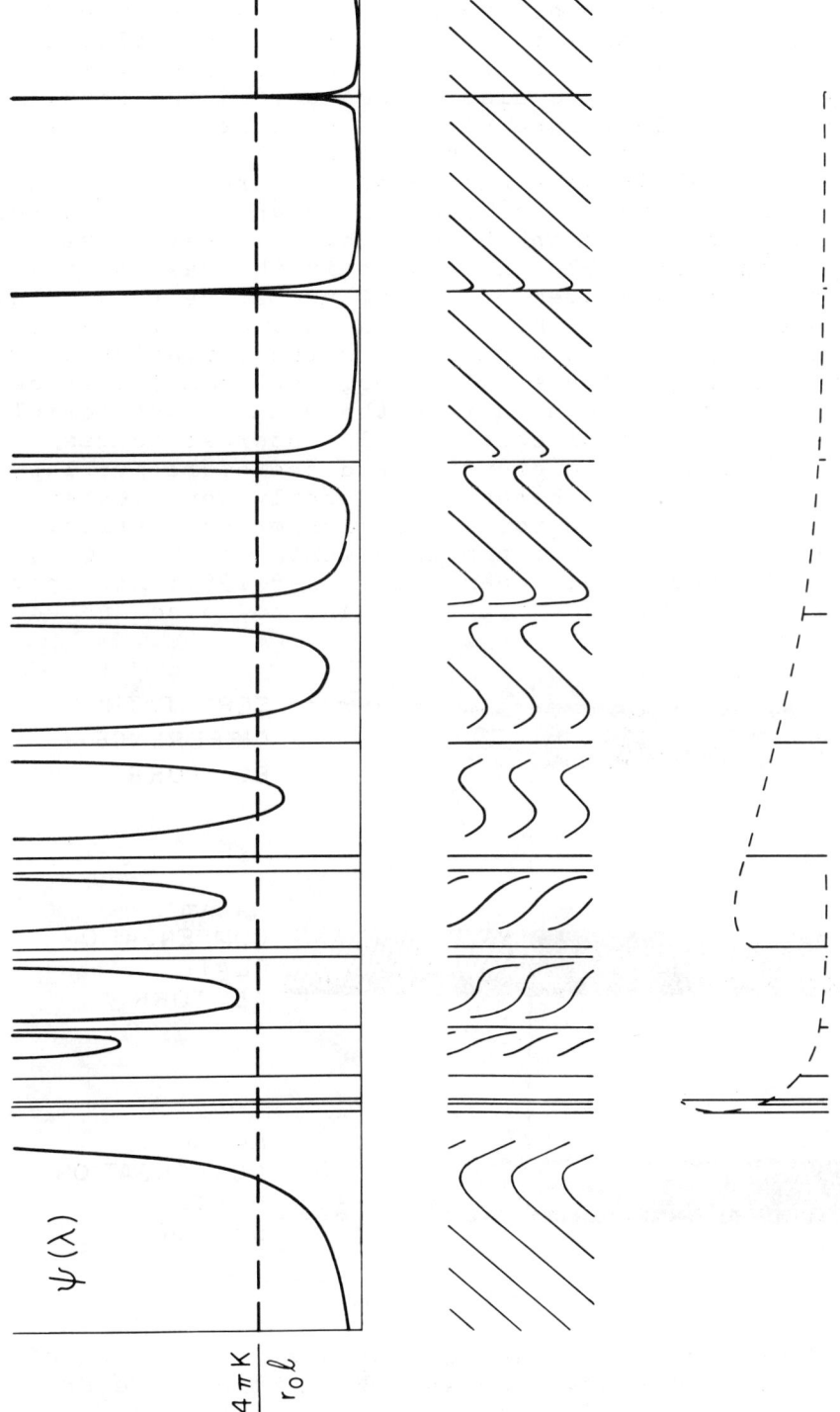

Fig. 10 Function $\psi(\lambda)$ and corresponding hook spectrogram for a molecular band exhibiting two branches. Wavelength and intensities of individual lines are indicated in bottom diagram by positions and lengths, respectively.

gible. In this example of two branches, as they would be formed by a $^1\Sigma$-$^1\Sigma$ transition of a diatomic molecule, hooks around individual lines are formed far away from the band head only. But a hook is also formed beyond the head, i.e., here on its left-hand side. The position of this latter hook is influenced by the crowded lines near the band head. If wavelengths and *relative* strengths of the individual lines within the band are known, one can determine Nfl just from this one hook. This is what Pery-Thorne and Banfield (1970) have done in the case of the (0,0) band of the γ-system of the nitric oxide (NO) molecule (Figure 11). Here the line spacing, especially near the band head, is much too narrow for the formation of hooks between single lines. However, the relative strength of the lines may be computed from the Hönl-London formulae. This method should be of considerable interest to gas diagnostics, because the crowding and therefore overlapping of lines near band heads considerably complicates measurements of Nfl from total emission or absorption. In fact, previous absorption measurements of the γ(0,0) band of NO (Bethke 1959; Weber and Penner 1957; Antropov, Sobolev, and Cheremisinov 1964) all had to be conducted

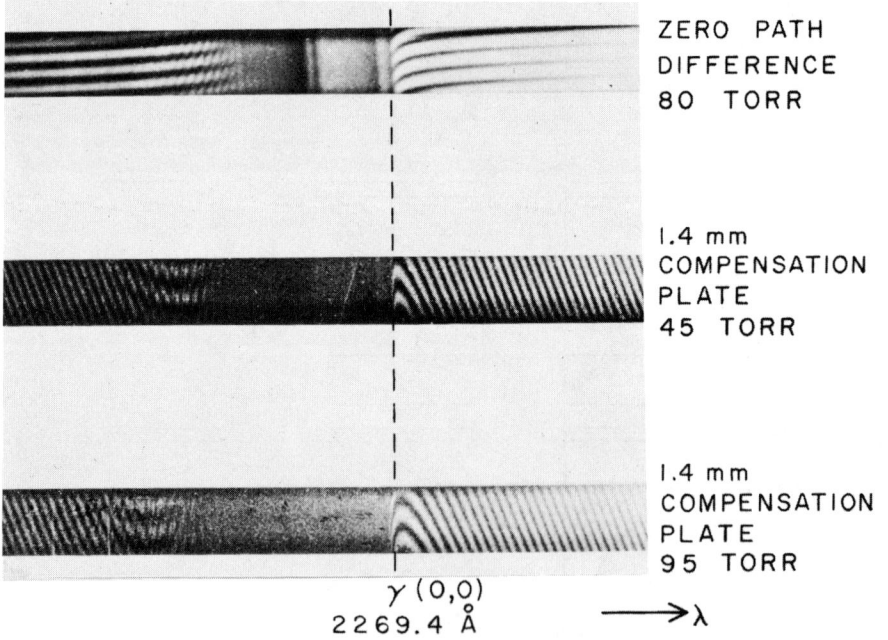

Fig. 11 Hooks formed at the head of the γ(0,0) band of nitric oxide. *(Courtesy Lady Anne Pery-Thorne and Dr. F.P. Banfield, Imperial College, London, England)*

at pressures high enough to smear out the rotational structure in order to avoid the subtle problems of overlapping line profiles. These measurements therefore depend on a long extrapolation back to zero pressure. By virtue of its insensitivity to line widths and saturation, the hook method avoids such difficulties.

In addition to these measurements of electronic bands, determinations of transition moments of infrared bands from refractivity measurements have been reported by Arcas and Hochard-Demollière (1967) for CO_2 and by Chamberlain (1967) for HCl as well as some solids and liquids. The former authors pointed out that the influence of line broadening could not be demonstrated within the limits of experimental error. Their experimental setup consisted of a spectrograph followed by a Michelson interferometer that measured fringes photoelectrically, either at a fixed wavelength by changing the gas pressure, or at a constant gas pressure by changing the wavelength (Alamichel, Assous, and Legay 1963). Chamberlain (1967) also gave an extensive discussion of numerical methods for the calculation of band strengths.

In conclusion, we should mention that Bradley, Gale, and Smith (1970) have reported the observation of dispersion delays in potassium vapor. Their experimental arrangement consisted of a dye laser with a 2 cm^{-1} bandwidth, an absorption cell, and a spectrograph equipped with a streak camera. Hence, according to linear dispersion theory (Brillouin 1960), the signal pulse delay τ far from resonance is

$$\tau = \frac{\pi N e^2 f \ell}{mc} \frac{1}{(\omega_o - \omega)^2} \approx \frac{N f \ell r_o}{4\pi c} \left(\frac{\lambda_o^2}{\lambda - \lambda_o} \right)^2$$

where $\omega = 2\pi\nu = 2\pi c/\lambda$ is the circular frequency, and e, m, and c are the electron charge and mass, and speed of light respectively. The delay is positive on both sides of the resonance, because the signal velocity is observed rather than the phase refractive index as is the case in the hook method. Bradley et al. (1970) also report reduced dispersive delays under different experimental conditions (lower density and higher peak power) where induced transparency (i.e., a coherency effect) becomes important.

SUMMARY

Refractivity measurements are shown to be advantageous for many applications in gas diagnostics. Most convenient for this purpose is the so-called hook method which, together with the required experimental equipment (Mach-Zehnder interferometer and stigmatic spectrograph), is discussed in detail. The principal advantages of the hook method are its large dynamic range and insensitivity to line profiles, notable assets in measurements on saturated lines. These and other applications of the hook method in particular and of refractivity measurements in general are illustrated by a number of reported experiments. Examples include simultaneous hook and absorption spectra, anomalous dispersion in the neighborhood of electronic as well as infrared molecular bands, and optical delays due to anomalous dispersion.

ACKNOWLEDGEMENTS

We wish to thank Dr. W.H. Parkinson for the vigorous support and frequent encouragement of the author's hook-method experiments. Discussions with Lady Anne Pery-Thorne, Drs. T.J. McIlrath, E.M. Reeves, and C.A. Forbrich were most helpful and are gratefully acknowledged. Thanks are also due to Prof. K. Dressler for supporting a sojourn at the Swiss Federal Institute of Technology in Zurich, where most of this manuscript was written.

The research performed at Harvard College Observatory was funded by the National Aeronautics and Space Administration through grant NGL-22-007-006.

REFERENCES

Alamichel, C., Assous, R., and Legay, F. 1963, Appl. Optics, $\underline{2}$, 495.

Alpher, R.A. and White, D.R. 1965, in *Plasma Diagnostic Techniques*, R.H. Huddlestone and S.L. Leonard, eds. (Academic Press, New York and London), p. 431.

Anketell, J. and Pery-Thorne, A. 1967, Proc. Roy. Soc. A, $\underline{301}$, 343.

Antropov, E.T., Sobolev, N.N., and Cheremisinov, V.P. 1964, Optics and Spectrosc., $\underline{16}$, 115.

Arcas, P. and Hochard-Demollière, L. 1967, C.R. (Acad. Sci., Paris) Series B, 264, 1258.

Ascoli-Bartoli, U. 1970, in *Physics of Hot Plasmas*, B.J. Rye and J.C. Taylor, eds. (Plenum Press, New York, and Oliver & Boyd, Edinburgh), p. 404.

Bershader, D. 1971, in *Modern Optical Methods in Gas Dynamic Research*, D.S. Dosanjh, ed. (Plenum Press, New York).

Bethke, G.W. 1959, J. Chem. Phys., 31, 662.

Borkenhagen, W.H., Ribe, F.L. and Sawyer, G.A. 1963, Los Alamos Sci. Lab., Univ. Calif., Rept. LA-2940.

Bradley, D.J., Durrant, A.J.F., Gale, G.M., Moore, M., and Smith, P.D. 1968, IEEE J. Quantum Electr., QE-4, 707.

Bradley, D.J., Gale, G.M., and Smith, P.D. 1970, Nature, 225, 719.

Brillouin, L. 1960, *Wave Propagation and Group Velocity* (Academic Press, New York and London), p. 121.

Brooks, R.G., Probert, S.D., and Maxwell, J. 1968, Meas. and Control, 1, T9.

Cary, B. and Nickels, W. 1963, General Electric, Missile and Space Division, Tech. Inform. Series R63SD82, DDC No. 423565.

Chamberlain, J.E. 1963, Ph.D. thesis, Dept. Phys., Imperial College, London. (For a short description, see also Pery-Thorne, A., and Chamberlain, J.E. 1963, Proc. Phys. Soc., 82, 133.)

Chamberlain, J.E. 1967, J. Quant. Spectrosc. Radiative Transfer, 7, 151.

Cordelle, J., Flamand, J., Pieuchard, G., and Labeyrie, A. 1970, in *Optical Instruments and Techniques* (Oriel Press, London), p. 117.

Day, R.A. 1967, Scientific Report No. 20, Shock Tube Spectroscopy Laboratory, Harvard College Observatory, Cambridge, Mass. 02138.

El-Wakil, M.M. and Jaeck, C.L. 1964, Am. Soc. Mech. Eng., Trans. Ser. C, J. Heat Transfer, 86, 464.

Forbrich, C.A. 1967, Ph.D. thesis, SU-IPR Rept. No. 191, Inst. Plasma Res., Stanford Univ., Stanford, Calif.

Forbrich, C.A. 1970 (private communication).

Fork, R.L. and Bradley, L.C. 1964, Appl. Optics, 3, 137.

Gebhart, B. and Knowles, C.P. 1966, Rev. Sci. Instrum., 37, 12.

Grasdalen, G.L., Huber, M., and Newsom, G.H. 1968, Rev. Sci. Instrum., 39, 886.

Hall, J.G. 1954, Rept. No. 27, Inst. Aerophys., Univ. Toronto.

Hariharan, P. 1969, Appl. Optics, 8, 1925.

Huber, M. 1966, J. Opt. Soc. Amer., 56, 1428.

Huber, M. and Tobey, F.L., Jr. 1968, Ap. J., 152, 609.

Johnstone, R.K.M. and Smith, W. 1965, J. Sci. Instrum., 42, 231.

Kaspar, J. 1965, Zpráva VZLU, No. 3, 31, Vyzkumný a Zkusebni Letecky Ustav, Prague.

Kopfermann, H. and Ladenburg, R. 1928, Z. Phys., 48, 51.

———. 1930, Z. Phys., 65, 167.

Labeyrie, A. and Flamand, J. 1969, Opt. Commun., 1, 5.

Ladenburg, R. and Bershader, D. 1955, "Interferometry in High Speed Aerodynamics and Jet Propulsion," in *Physical Measurements in Gas Dynamics and Combustion,* vol. IX, High Speed Aerodynamics and Jet Propulsion (Princeton, Princeton University Press).

Ladenburg, R. and Kopfermann, H. 1928, Z. Phys., 48, 26.

Ladenburg, R. and Levy, S. 1930, Z. Phys., 65, 189.

Lee, T.S. and Park, S.D. 1964, Tokyo University, Japan, 7 p. research report (Clearinghouse f. Sci. and Tech. Infor. acc. no. N65-15449).

Marlow, W.C. 1967, Appl. Optics, 6, 1715.

Meroz, I. 1962, ed., *Optical Transition Probabilities, A Collection of Russian Articles, 1924-1960*, Israel Program for Sci. Transl., Jerusalem (available from Office of Tech. Services, U.S. Dept. Commerce, Washington, D.C.).

Meroz, I. 1963, ed., *Optical Transition Probabilities, A Collection of Russian Articles, 1932-1962*, Israel Program for Sci. Transl., Jerusalem (available from Office of Tech. Services, U.S. Dept. Commerce, Washington, D.C.).

Penkin, N.P. 1964, J. Quant. Spectrosc. Radiative Transfer, 4, 41. (Translated by J.L. Tech in Astron. Papers Transl. from the Russian, No. 3, 1964, Smithsonian Ap. Obs., Cambridge, Mass.).

Penkin, N.P. and Shabanova, L.N. 1965, Optics and Spectrosc., 18, 504.

Pery-Thorne, A. and Banfield, F.P. 1970, J. Phys. B (Great Britain), 3, 1011.

Peterson, O.G., Tuccio, S.A. and Snavely, B.B. 1970, Appl. Phys. Letters, 17, 245.

Prokof'ev, V.K. 1924, Trudy Gosudarstvennogo Opt. Inst. (Leningrad), 3, 1; transl. in Meroz 1962, op. cit., p. 1.

Prowse, D.B. 1967a, Appl. Optics, 6, 773.

____. 1967b, Tech. Note 100, Dept. Supply, Austral. Def. Sci. Service, Def. Stand. Lab., Ascot Vale, Victoria.

Rowe, R.L. 1966, Instrum. Soc. Am., Trans. 5, 44.

Rozhdestvenskii, D.S. 1912, Ann. Phys., 39, 307.

____. 1951, *Works on Anomalous Dispersion in Vapors of Metals*, S.E. Frish and N.P. Penkin, eds., Akad. Nauk, USSR. (Translated in part by L.G. Robbins, U.S. Army Engineer Research and Development Laboratories, Ft. Belvoir, Va., T-1024, 1961.)

Sawyer, R.A. 1963, *Experimental Spectroscopy*, 3rd. ed. (Dover Publ., New York).

Shukhtin, A.M. 1965, Optics and Spectrosc., 19, 457.

Slavenas, I.-Yu. Yu. 1966, Optics and Spectrosc., 20, 264.

Snavely, B.B. 1969, Proc. IEEE, 57, 1374.

Spurk, J. 1971, in *Modern Optical Methods in Gas Dynamic Research*, D.S. Dosanjh, ed. (Plenum Press, New York).

Stein, A. and Shultz, T. 1965, Appl. Optics, 4, 1510.

Tanner, L.H. 1968, in Tech. Rept. for the year 1956, Aeronaut. Res. Council (London: Her Majesty's Stationary Office), p. 595.

Tech, J.L. 1964, Astron. Papers Transl. from the Russian, No. 3, Smithsonian Ap. Obs., Cambridge, Mass.

Tumakaev, G.K. and Lazovskaya, V.R. 1965, Soviet Phys. Tech. Phys., 9, 1449; see also Dunayev, U.A., Maslennikov, V.G., Mishin, G.I., Sistchikova, M.P., and Tumakayev, G.K. 1967 in *Proc. 7th Intl. Congr. High Speed Photogr.*, Zurich, Sept. 1965, O. Helwich ed. (O. Helwich, Darmstadt and Wien), p. 588.

Weber, D. and Penner, S.S. 1957, J. Chem. Phys., 26, 860.

Wheaton, J.E.G. 1964, Appl. Optics, 3, 1247.

Wiese, W.L. 1968, "Atomic and Electron Physics, Atomic Interactions, Part A," in *Methods of Experimental Physics*, vol. 7, B. Bederson and W.F. Fite, eds., p. 117.

Wiese, W.L., Smith, M.W., and Glennon, B.M. 1966, *Atomic Transition Probabilities*, vol. I, "Hydrogen through Neon," Natl. Stand. Ref. Data Series, Natl. Bur. Stand. 4 (U.S. Govt. Print. Off., Washington, D.C. 20402).

Wiese, W.L., Smith, M.W., and Miles, B.M. 1969, *Atomic Transition Probabilities*, vol. II, "Sodium through Calcium," Natl. Stand. Ref. Data Series, Natl. Bur. Stand. 22 (U.S. Govt. Print. Off., Washington, D.C. 20402).

Wygnanski, I. 1963, Tech. Note 63-1, Supersonic Gasdynamics Res. Lab., Mech. Eng. Res. Lab., McGill Univ., Montreal.

HOOK INTERFEROMETRY USING A SINGLE PLATE INTERFEROMETER

Joseph H. Spurk

Ballistic Research Laboratories, USAARDC

Aberdeen Proving Ground, Maryland 21005

The hook method of Roschdestvensky[1] has been used in the past to study the anomalous dispersion of gases and has recently received much attention as a tool to measure the Nf values of a variety of gases in conjunction with the shock tube[2]. The method is generally more accurate than other methods[3] and may also be used when the line in question is substantially broadened. The method is in fact self-calibrating, but it entails the complexity of an interferometer. Most of the recent work has been done using a Mach-Zehnder interferometer. This instrument allows considerable freedom in adjustments, such as fringe spacing and fringe orientation, and also allows a wide beam separation. Of these advantages only the wide beam separation can be important in the hook method; the fringe orientation, by contrast, is always fixed in this application and the fringe spacing is largely, but not solely, determined by the order of the fringes being observed, i.e., by the difference in optical path in the two legs of the interferometer. This path difference, which is purposely introduced by inserting a parallel plate in one leg, is the essential difference of the hook method of Roschdestvensky and the method by Puccianti. As a result of the additional path the fringes are inclined in the spectrum.

The Mach-Zehnder interferometer in this use comprises a number of optically high quality surfaces and is in practice difficult to adjust. The difficulties in adjustment and the requirement on flatness of the plates are compounded if work is attempted in the vacuum ultraviolet region of the spectrum, where lie the major resonance lines of the practically important gases such as the constituents of air. The material for interferometers in this spectral region is quite expensive and homogeneity of the crystalline material is a critical consideration. The material most attractive from the standpoint of transmission properties in the VUV region, lithium fluoride, does not polish well and is easily scratched. These considerations require that the number of transmissive plates in the interferometer be kept to a minimum.

A simple interferometer which consists of a flat plate whose surfaces may be inclined under a small angle to form a wedge is considered here. If this wedge is illuminated by parallel light coming from an extended source made up of incoherent point sources, fringes can be seen only (in quasi-monochromatic light) in the vicinity of the plane of localization. This plane is the locus of the intersection of pairs of rays derived from single incident rays[4] as shown in Figure 1. The fringes in this plane are equidistant and parallel to the apex of the wedge. It is apparent that one ray may be used as the reference beam and the other as the test beam of an interferometer, where the additional path needed in the hook method is now seen to be an integral part of the interferometer. Thus, when the plane of localization is viewed through a spectrograph in white light, fringes will be seen in the exit of the spectrograph which are inclined under an angle determined by the wedge angle and by the order of the fringes being observed.

The beam separation is determined by the thickness of the plate (just as in a Jamin interferometer) and the thickness is restricted by the resolving power of the spectrograph. It may be shown that one must have $\lambda/\Delta\lambda > m$ in order to see fringes in the

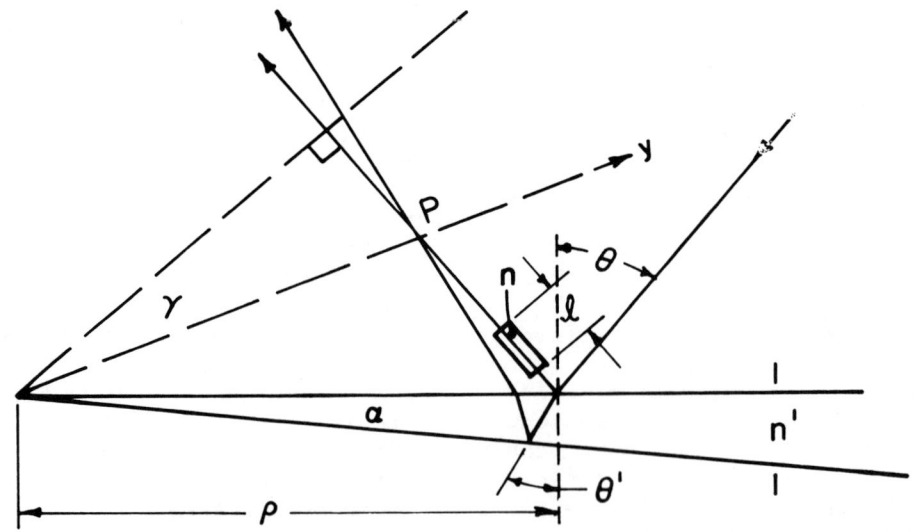

Figure 1. Schematic of single plate interferometer.

exit of the spectrograph. Here $\lambda/\Delta\lambda$ is the resolving power of the spectrograph and m the order of the fringes. The largest admissible optical path difference is then given by $\Delta S_{max} = \lambda m_{max} = \lambda(\lambda/\Delta\lambda)$. In the setup shown in Figure 1 the maximum separation σ_{max} of the two beams is then seen to be

$$\sigma_{max} = \tan^2 \theta' \cot \theta \, \Delta S_{max}$$

With $\tan^2 \theta' \cot \theta \sim 1$ for simplicity and using a spectrograph with a resolving power of 120,000 say, the maximum separation is $\sigma_{max} = 1.2$ cm for a wavelength of 1000 Å. This separation is sufficient in shock tube work, where the shock wave is straddled and when using a light source of sufficiently short duration. The actual order m may have to be considerably less than the above upper limit; this is due to the fact that the inclination of the fringes leads to an effective reduction of fringe spacing from the spacing in monochromatic light by the factor $[1 + (m/b\alpha)^2 (d\lambda/d\ell)^2]^{-1/2}$ so that the fringes may become too fine to be easily seen. Here $(d\lambda/d\ell)$ is the inverse linear dispersion and b is a function of θ and n' only.

A larger beam separation may be achieved without increasing the optical path by choosing a very thick wedge plate and then compensating for the too large optical path by inserting an additional flat plate in the test beam. A larger beam separation can also be gotten using only reflective optics, in which case the wedge plate is the only transparent element in the system.

The feasibility of the single plate interferometer has been tested by observing the anomalous dispersion of sodium vapor in the vicinity of the D_1 and D_2 lines. The experimental setup is shown in Figure 2. In this arrangement the intersection of the

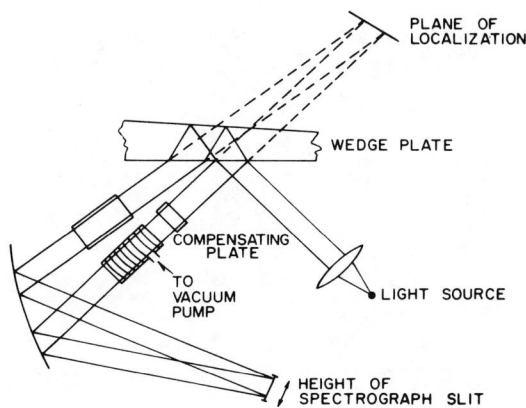

Figure 2. Sketch of experimental setup.

two rays derived from one incident ray is a virtual intersection and the plane of localization is imaged on the entrance slit of the spectrometer by a spherical mirror. The "monochromatic" spacing is now affected by the f-number of the mirror and it is, in fact, possible to use a flat plate instead of a wedge-like plate. As is indicated in Figure 2 this setup also has provision to use a compensating plate in the test beam in order to increase the allowable beam separation.

The sodium vapor in this test was produced by heating a circular steel tube into which small pieces of sodium had been introduced. Figure 3 shows samples of spectra obtained in this setup. The quality of the spectra is quite comparable to spectra obtained using a Mach-Zehnder interferometer. The cost of this simple interferometer is, of course, vastly smaller. The versatility of

Figure 3. Examples of spectra in the vicinity of the sodium D lines.

the Mach-Zehnder type arrangement is rarely needed in the hook application and it appears that the single plate interferometer can be used with profit in this particular application. An interferometer using a lithium fluoride plate and reflective optics only has been made and work in the VUV region of the spectrum is now in progress.

REFERENCES

1. D. Roschdestvensky, Ann. Physik 39, 307 (1912), "Anomalous Dispersion in Sodium Vapor."
2. Yu. A. Dunaev, G. K. Tumakaev and A. M. Shukhtin, Soviet Phys.--Tech. Phys. 6, 815 (1962).
3. W. C. Marlow, Appl. Optics 6, 1715 (1967); also Lockheed Missile and Space Division LMSC A034392 (1965).
4. M. Born and E. Wolf, Principles of Optics (Pergamon Press, New York, 1959), p. 296.

SOME OPTICAL DIAGNOSTIC TECHNIQUES INVOLVING HIGH POWER LASERS

A.J. Alcock

Physics Division, National Research Council of Canada

Ottawa 7, Canada

1. INTRODUCTION

Although lasers have permitted the development of several completely new diagnostic techniques such as Holographic Interferometry and Thomson Scattering they have also brought about significant improvements in other more conventional methods, and in the present paper some examples of optical interferometry and Schlieren photography involving high power lasers are described.

Since about 1958 interferometric and schlieren techniques at optical wavelengths have been widely applied due to the increased interest in plasmas with electron densities exceeding $10^{14} cm^{-3}$. Additional impetus was provided by the development of lasers in the early sixties and the scope of optical diagnostic methods was greatly increased by these new and extremely powerful light sources. In particular, giant pulse lasers, capable of generating pulses with peak powers exceeding 1 MW and having pulse durations ranging from a few tens of nanoseconds down to several picoseconds, offer a number of important advantages over conventional light sources.

These are:-

1) a high degree of spectral purity
2) extremely low values of beam divergence
3) high intensity
4) extremely short pulse durations.

The high degree of temporal and spatial coherence responsible for (1) and (2) permits interferometric studies to be carried out with relatively crude instruments and, in some cases, completely eliminates the need for compensating cells or plates within an interferometer. In addition, the high intensity and coherence greatly alleviate the problem of overcoming the plasma luminosity, both directly due to the brightness of the source and also by the ease with which unwanted radiation can be eliminated by means of spectral and spatial filtering. The short pulse duration offers the advantage of extremely good time resolution although it should be remembered that the optimization of this parameter may degrade the spectral purity sufficiently to compel a compromise between different requirements of the system.

Many of the above features are exemplified by the Q-switched ruby laser which has the additional advantage of operating at a visible wavelength. Indeed, such a laser was used, in two of the earliest applications of lasers in optical diagnostics, for the

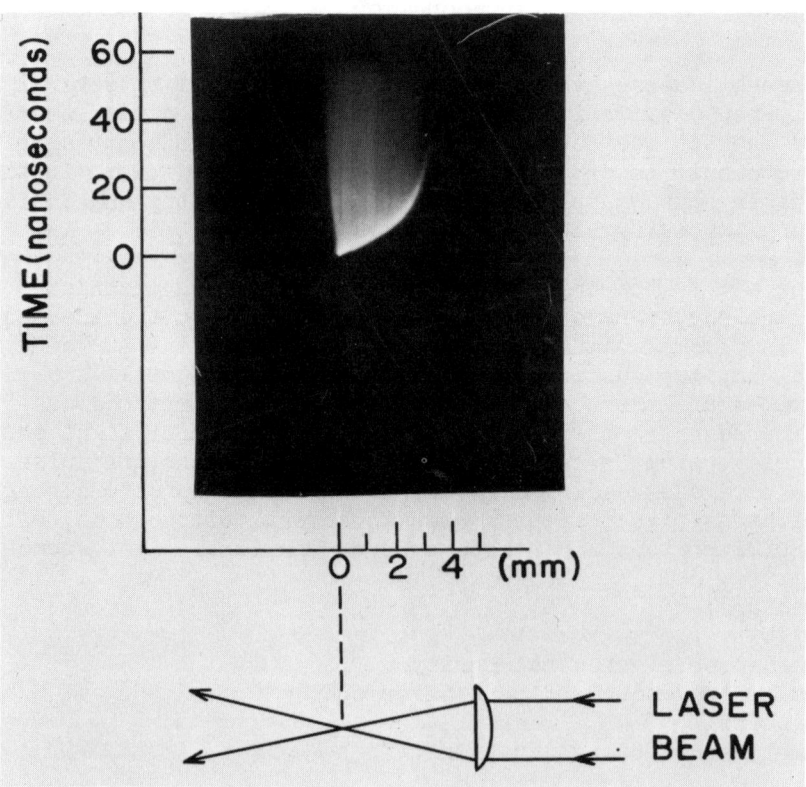

Fig. 1. Streak photograph of a typical laser spark in air at atmospheric pressure.

investigation of theta-pinch plasmas by means of optical interferometry[1] and schlieren photography.[2] However, another type of plasma, which has been extensively studied by such techniques, is the spark produced by focusing the radiation from a high power laser within a gas. In many cases the resulting plasmas are characterized by extremely small dimensions (<1cm), high electron densities (~ 10^{19}cm^{-3}), high electron temperature (10^5-10^6 °K) and rapid expansion along the laser beam axis (Fig. 1). Thus the detailed investigation of their structure has created a unique demand for improved diagnostic techniques. Some of these have already been employed in other situations and it is hoped that more widespread applications will be realized in the future.

2. LASER-PRODUCED PLASMA STUDIES IN THE NANOSECOND REGION

The development of diagnostic techniques applicable to the investigation of laser sparks has been closely associated with the

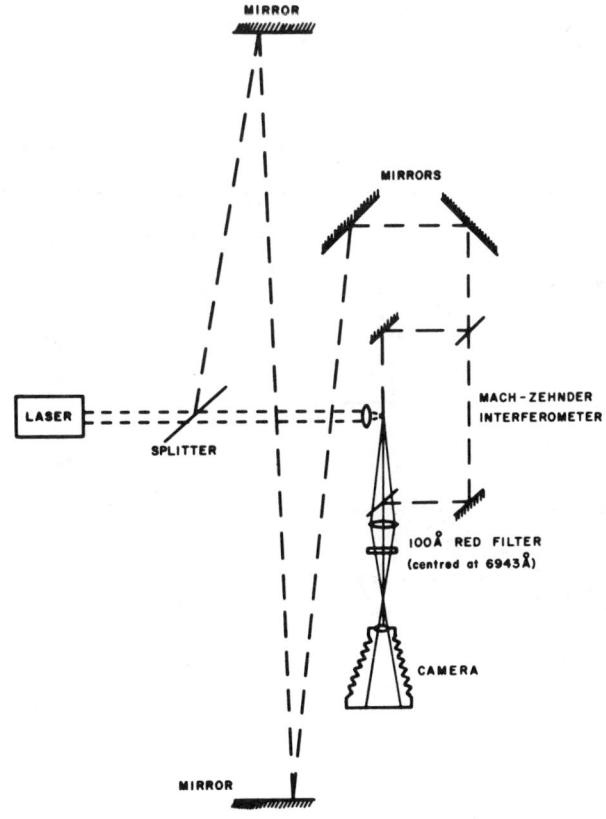

Fig. 2. Spark interferometry using one laser.

evolution of laser technology over the last ten years. Thus, when the first sparks were produced by lasers having pulse durations in the 30 - 50 nanosecond range, it was natural to use the same lasers to investigate their characteristics. The experimental arrangement used to obtain the first interferograms of laser sparks in air[3] is shown in Fig. 2. It can be seen that part of the beam used to produce the spark is reflected by a beam splitter into an optical delay line and eventually enters a Mach-Zehnder interferometer. Although such a system has the advantage of eliminating the problem of synchronization between the light source and the production of the plasma it becomes inconvenient when long delay times or shorter exposure times are required. An example of a more sophisticated arrangement, involving two lasers, is shown in Fig. 3. In this case the second laser, which is Q-switched by means of a Pockels cell, is synchronized with the laser producing the spark by deriving a 'prompt' pulse from the rotating prism Q-switch of the latter. With this system a jitter of ~25 nanoseconds was obtained and the insertion of a variable electronic delay permitted the development of sparks to be followed for several tens of microseconds after the occurrence of breakdown. An example of the interferograms obtained is shown in Fig. 4 and clearly illustrates the development and propagation of the blast wave produced by the spark. The minimum

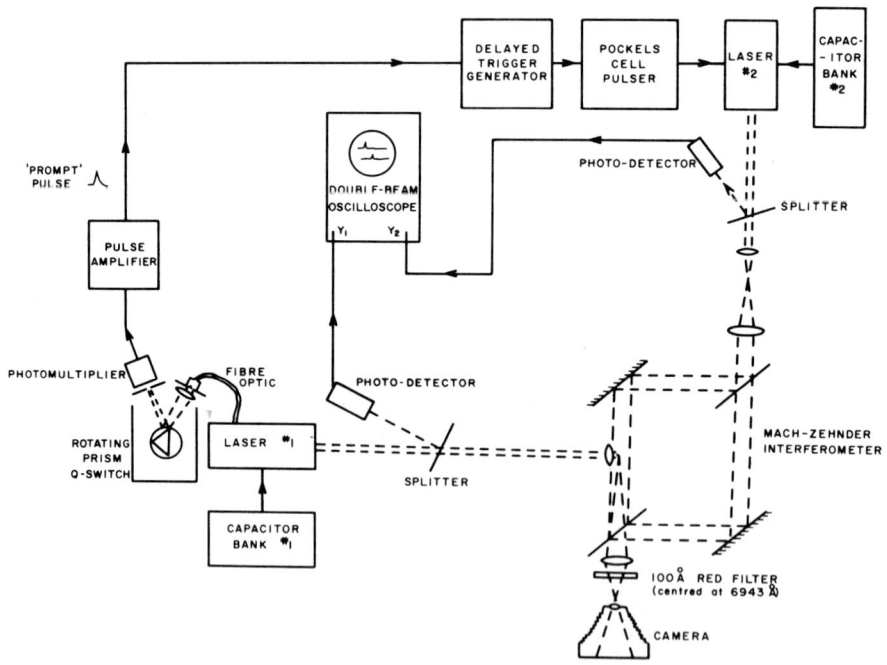

Fig. 3. Spark interferometry using two lasers.

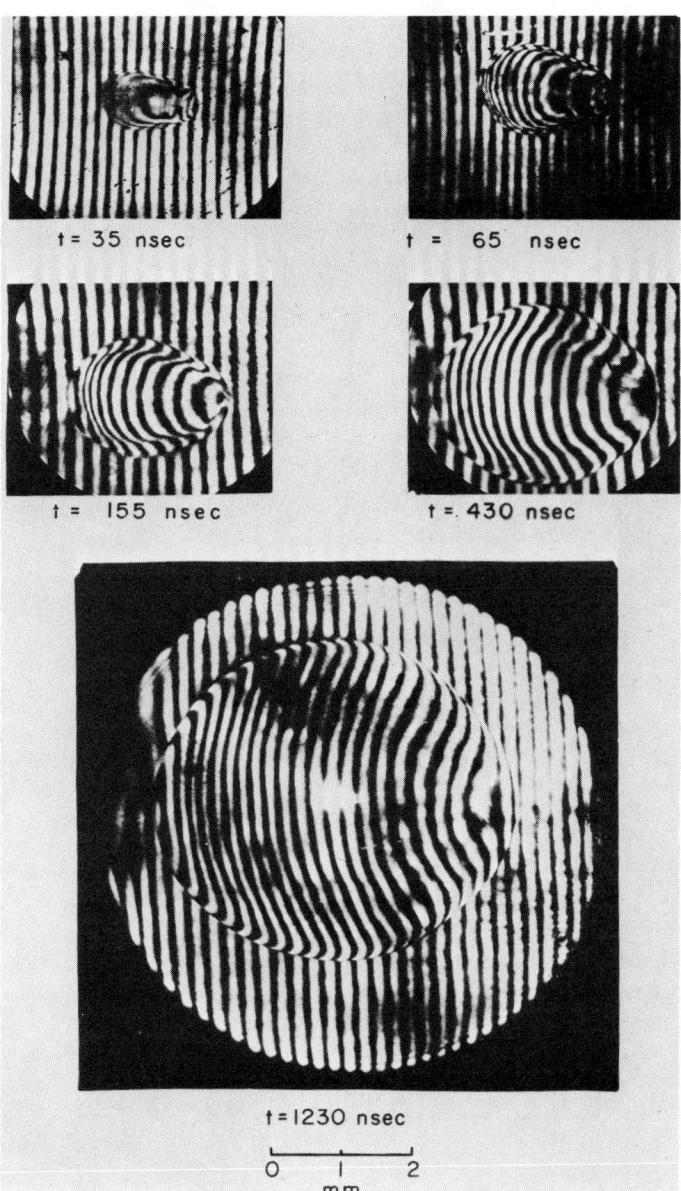

Fig. 4. Photographs of fringe pattern at various times after breakdown. The laser beam is incident from the left, and the spark was produced by focusing a 0.2 joule, 30 nsec, laser pulse with a 2.8cm focal length lens.

usable delay time of ~35 nsecs was determined primarily by the length of the illuminating pulse which was in fact reduced to ~6 nanoseconds by the addition of a Kerr cell shutter between the laser and the interferometer.

Although the time-resolved, two dimensional record provided by these interferograms permitted the development of a theoretical model for the asymmetrical blast wave produced by a laser spark,[4] a more detailed interpretation of the interferograms was precluded due to the uncertain contribution of neutral atoms to the fringe shifts.

It is well known that the measured fringe shift, S, over a path length, L, is given by

$$S = \frac{1}{\lambda} \int_0^L \delta(\mu-1)d\ell \quad (1)$$

where λ is the illuminating wavelength and $\delta(\mu-1)$ the total change in refractivity. In the absence of a magnetic field, one has for electrons with a density of N_e cm^{-3}

$$\delta(\mu - 1)_e \simeq -4.46 \times 10^{-14} \lambda^2 N_e \quad (2)$$

while, in the visible region, the contributions of other species depend only weakly on the wavelength. Thus, as originally pointed out by Alpher and White[5], it is necessary to obtain simultaneous interferograms at more than one wavelength if direct measurements of electron density are required. Such measurements can now be made with relative ease since the high intensity of Q-switched laser radiation permits the generation of additional wavelengths by means of a number of nonlinear processes. Of these, harmonic generation is particularly convenient since it is capable of converting approximately 10 - 20% of the laser radiation to the second harmonic wavelength, and the use of ruby laser light and its second harmonic for plasma investigations was first reported by Martellucci and Mazzucato[6], who obtained Schlieren photographs of a theta-pinch at these two wavelengths. At the same time interferometric studies of laser sparks were extended to two wavelengths[7] using the experimental arrangement shown in Fig. 5. Apart from the ADP harmonic generator crystal, it can be seen that the only additional components are a second camera, a beam splitter and narrow band filters to separate the two wavelengths. Using the relation

$$S_1\lambda_1 - S_2\lambda_2 \simeq -4.46 \times 10^{-14}(\lambda_1^2 - \lambda_2^2)\bar{N}_e \ell \quad (3)$$

where $\bar{N}_e$ is the average electron density over the path length ℓ, the simultaneous interferograms obtained at 6943Å and 3472Å (Fig. 6) permitted the determination of the average electron density during

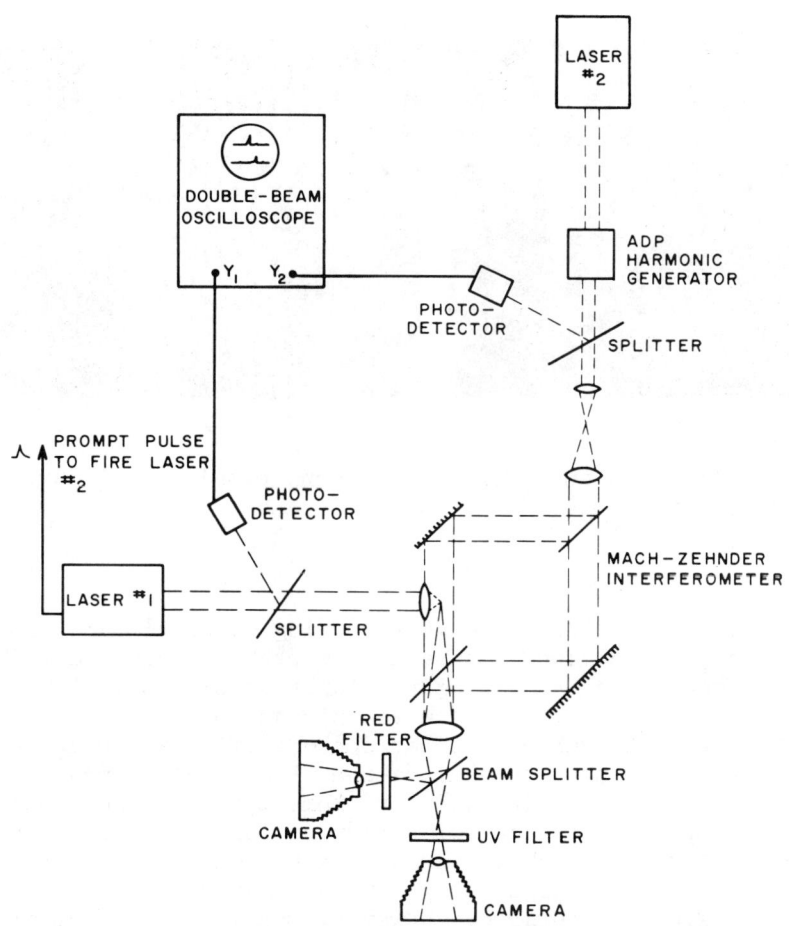

Fig. 5. Experimental arrangement used for two wavelength interferometry.

the first 300 nanoseconds after breakdown, (Fig. 7).

Two wavelength laser interferometry has subsequently been employed by Panarella for the investigation of the cylindrical converging shock waves generated in a small "high pressure" theta-pinch[8], and by Oertel and Spurk for the study of hypersonic ionized flows[9].

3. LASER-PRODUCED PLASMA STUDIES WITH SUBNANOSECOND RESOLUTION

As can be seen from Fig. 1 the initial velocity, at which a laser spark expands towards the source of energy, is of the order of 10^7 cm/sec. Thus the investigation of this rapid expansion phase requires diagnostic pulses with durations of one nanosecond or less if a spatial resolution of 0.1 mm is required.

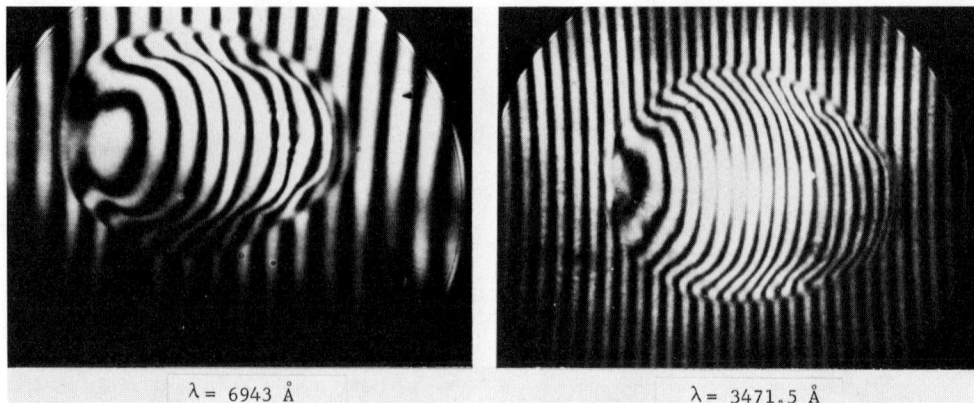

Fig. 6. Simultaneous interference patterns obtained 170 nanoseconds after breakdown, at $\lambda = 6943$Å and $\lambda = 3472$Å.

The achievement of passive mode-locking of ruby[10] and neodymium:glass[11] lasers, by means of saturable absorbers, yielded a light source which generates a train of regularly spaced, high intensity, pulses each having a duration of a few picoseconds. By suitable adjustment of the saturable absorber concentration and pumping level, trains of pulses having a total duration of several hundred nanoseconds could be obtained (Fig. 8) and this type of laser appeared potentially useful for the investigation of laser produced, or other ultra-fast, plasma phenomena, even though the duration of the individual pulses was considerably shorter than the value of one nanosecond originally required. However, the lack of spectral purity, and the difficulty involved in separating the interference patterns produced by successive mode-locked pulses prevented the immediate application of such a laser to the illumination of an interferometer. Thus, in spite of its inability to provide as much quantitative information, a schlieren system illuminated by a mode-locked neodymium:glass laser was employed.[12] A schematic diagram of the experimental arrangement is shown in Fig. 9. The mode-locked laser consisted of a Brewster-ended neodymium:glass rod, two dielectric mirrors having reflectivities of 75% at 1.06μ, and a Brewster angle cell containing Kodak 9740 saturable dye. Two-photon fluorescence measurements indicated that the mode-locked pulses had a duration of approximately 5 picoseconds and appropriate adjustment of the operating conditions yielded pulse trains with durations ranging from 0.4-1.0 microseconds. The use of two partially transmitting mirrors resulted in an output being available at each end of the laser cavity. One beam was focussed onto one of the electrodes of a spark gap while the other passed through an ADP harmonic generator crystal. Approximately 1 percent of the incident

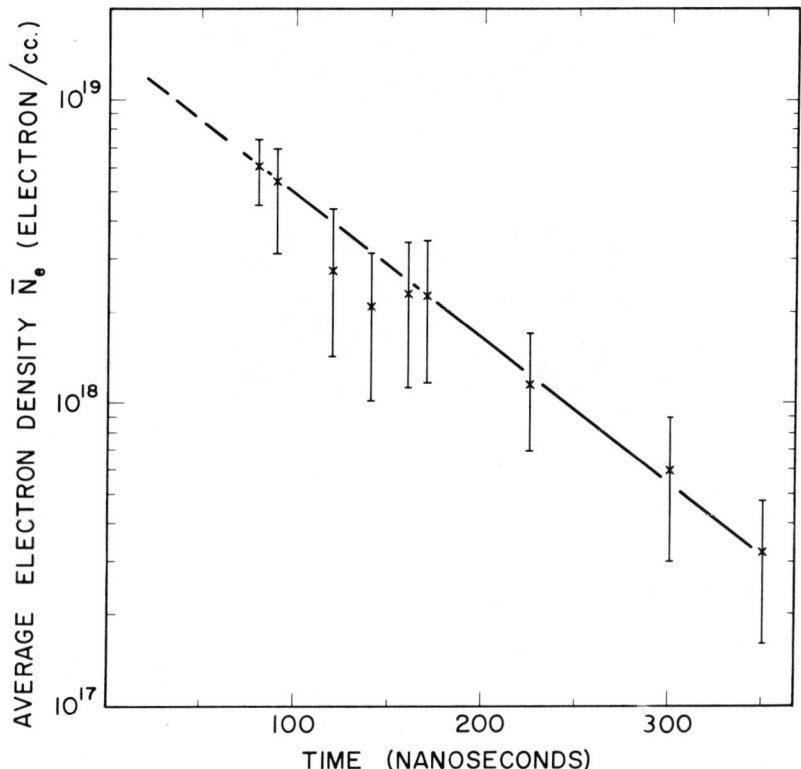

Fig. 7. Plot of average electron density vs. time for laser spark in air, at initial pressure of one atmosphere.

radiation was converted to the second harmonic wavelength of 5300Å after which it passed through a cell containing copper sulphate solution to absorb the fundamental. The sparks investigated were produced by the focused radiation from a ruby laser which was Q-switched by means of a Pockels cell. In order to synchronize the two lasers, the spark gap, triggered by the mode-locked laser, was used to switch the Pockels cell and with this arrangement the ruby laser pulse occurred between 200 and 400 nanoseconds after the start of the neodymium laser emission. The actual delay was determined by the length of the mode-locked pulse train but for a given pulse the jitter was not greater than 20 nanoseconds.

The second harmonic radiation from the mode-locked laser passed through the breakdown region in a direction perpendicular to the ruby laser axis and was then focused by the schlieren lens onto a

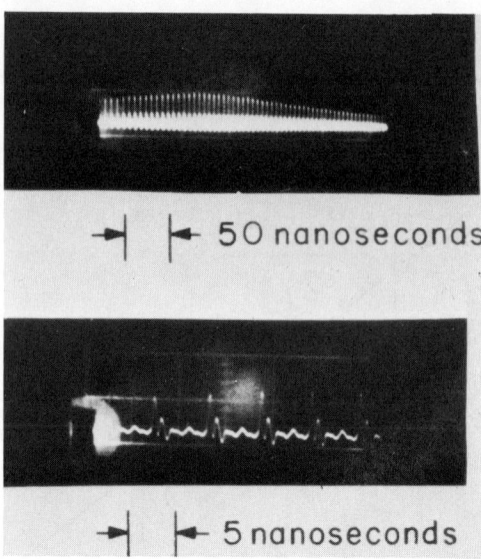

Fig. 8. Oscilloscope traces showing mode-locked pulse train. The time scale on the upper trace is 50 nsec/div; the lower trace shows the same pulses on an expanded scale of 5 nsecs/div.

knife edge or round obstacle. An additional lens was used to image the spark onto the film while a narrow band intereference filter and a polarizer were used to eliminate light from the spark. Typical Schlieren photographs of sparks in argon and neon are shown in Fig. 10 and in each case it can be seen that the central part of the photograph consists of a number of well-defined contours. Additional measurements of the longitudinal expansion of the plasma, made by means of an image converter streak camera, confirmed that each contour was produced by a single pulse from the mode-locked laser and corresponded to the boundary of the highly ionized core of the spark. Thus a very precise two-dimensional record of the plasma expansion velocity, during the first 20 to 30 nanoseconds after breakdown, was obtained (Fig. 11). Since the mode-locked pulse train continues for some hundreds of nanoseconds the later stages of the spark development are also recorded. However the outer region on the photograph is more difficult to interpret due to the much lower expansion velocity and the combined effect of the ionized core and the shock wave moving into the neutral gas.

Although mode-locked laser illumination can provide extremely good temporal resolution there are situations where the use of such a laser is inconvenient or extremely difficult. Such a case arose during an investigation of sparks produced by means of a giant-pulse ruby laser operating in a single axial and transverse mode. Since the axial mode selection was achieved by using a resonant reflector and saturable absorber Q-switch, low jitter triggering of the laser

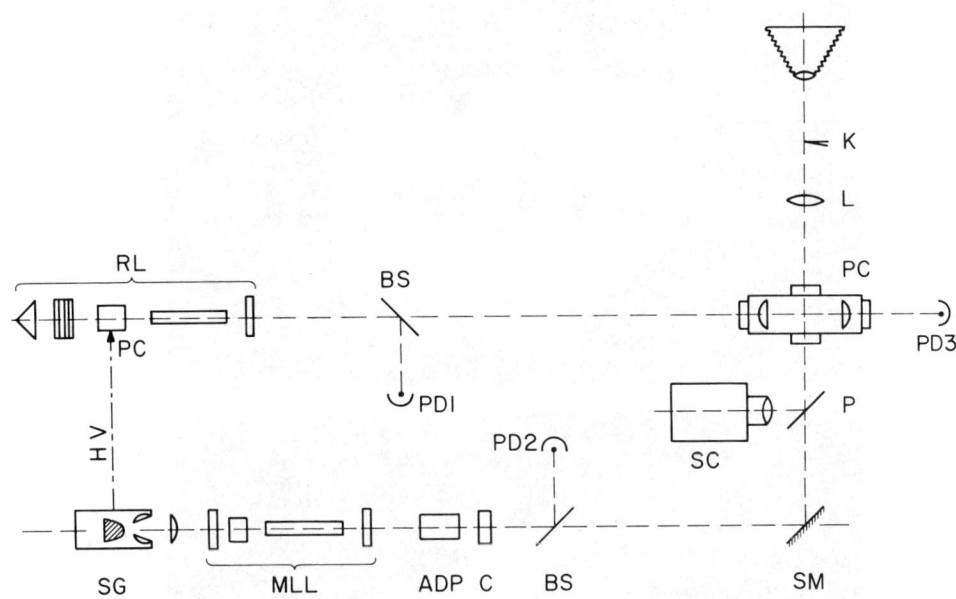

Fig. 9. Schematic diagram of mode-locked laser Schlieren setup;
MLL - mode-locked laser; RL - ruby laser; PC - Pockels cell;
SG - spark gap; ADP - second harmonic generator; C - copper
sulphate cell; BS - beam splitter; SM - silver mirror;
PD - photodetector; L - lens; K - knife edge; P - pellicle;
SC - streak camera.

pulse was not possible. Since this has not been achieved either in
the case of a laser mode-locked by means of a saturable absorber, an
alternative method for the generation of a subnanosecond illumination
pulse had to be developed. The technique which was used[13] permitted
a subnanosecond segment to be chopped from the 10 nanosecond single-
mode laser pulse and a diagram of the experimental arrangement is
shown in Fig. 12. Approximately 10% of the output from the single-
mode laser is reflected by a beam splitter into an amplifier rod
with a gain of ~8. A 50% beam splitter selects half of this amplified pulse which is then focussed on to the positive electrode of a
high pressure nitrogen spark gap[14]. The spark gap acts as the switch
in a 50Ω transmission line pulse generator capable of delivering,
with minimal delay and jitter, 10 Kv, rectangular, pulses having a
rise and fall time <300 psecs and a duration ranging from ~700 psecs
to several tens of nanoseconds. The resultant voltage pulse at the
terminated end of the output cable is approximately equal to the half
wave voltage of a small, low capacitance, Pockels cell which has an
aperture of 2mm and is situated between a pair of Glan-air polarizers.

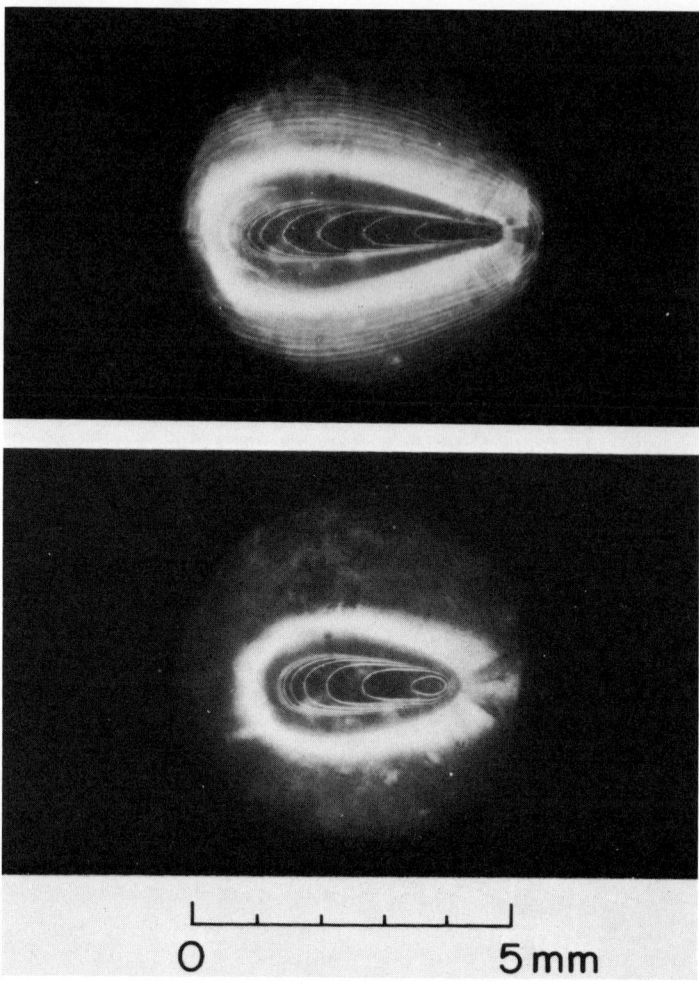

Fig. 10. Schlieren photographs of laser-induced sparks. The laser beam is incident from the left. Upper photograph - argon; lower photograph - neon: In both cases the pressure was 1010 torr and the ruby laser radiation was focused by a 3.5 cm focal length lens.

The remainder of the amplifier output is delayed by 6 nanoseconds to ensure that the high voltage pulse and the peak of the laser pulse arrive at the Pockels cell simultaneously. Since the intrinsic delay between the breakdown of the spark gap and the peak of the laser pulse is ~1 nanosecond, the use of a 700 psec voltage pulse permits the selection of a subnanosecond segment from the amplifier output. The performance of this extremely fast electro-optical

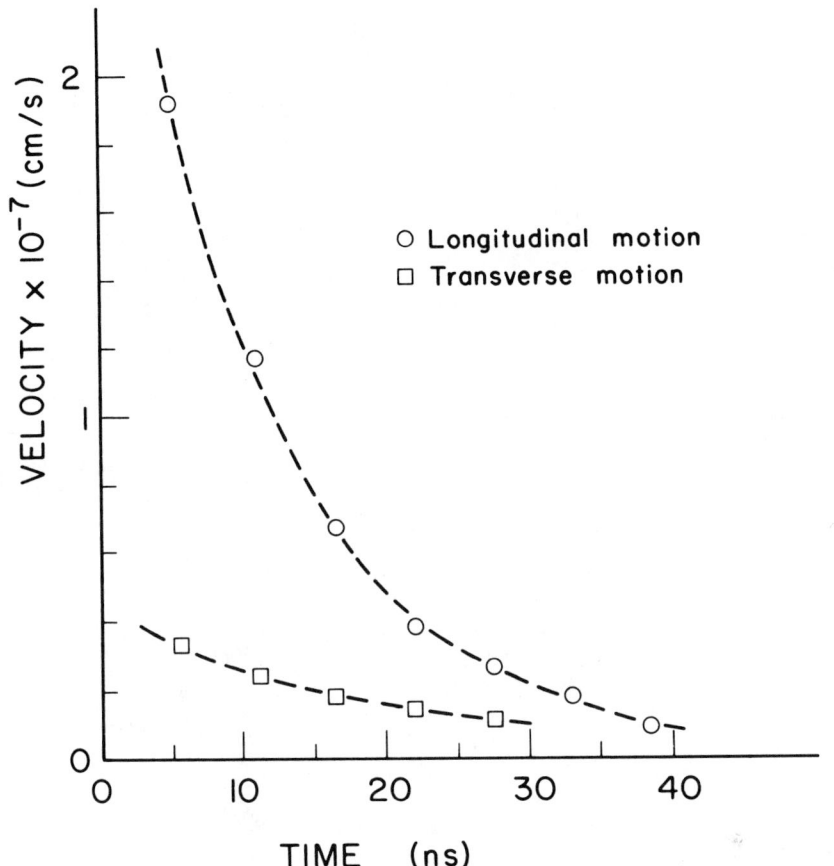

Fig. 11. Temporal variation of the longitudinal and transverse velocity of a spark in air at atmospheric pressure. Focal length of lens = 3.5 cm; peak laser power = 100 MW.

shutter is illustrated in Figs. 13(a) and 13 (b). The oscilloscope trace of the transmitted pulse, Fig. 13 (a), shows rise and fall times equal to the 450 psec response time of the fast photodiode (ITT F4000) - Tektronix 519 detection system. The pulse width at half peak intensity is ~700 psec. The efficiency of the shutter is demonstrated by the oscilloscope trace of the laser light rejected by the second polarizer, Fig. 13(b), which shows that ~100% of the peak power is transmitted in the selected subnanosecond pulse. Using these pulses to illuminate an interferometer the filamentary structure of sparks, produced by means of the single-mode ruby laser[15], is being investigated.

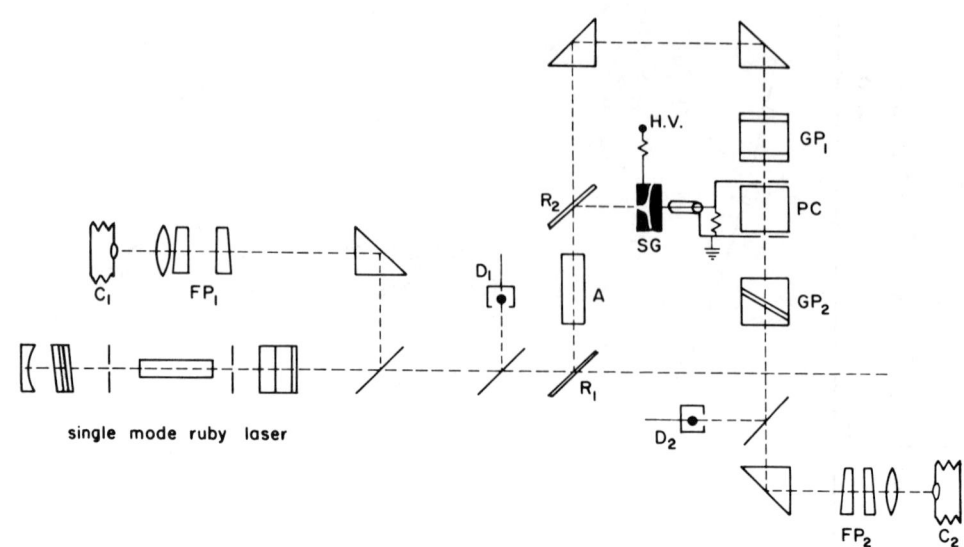

Fig. 12. Experimental arrangement used for the generation of sub-nanosecond light pulses. D_1, D_2 - photodiodes; R_1, R_2 - beam splitters; A - amplifier; SG - spark gap; PC - Pockels cell; GP_1, GP_2 - polarizing prisms.

4. RECENT DEVELOPMENTS

Although the techniques described above were developed for the investigation of laser-produced plasmas a number of additional applications may be permitted as a result of recent advances in laser technology. One important example is the achievement of single-mode operation in a laser Q-switched by means of a Pockels cell[16]. Reduction of the pulse duration by means of an optical gate, such as that described above, would yield a triggerable illuminating pulse with a duration which could be varied between 1 nanosecond and several tens of nanoseconds.

A number of other developments may well result in the extension of interferometric methods to the picosecond region. One approach is the selection of a single pulse from the output of a mode-locked laser, a technique which was first demonstrated by Penney and Heynau[17]. However, with this and other schemes which have since been developed, the rejection of all pulses preceding and following the selected pulse is not 100% efficient and thus some loss of contrast must be expected due to background radiation reaching the film. An alternative approach to the same objective involves the use of a gated image converter or image intensifier which would reject background radiation much more effectively. Recently

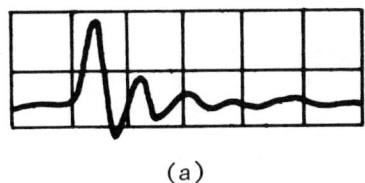 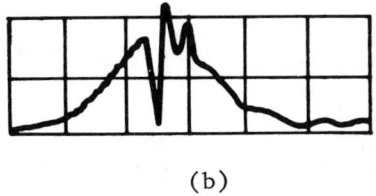

(a)   (b)

Fig. 13. (a) Oscilloscope trace showing pulse transmitted through second polarizing prism. Horizontal scale - 2 nsecs/div. Vertical scale - 1.0 MW/div.
(b) Oscilloscope trace showing laser radiation rejected by second polarizer. Horizontal scale - 5 nsecs/div. Vertical scale ~1.0 MW/div.

Bradley et al.[18] have reported a spatial resolution of 8 line pairs per millimetre for an image intensifier system illuminated by a single picosecond pulse, while another interesting development is the successful operation of a broadband picosecond shutter[19] employing the optical Kerr effect induced in an organic liquid by a mode-locked laser pulse.

Advances in dye laser technology may also have important implications since it may be possible to tune these high intensity sources to resonance with an optical transition. This would not only permit a significant improvement in the sensitivity of optical interferometry but might also lead to some interesting extensions of multiple wavelength interferometry as has been suggested by Measures[20].

Finally it is necessary to mention the important developments in pulsed CO_2 lasers which have been made during the last few months [21]. These have resulted in the availability of a relatively simple, high intensity source of submicrosecond pulses at a wavelength of 10.6 microns, thus greatly improving the potential sensitivity of interferometric plasma investigations. Although operation at such a wavelength introduces some difficulties in the recording of interference patterns the development of special image converter tubes[22] and up-conversion techniques[23] offer a number of interesting possibilities and it is likely that interferometric studies will be extended to the infra-red region in the not too distant future.

Many of the advances mentioned above may be applied equally well to the technique of holographic interferometry and thus, whether conventional or more recently developed methods are applied, lasers are likely to play an important role in the continuing development of optical diagnostic techniques.

ACKNOWLEDGMENTS

In addition to the participation of the many colleagues whose names appear in the list of references it is a pleasure to acknowledge the contribution of Professor S.A. Ramsden who initiated the application of optical diagnostic techniques to laser-produced plasma investigations.

REFERENCES

1. E.M. Little, W.E. Quinn and G.A. Sawyer, Phys. Fluids $\underline{8}$, 1168 1965.

2. U. Ascoli-Bartoli, S. Martellucci and E. Mazzucato, Nuovo Cimento, $\underline{32}$, 298 (1964).

3. A.J. Alcock, E. Panarella and S.A. Ramsden, Proc. Seventh International Conference on Phenomena in Ionized Gases, Belgrade 1965, p.224.

4. E. Panarella and P. Savic, Canadian Journal of Physics, $\underline{46}$, 183 (1968).

5. R.A. Alpher and D.R. White, Phys. Fluids $\underline{2}$, 162 (1959).

6. S. Martellucci and E. Mazzucato, 7th International Congress on High Speed Photography, Zurich 1965.

7. A.J. Alcock and S.A. Ramsden, Appl. Phys. Letters $\underline{8}$, 187 (1966).

8. E. Panarella, CASI Transactions, $\underline{2}$, 21 (1969).

9. F.H. Oertel and J.H. Spurk, ICIASF '69 Record p. 229.

10. H.W. Mocker and R.J. Collins, Appl. Phys. Letters, $\underline{7}$, 270 (1965).

11. A.J. DeMaria, D.A. Stetser and H. Heynau, Appl. Phys. Letters, $\underline{8}$, 174 (1966).

12. A.J. Alcock, C. DeMichelis, K. Hamal and B.A. Tozer, IEEE Journal of Quantum Electronics, $\underline{QE-4}$, 593 (1968).

13. A.J. Alcock and M.C. Richardson, Optics Communications $\underline{2}$, 65 (1970).

14. A.J. Alcock, M.C. Richardson and K. Leopold, Rev. Sci. Instr., **41**, 1028 (1970).

15. A.J. Alcock, C. DeMichelis and M.C. Richardson, IEEE Journal of Quantum Electronics, **QE-6**, 622 (1970).

16. J.P. Budin and J. Raffy, IEEE Conference on Laser Engineering and Applications, paper 12.6, Washington, June 1969.

17. A.W. Penney Jr. and H.A. Heynau, Appl. Phys. Letters, **9**, 257, (1966).

18. D.J. Bradley, J.F. Higgins and M.H. Key, Appl. Phys. Letters, **16**, 53 (1970).

19. M.A. Duguay and J.W. Hansen, Appl. Phys. Letters, **15**, 192 (1969).

20. Annual Progress Report, University of Toronto Institute for Aerospace Studies, p.110 (1969).

21. A.J. Beaulieu, Appl. Phys. Letters, **16**, 504 (1970).

22. S. Spinak, P.P. Barron, S. Karp, R.B. Hankin and R.H. Meier, Appl. Optics, **7**, 17 (1968).

23. J. Warner, Appl. Phys. Letters, **13**, 360 (1968).

PULSE LASER HOLOGRAPHIC INTERFEROMETRY

F. C. Jahoda

Los Alamos Scientific Laboratory, University of

California, Los Alamos, New Mexico

In the expectation that other speakers at this symposium will have thoroughly discussed the physical information that refractive measurements provide in the study of gas dynamics, the emphasis in the present contribution is entirely on the holographic techniques by which refractivity can be measured. The interpretation of holographically produced interference fringe patterns is, generally, equivalent to that of the fringe patterns produced by a conventional interferometer, although some exceptions are noted later. The primary justification for these techniques is the greater experimental simplicity and/or flexibility thereby achieved over older methods.

Since practical holography depends on the coherence of laser sources, it also seems pertinent to remark at the outset that the laser by itself, without holography, has greatly simplified conventional interferometry. It does this by relaxing the tolerances with which the spatial and temporal overlap of recombining wavefronts must be achieved. (This is no minor matter and has enabled non-specialists to apply interferometry routinely, while one can only marvel at the virtuosity of the pre-laser day experts.)

The great general interest in holography, presumably because of the spectacular nature of three-dimensional reconstructions of ordinary scenes, has resulted in wide dissemination recently of the basic principles of wavefront reconstruction. None-the-less I consider it purposeful here to briefly resummarize these. While all my later examples deal with pulsed ruby laser holography of transient plasmas, much of what follows is applicable to holography with steady state lasers as well.

PRINCIPLES OF HOLOGRAPHY

A hologram stores both the amplitude and phase of an optical field incident on the hologram plane in a manner that permits a faithful recreation at later times of the original phase and amplitude distribution. (Fig. 1). The reconstructed wavefront may then be used instead of the original to determine, for instance, the spatial phase variations.

The hologram exposure is an interference pattern between 1) a "scene" beam reaching the hologram plane (photographic plate) by reflection from or, as in the case of plasma, transmission through the object and 2) a reference beam, originating in the same source, and reaching the hologram plane uninfluenced by the object, often by a path that circumvents the object.

Expressed analytically the intensity at the plate is

$$I = \left(E_r e^{i\alpha x} + E_s e^{i\varphi(x)}\right)\left(E_r^* e^{-i\alpha x} + E_s^* e^{-i\varphi(x)}\right)$$

$$= |E_r|^2 + |E_s|^2 + E_r^* E_s e^{i\varphi(x)-i\alpha x} + E_r E_s^* e^{-i\varphi(x)+i\alpha x}. \qquad (1)$$

E_r is the amplitude of the reference beam, E_s is the amplitude of the scene beam, * indicates complex conjugate, and x is the coordinate in the hologram plane (taken one dimensional for notational convenience only). The amplitudes E_s and E_r of both beams are assumed constant, while the phase φ of the scene beam is an unspecified function of position, and the phase of the reference beam, assumed a plane wave incident at angle $\theta = \sin^{-1}(\lambda/2\pi)\alpha$, is linearly proportional to the hologram coordinate. The time factors have been omitted because oscillations at optical frequencies are not resolved by any detectors and all our considerations apply to the

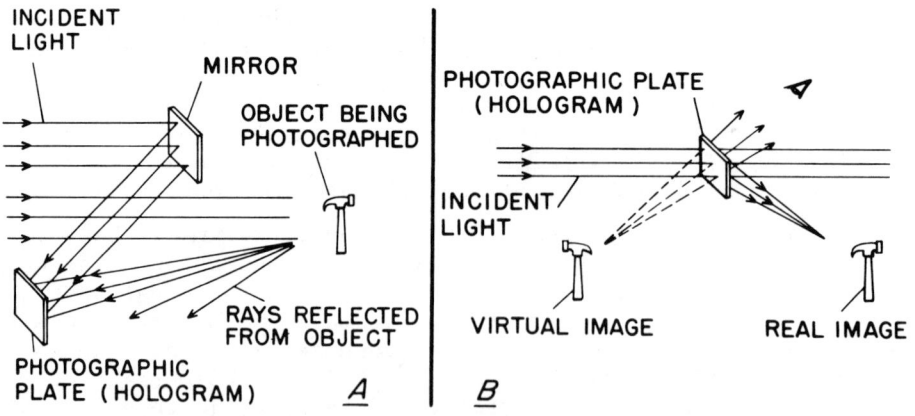

Fig. 1. Schematic diagram of hologram recording, left, and hologram reconstruction, right.

envelopes of amplitude or intensity.

For incoherent beams the cross-product terms would average to zero over any realizable observation time due to additional randomly varying relative phases between the two beams; thus the desirability of laser sources. The greater the temporal and spatial coherence of the laser the less restrictive, respectively, the requirements on path equality of the two beams and on accurate point-for-point overlay of the two wavefronts at the photographic plate. These are just the same requirements as those encountered in conventional interferometry. However, because the mean angle between the two beams will usually be several degrees, the fringe structure is exceedingly fine. In consequence, the emulsion recording the hologram must be capable of high resolution. (It is important to distinguish this fine scale intensity modulation on the photographic plate, produced by conventional interference in the construction of any one hologram, from the gross interference pattern due to the beating together of two similar holograms produced by two consecutive exposures that is the basis of holographic interferometry.)

The recorded photographic density is linearly proportional to the intensity I for variations small compared to the constant background term E_r^2 ($E_r \gg E_s$) and the amplitude transmittance of the developed emulsion becomes $T(x) = T_o - kI(x)$ where k is a proportionality constant determined by the film. When the hologram is reconstructed by illuminating it with a plane wave reconstruction beam the transmitted amplitude contains a term, among others, proportional to the third term in Eq. (1). This term itself is proportional to the original subject waveform deflected by the linear phase term $e^{-i\alpha x}$ (analogous to ordinary grating deflection into 1st order) in the direction θ. Thus an observer looking through the hologram toward the reconstruction beam, as through a window, will see a virtual image of the original scene angularly separated from the reconstruction beam direction. There is another term in the negative θ direction which duplicates the original subject phase distribution except for reversal in sign. This is equivalent to the original wavefront traveling with reversed sign in time and thereby converging to a real image in front of the plate. Since the absolute sign of the phase is of no concern, either the real or virtual reconstructions can be used according to convenience. The use of other than plane parallel reference and reconstruction beams or a change of wavelength between hologram formation and reconstruction changes the image locations and magnifications, so that more properly one should designate the images as conjugate and primary rather than real and virtual.

The same results are obtained more fundamentally from a diffraction theory point of view and the interested reader seeking a comprehensive overview of the subject is referred to two books: "Principles of Holography" by Howard M. Smith, Wiley-Interscience,

1969 and "Introduction to Fourier Optics" by Joseph W. Goodman, McGraw-Hill, 1968.

In practice, a great deal of latitude in relative beam intensities and processing procedure is permissible. (This is particularly so in the interferometric applications discussed below, for which only the relative differences between two exposures are of primary interest.) If the hologram exposure falls outside the range of linear film response, the non-linearities primarily generate higher order terms that are deviated into higher multiples of the angle θ.

The most restrictive requirements in making a hologram are those on source coherence and detector resolution. For the former, even giant pulse ruby lasers without special mode selection are adequate if the reference and scene beams are kept equal in length to within a few centimeters, and if a small pinhole which may be external or internal to the laser is used to improve the spatial coherence. The pinhole, however, decreases the energy density reaching the film plane sufficiently (especially after the beam is expanded to a useful working aperture) to make it desirable to use a faster emulsion than the extremely slow, ultra-high resolution Kodak 649 emulsion often used for CW holography. The best speed with only a slight decrease in resolution is currently obtained from Agfa-Gevaert 10E75 plates.

HOLOGRAPHIC INTERFEROMETRY

If two sequential hologram exposures are made on the same photographic emulsion the wavefront amplitude and phase structure of the scene beam from both exposures is recorded. When the double exposure hologram is reconstructed, the wave fields of the two exposures are reproduced simultaneously. The reconstructions are coherent since they originate in a common source and the two images will interfere with each other according to the phase differences between the two exposures. The result is analogous to a conventional interferogram. However, the wavefronts being compared originally existed at different times rather than, as in a conventional interferometer, existing simultaneously and requiring first a spatial separation and subsequent recombination. Since these wavefronts traverse a common path (aside from the difference purposely introduced between exposures, e.g., plasma) the experimental requirements are greatly simplified. In holographic interferometry there do not exist the needs of conventional interferometry for a) precise alignment to achieve recombination of the beams and b) for extremely high quality optical flats to prevent phase differences between the beams other than those one desires to measure in the plasma. Any kind of windows, in particular transparent walls of arbitrary curvature in the plasma vessel, are sufficient providing the optical distortions of the windows are taken into account in the analysis. The complexity associated with a conventional interferometer capable of spanning the plasma discharge is entirely dispensed with.

The double exposure holographic interferogram described so far is actually the analog of the conventional interferometer adjusted for a uniform field, the case sometimes called the infinite fringe spacing. This adjustment is not generally desirable because it is not possible to detect phase disturbances small compared to a full fringe nor to recognize regions where reversals of the fringe order numbers may occur. Usually the conventional interferometer is adjusted for a straight line fringe background by introducing a slight angle between the recombined beams with a small tilt of at least one mirror or beam splitter. In the holographic case the background fringes are easily produced by purposely making a linear phase change between exposures. This is readily done either with a hollow wedge inserted in the optical path of the scene beam whose filling gas is changed between exposures or by a mechanical motion of one mirror between exposures. Figure 2 illustrates this.

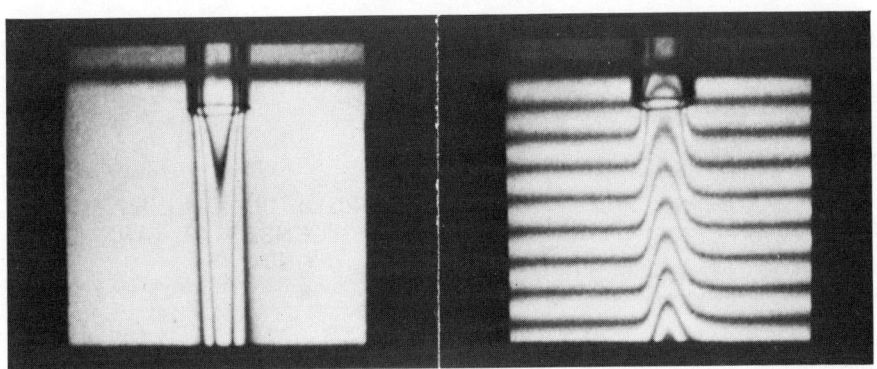

Fig. 2. Reconstructions of double exposure with and without argon gas stream, left, and the same with a change in gas filling of wedge in addition, right.

These are two reconstructions of double exposures in which in each case an argon gas flow was turned on for one exposure and on the right-hand side, in addition, a wedge in the scene beam had its filling gas changed between exposures. This illustration is useful to stress two other points a) at the top the fringes can be followed right into the glass capillary and b) while all my other examples will be concerned with electron refractivity, any physical parameter that manifests itself as a sufficient refractivity change can be measured.

This deliberately introduced background serves another extremely useful purpose. If one does not have good antivibration mounting of the entire holographic optical system, spurious phase shifts are introduced by relative motions between various components that have occurred during the time between the two exposures. If the deliberately introduced change produces a large number of background fringes compared to the number of fringes produced by the random vibration, the vibration produced phase changes - which are generally linear - only slightly alter the orientation and spacing of the background fringes. Without the controlled background the random motions would produce a background of wholly random spacing and orientation which might not even be recognized as spurious and the extra phase changes would then be mistakenly added to the phase changes attributed to the plasma.

Figure 3 is a schematic of the experimental arrangement used in the application of this technique to a plasma generated by a plasma gun located in the center of a 4-ft diameter vacuum tank.

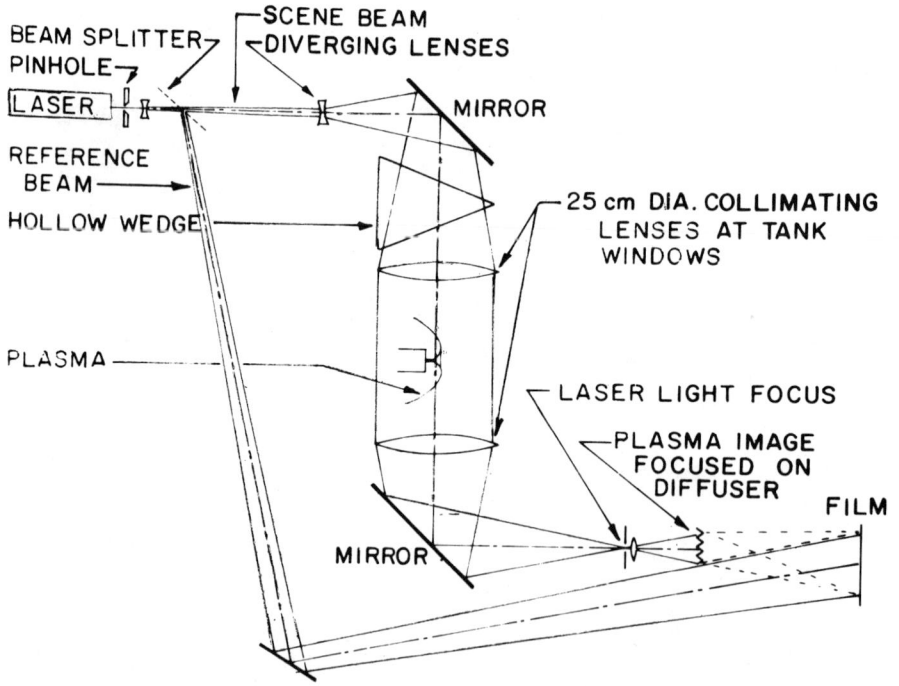

Fig. 3. Schematic of holographic interferometer to measure plasma density.

The tank could not readily have been spanned by a conventional interferometer. The scene beam is expanded to cover 10-in. windows (also an extreme dimension for conventional interferometry), and is again focused down to pass through an aperture which prevents most of the plasma self-luminosity from getting through. An auxiliary lens images the plasma location either onto the hologram plane or, as shown here, onto a diffuser. This imaging counteracts the refractive bending effects of the plasma which otherwise would degrade the fringes in the region of the plasma boundary. The reference beam is made of "equal" length to the scene beam as measured with a meter stick. Two ruby giant pulse laser exposures are made separated in time by a minute or two, the first without plasma present and air in the wedge, the second synchronized with a plasma discharge and helium in the wedge. An advantage of having the diffuser is that each point of its surface becomes a secondary source which illuminates all points of the hologram plane. In consequence, reconstruction of any small portion of the hologram will reproduce the entire scene as projected onto the diffuser. Equivalently, even if both beams overlap over only a small area of the hologram the complete information is encoded there, and the reconstruction is completely independent of any artifacts such as dust or scratches associated with the photographic emulsion.

Typical results obtained with a processed doubly exposed hologram of a neon plasma leaving the muzzle of a coaxial gun at various times after gas breakdown are shown in Fig. 4. The pictures were recorded on film placed at the real image location during the reconstruction of the doubly exposed holograms with CW-laser light. Each frame corresponds to a different discharge, and the slightly different slants of the background fringes illustrate the result of slight uncontrolled motions discussed above.

The diffuser may also be placed in the scene beam before it traverses the plasma instead of after. The plasma is then illuminated over a whole range of angles and the double exposure holographic recording will give changing fringe patterns (and thus integrated density distributions) corresponding to different viewing directions in the play-back. If the geometry permits viewing at sufficiently different angles this allows some inference about the depth dimension. In this case it is essential to restrict the viewing angle for any given direction in the reconstruction to prevent complete smearing out of fringes due to overlapping views. One hologram is then equivalent to simultaneous interferograms taken over a range of viewing directions.

An advantage of operating without the diffuser is that the fringes are not localized in space and will appear on the hologram itself when it is viewed at glancing angle in incoherent nonmonochromatic light. (It is instructive in this case to recognize that the dark fringes are just the moiré beats produced where the

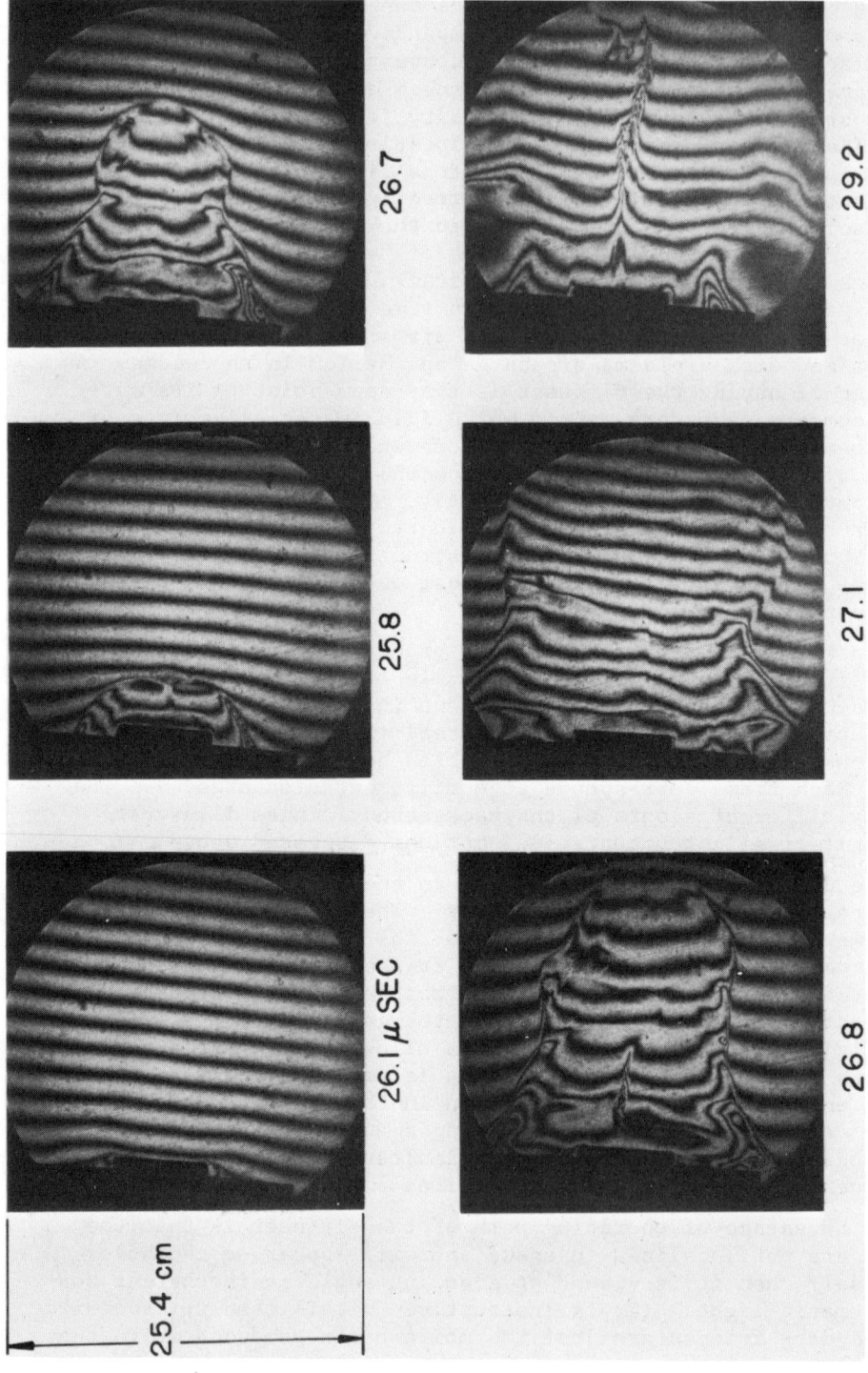

Fig. 4. Holographic interferograms of coaxial gun neon plasma at different times after breakdown.

fine scale grating of one exposure is out of registry with the grating structure of the other exposure; (see Fig. 5) By avoiding the speckle effect produced by reconstruction in laser light better spatial resolution is achieved in cases where fringes crowd together. Also by overlapping corresponding portions of the reference and scene beams on the hologram the spatial coherence requirement on the laser source is further reduced.

Figures 6 and 7 show, respectively, symmetric and non-symmetric integrated density distributions for a 3-meter long theta pinch, obtained by Dr. R. E. Siemon at Los Alamos, without a diffuser in the scene beam. To produce the slide in this case merely involves a white light source and a lens-pinhole combination that images the hologram onto the film in one non-zero diffraction order. The visibility of the fringe pattern on the holographic plate is greatly enhanced by bleaching the developed emulsion. Again we do not pause here for a discussion of the plasma physics except to note that quantitative electron density profiles are readily derived from these interferograms.

For plasmas which are not fully ionized, one can separate the electron contribution to the refractivity, which is proportional to wavelength squared, from that of the neutral component, approximately constant with wavelength, by making two interferograms at different wavelengths. Holography is particularly convenient in this case. The ruby laser fundamental wavelength and its second harmonic, produced in a KDP crystal, can be used simultaneously. If common scene and reference beams each contain both colors, separation of the two interferograms is obtained automatically during the reconstruction with a single wavelength, since the average grating spacings produced by the two colors at common angle of incidence are different.

As already mentioned holographic interferometry is experimentally simpler to implement than the conventional techniques, given a suitable laser source. Its sensitivity range spans at least the same range as the Mach-Zehnder interferometer. The number of fringes, p, is related to the refractive index, μ, the wavelength λ, and the pathlength ℓ by $p = (\mu - 1)\ell/\lambda$. For electrons $\mu = (1 - \omega_p^2/\omega^2)^{\frac{1}{2}}$, where $\omega = 2\pi c/\lambda$ is the angular frequency of the illumination, and ω_p, the plasma frequency, is given in terms of the unit charge e, electron mass m, and electron density n, by $\omega_p = (4\pi ne^2/m)^{\frac{1}{2}}$. If we arbitrarily assume that fringe shifts between 0.1 and 30 fringes are readily detectable, at the ruby wavelength this encompasses density-path length products between 3×10^{16} and 1×10^{19} cm^{-2}.

The flexibility of holographic systems makes it feasible moreover to extend this range at both limits. For instance, when fringes crowd together the region of interest can be magnified by ordinary optics. In a conventional interferometer this would be

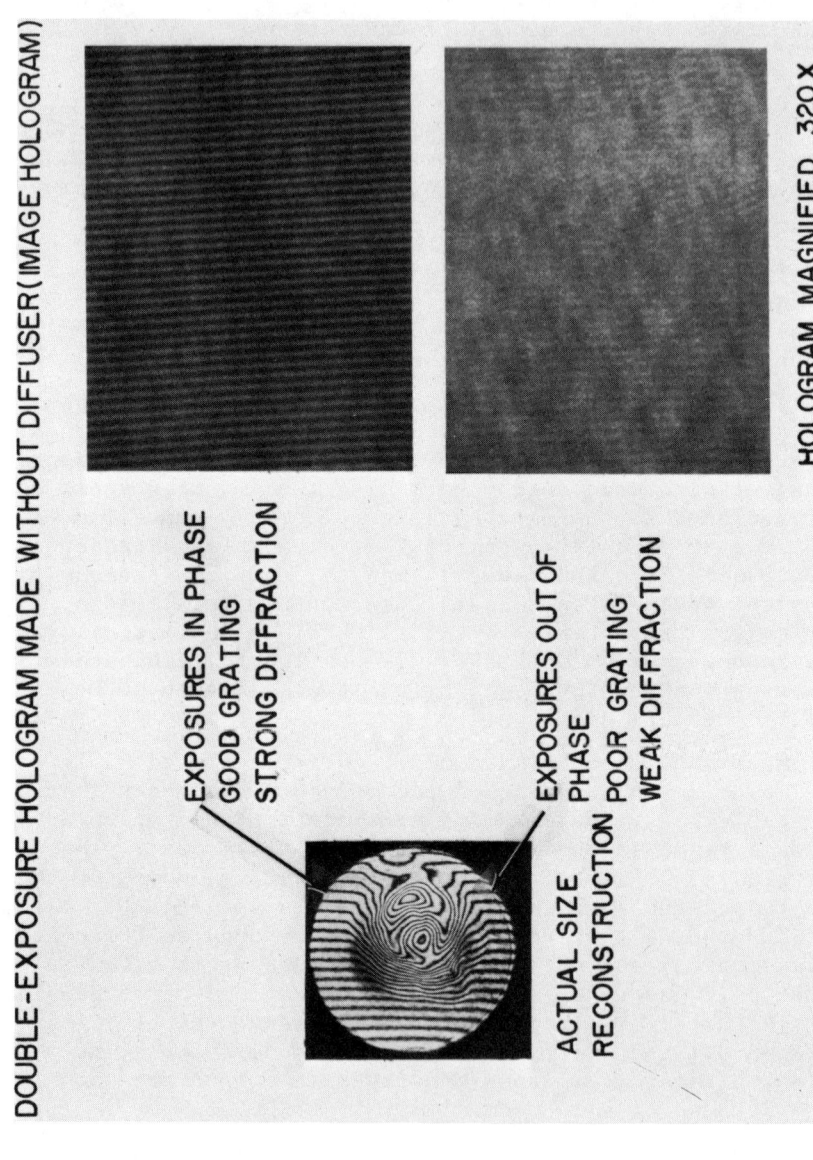

Fig. 5. Magnified sections of hologram made without diffuser.

HOLOGRAPHY 147

Fig. 6. Holographic interferogram of theta pinch plasma at peak magnetic field.

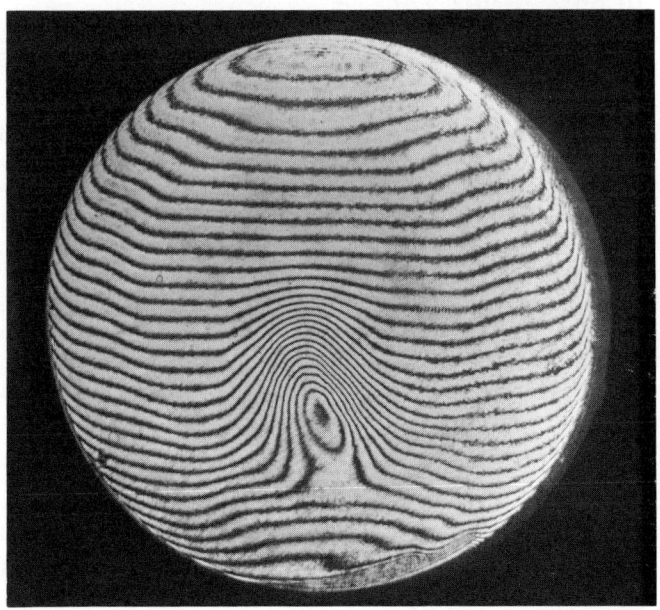

Fig. 7. Holographic interferogram of theta pinch plasma at later time after onset of instability.

impractical because of the requirement of phase matching optics in
the comparison beam. The limitation is set not by the holographic
technique, but by refractive bending due to density gradients. To
some extent this can be corrected for in the analysis, and since
the fractional error is proportional to the square of the length,
for short path lengths through the subject this limit is not
reached until much higher densities. By programming the two exposures to occur at two times during the event, rather than before
the event and at one chosen time during the event, a time differential interferogram is produced, and this obviously enables one
to decrease the number of fringes that must be resolved. (This is
the first instance where the interpretation of the final product
is quite different from a conventional interferogram.)

The sensitivity can also be decreased if the two hologram
exposures are made simultaneously in different wavelengths instead
of sequentially with the same wavelength, and the reference beams
in the two colors are incident at appropriately related angles so
that the two reconstructions (at a common wavelength) occur in the
same direction. Interference fringes will be formed with a sensitivity to electron density that is a function of the wavelength
difference. Various wavelength differences might be obtained, for
instance, by stimulated Raman emission from appropriate materials.
Small inaccuracies in the angular settings of the reference beams
would cause a background fringe pattern of the type one seeks to
add purposely anyway as described above. (The principle has been
demonstrated in holographic contour mapping, but not yet applied
to plasmas, to our knowledge.) Again the quantitative interpretation of the resultant fringe pattern is modified from the conventional mode.

To increase the sensitivity the most direct means would be
longer wavelength holography. The CO_2 laser is an obvious candidate for a 15-fold increase in sensitivity to electron refraction,
but practical high-spatial resolution detectors to replace photographic film are not yet available. Some promising methods, however, are on the technological horizon and this can definitely be
regarded as a future prospect. In the meantime various multiple
path geometries through the test volume can be used. Here again,
the holographic freedom from having to phase match the test path
with a spatially separated comparison path is the practical key
that opens a variety of possibilities. For example, we have used
two partially reflecting mirrors forming a lossy resonant cavity
around the plasma, slightly misaligned in order that a lens-pinhole
combination can select the particular emergent beam corresponding
to the desired number of traversals. The limitation on number of
traversals is set by the decreasing intensity of the beam. A
proper relationship between the angular tilts of the mirrors to
keep the maximum lateral displacement as small as possible is
easily derived. Thus we have achieved a five traversal system
(five-fold increase in sensitivity) with better than 1-mm spatial

resolution within a 1-meter test cavity. However, in order that fractional fringe measurements in the enhanced sensitivity geometry are themselves meaningful, and a true gain in lower limit is achieved, it becomes necessary to use very high-quality optical components, thereby negating one of holography's advantages. The reason is that any spatially varying phase change implies some deflection, however minute, due to spatial gradients, and schlieren striations within the optics may cause intensity modulation effects in coherent light that tend to mask small or irregular true plasma interference effects resulting from the two consecutive exposures.

CINE-INTERFEROMETRY

The methods so far described are limited to a single interferogram at a chosen moment during the event. If the individual events are expensive or are not accurately reproducible it is often desirable to obtain a time sequence of interferograms during a single event. In conventional interferometry this can be achieved with some form of image transport- e.g., streak or framing camera, whereas in holographic interferometry the need for accurate registry of two exposures, one before and one during the event, precludes these methods. The major difficulty with the methods that can be used, described below, is a scarcity of laser sources that can illuminate the entire duration of the event. The same problem, of course, is encountered in conventional interferometry with laser sources, so that in this case the merits of the holographic techniques must be evaluated by a comparison with the conventional techniques using conventional (sic! e.g., argon flash) sources.

One obvious possibility, is to use multiple giant pulse laser sources. With the inclusion of a diffuser in the scene beam, one can simply utilize a hologram's ability to store independently several (double) exposures by separating from one another the relatively small areas of the emulsion exposed by the individual reference beams. The light scattered by the diffuser from the superposed scene beams of all the other lasers, pulsing at different times, is only an incoherent background that does not affect the individual holograms. For a sufficiently thick emulsion the reference beam separation can be angular rather than spatial. An equivalent, less expensive, means is to use portions of a single giant laser pulse after storing them in variable optical delays. This method, however, is limited in practice to total delay times considerably less than 1 μsec, and even then requires rather complex experimental configurations.

With a laser having a pulse width spanning the total duration of the event, consecutive real time interferograms can be obtained with an adaptation of the "live fringe" method sometimes used in CW laser holography. In this case, instead of two consecutive exposures of the hologram emulsion that yield a permanent fringe pattern, the hologram is accurately repositioned after photographic processing following a single exposure. Whenever the laser illumination source is again turned on interference can be observed in real time between a) the stored virtual image of the scene, reconstructed by the beam that was the reference beam in the holographic recording, and b) the current status of the scene. Consecutive interferograms can then be photographed by image converter cameras gated on at pre-selected times.

Interference fringes will occur both because of those phase changes in the scene that one wishes to measure and because of inexact repositioning of the hologram. With CW lasers the positioning error can be eliminated by fine adjustment during observation of the unaltered scene through the hologram. With ruby lasers observations can only be made by camera monitoring of discrete pulses (the reconstruction must be in approximately the same wavelength as the construction to provide "live" fringes) and the fine adjustment procedure is impractical. However, by mounting photographic plates in a repositionable jig that goes through the development process, the residual phase shift due to repositioning error is slight. It is, moreover, small compared to the linear phase shift deliberately introduced into the scene between exposure and reconstruction to give a background of straight fringes, and again only alters slightly the orientation and spacing of the background in a harmless manner.

The minimum framing time of the image converter and thus the ultimate time resolution is limited by the combination of laser intensity, hologram diffraction efficiency, and photocathode quantum yield of the image converter.

Figure 8 shows interference fringes obtained with two image converter cameras at the indicated times after initiation of an air spark. This spark is created by dissipating 1.5 J between point electrodes spaced 2-mm apart on the 1-cm-diameter insulator seen in shadow projection. Two Beckman-Whitley single-frame image converters were used. The shorter exposure was obtained with an "S-11, extended red" photocathode, whereas the longer exposure camera had only the regular S-11 photocathode. There is a mirror inversion due to a beam splitter which enabled both cameras to view the object in the collimated laser beam. The reconstruction laser power was approximately 5 kW, obtained from a chilled free-running ruby laser, with mirrors coated directly on the ruby, giving 30% modulation but no dead time over a 50-μsec period. This laser has rather limited coherence quality when integrated over the entire pulse length, and is inadequate to produce high-efficiency holograms. Therefore the hologram exposure was actually made with a Q-switched laser of better coherence. This change of lasers is feasible if no diffuser is used and the reference and scene beams recombine at the hologram plane approximately point for point in the way they originate at the beam splitter. The processed plate was bleached in a chromium intensifier solution

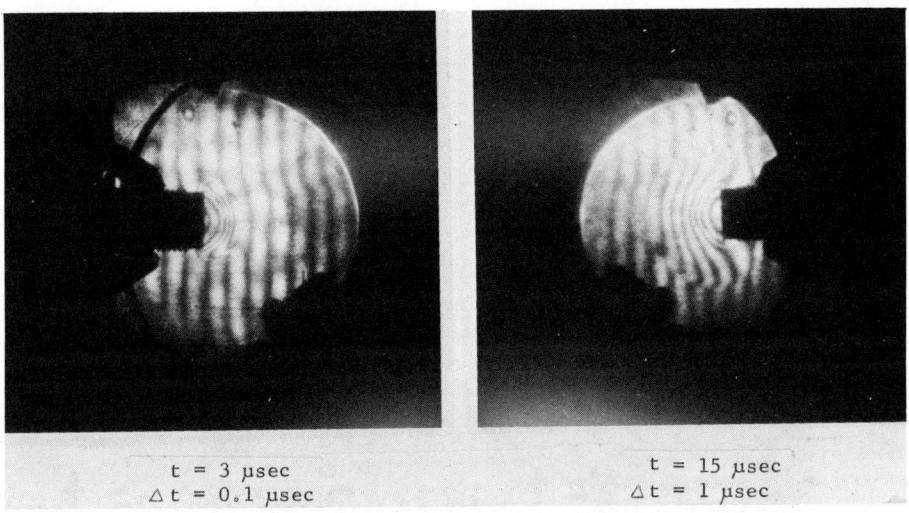

Fig. 8. Two consecutive holographic interferograms of a single spark discharge.

to enhance the brightness of the reconstructed image. Various improvements in the system subsequent to the time the last figure was produced have achieved a reduction in the limiting framing time to 5 nanoseconds. This means that if a reliable image converter with S-20 photocathode and efficient output coupling between phosphor and recording film were available, present technique would already only be limited by the electronic switching capability of the image converter.

For this technique the use of dry-processing photopolymer material developed by Hughes Research Laboratory, already competitive with holographic film in sensitivity and resolution and capable of higher diffraction efficiency, appears particularly promising, since it eliminates both the necessity of accurate repositioning and the stability requirements on the entire set-up during the lengthy photographic processing. It also has a convenience value for the single frame two-exposure holographic techniques, that is best appreciated by considering the difference between polaroid oscilloscope cameras and ordinary film scope cameras. Its practical implementation, in particular to the point of absolute reliability that potential data will not be lost, is being actively pursued. The final slide (Fig. 9) shows an interferogram made with the photopolymer, and transferred onto polariod film, in situ, immediately after the second exposure.

Fig. 9. Holographic interferogram produced on photopolymer material.

BIBLIOGRAPHY

The preceeding text was written in a format primarily intended for oral presentation at the symposium, without specific citations. For the record and the interested reader some references, weighted toward the author's work and without regard for completeness or priority are given in this section.

Numerous popular and scientific journal articles on the principles of holography have appeared. Besides the two books given distinction by inclusion in the text above, several others are known to the author:

G. W. Stroke, An Introduction to Coherent Optics and Holography, Academic Press, Inc., New York and London (2nd edition, 1969).

J. B. De Velis, and G. O. Reynolds, Theory and Applications of Holography, Addison-Wesley (1967).

H. Kiemle and D. Röss, Einführung in die Technik der Holographie, Akademische Verlagsgesellschaft (1969); In German.

M. Francon, Holographie, Masson and Cie (1969). In French.

Yu. I. Ostrovskii, Holography, Leningrad (1970). In Russian.

The concept of diffuse illumination was introduced and lucidly explained in

E. N. Leith and J. Upatnieks, J. Opt. Soc. Am. 54, 1295 (1964).

Pulsed laser holographic interferometry was first discussed in a pioneer paper by:

L. O. Heflinger, R. F. Wuerker, and R. E. Brooks, J. Appl. Phys. 37, 642 (1966).

The background fringe idea is introduced by:

F. C. Jahoda, R. A. Jeffries, G. A. Sawyer, Appl. Optics 6, 1407 (1967)

and the first application to plasmas in

T. D. Butler, I. Henins, F. C. Jahoda, J. Marshall, and R. L. Morse, Phys. Fluids 12, 1903 (1969).

Two wavelength holographic interferometry of plasmas has been reported by:

I. I. Komisarova, G. V. Ostrovskaya, L. L. Shapiro, and A. N. Zaidel, Phys. Letters 29A, 262 (1969) and

R. A. Jeffries, Phys. Fluids, 13, 210 (1970).

Holographic contour mapping is explained in

B. P. Hildebrand and K. A. Haines, J. Opt. Soc. Am. 57, 155 (1967).

Possibilities that may lead to CO_2-laser holographic interferometry are indicated by:

J. J. Amodei and R. S. Mezrich, Appl. Phys. Letters 15, 45 (1969).

T. Izawa and M. Kamiyama, Appl. Phys. Letters, 15, 201 (1969).

W. A. Simpson and W. E. Deeds, Appl. Optics 9, 499 (1970).

The use of delayed portions of a single giant pulse laser for multiple interferograms is reported in

A. Kakos, G. V. Ostrovskaia, Yu. I. Ostrovsky, and A. N. Zaidel, Phys. Letters, 23, 81 (1966) and

J. C. Buges, C. R. Acad. Sci. Paris, B268, 1624 (1969),

and the adaptation of the "live fringe" method by:

F. C. Jahoda, Appl. Phys. Letters, 14, 341 (1969).

The photopolymer recording of holograms is published by:

D. H. Close, A. D. Jacobson, J. D. Margerum, R. G. Brault and F. J. McClung, Appl. Phys. Letters 14, 159 (1969).

Finally, we would like to refer to a review chapter on

"Optical Refractivity of Plasmas" by F. C. Jahoda and G. A. Sawyer, which will appear in the Academic Press series on "Experimental Methods in Physics," Vol. 9B, R. Lovberg, editor (in press).

This work has been performed under the auspices of the U. S. Atomic Energy Commission.

THE ANALYSIS OF PLASMA BY LIGHT SCATTERING

Ralph H. Lovberg

University of California, San Diego

I. INTRODUCTION

The unprecedented brightness of laser sources has made practical for the first time the use of Thomson scattering of light as a plasma diagnostic tool. In addition to the simple low-density process, where scattering intensities from the individual plasma electrons are summed, it has become possible to exploit the theory of plasma fluctuations in order to extend the analysis of scattering data to densities where collective effects predominate.

The theory required to interpret scattering data in the high density "collective" regime was set forth a decade ago by Saltpeter[1] and Fejer[2], and has been reviewed and elaborated by many other workers since that time[3,4]. A particularly good review of the theory has been given by Bernstein, et al[5].

Extremely comprehensive reviews of the whole subject of light scattering in plasmas, both theoretical and experimental, are given by Kunze[6] and by Evans and Katzenstein[7]. We draw rather heavily on the latter reference in this discussion, as well as on recent experimental work described in a thesis of R. E. Siemon[8].

II. ELEMENTS OF THE THEORY

a. <u>Low Density</u>

For plasmas of density sufficiently low that the number of electrons in a cubic wavelength of light is much less than one, the scattering problem is very simple. Individual electrons absorb and re-radiate energy from an incident beam according to classical electrodynamics. Each exhibits the Thomson cross section

1) $$\sigma_T = \frac{8\pi}{3} r_e^2$$

for total scattering, and a differential cross section

2) $$\frac{d\sigma}{d\Omega} = r_e^2 \sin^2 \phi$$

where $r_e = e^2/mc^2$ is the classical electron radius, and $d\sigma/d\Omega$ is the cross section for scattering energy into a unit solid angle at a polar angle ϕ with respect to the direction of the electric vector in the incident beam.

For the special case of unpolarized radiation, the differential cross section in terms of the scattering angle θ (the angle between the incident propagation $\vec{k}_o$ and the final scattered wave vector $\vec{k}_f$) is given by

3) $$\frac{d\sigma}{d\Omega} = \frac{r_e^2}{2} (1 + \cos^2 \theta)$$

In general, equation 2) is more useful for laser scattering, since the incident beam is usually plane polarized.

The essential point of low-density scattering is that, by virtue of their large spacing as compared with λ_o, the incident wavelength, the individual electrons make their contributions to the detector in random phase i.e., their phases are entirely uncorrelated. Hence, the total received intensity is the sum of the scattered intensities.

We may make a simple estimate of the scattered power in this limit for a typical experiment. The intensity at the detector will be

4) $$I_s = I_o \cdot \frac{1}{r^2} \cdot \frac{d\sigma}{d\Omega} \cdot N$$

where N is the number of electrons in the scattering volume as observed by the detector. Setting the differential cross section to about r_e^2, and letting

$$N = 10^{16}$$
$$r = 30 \text{ cm},$$

we get

$$\frac{I_s}{I_o} \simeq 10^{-12}.$$

LIGHT SCATTERING

This ratio makes it very clear that, prior to the advent of the pulsed laser, Thompson light scattering was not a practical diagnostic scheme.

Rather than pursue a separate discussion of the low density limit in more detail, we will set forth a more general description of the problem, out of which the pure Thomson scattering will emerge as a special case.

b. The General Scattering

Suppose that a beam consisting of ideal monochromatic plane waves is incident upon a volume of plasma and scattered into various angles by the electrons. (We assume, in all of what follows, that the attenuation of the beam due to the scattering is negligible). In the present problem as illustrated schematically in Figure 1, we suppose the incident beam to have a propagation vector $\vec{k}_o$, and we consider a particular scattered component $\vec{k}_f$

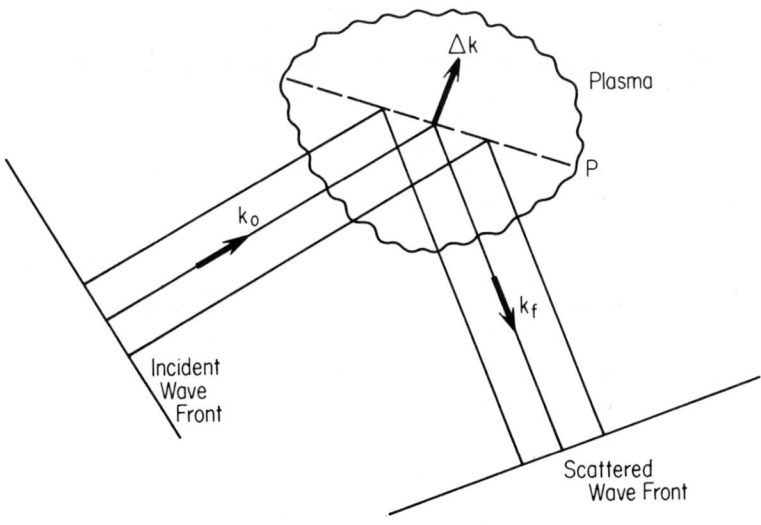

Fig. 1. Planes of constant phase (P) are normal to $\vec{\Delta k}$ where $\vec{\Delta k} = \vec{k}_o - \vec{k}_f$.

Light scattered along $\vec{k}_f$ can have many phases; however, all the electrons which lie in a certain plane P contribute identical phases to the scattering, since they form a virtual plane mirror which has the property of converting an incident plane wave into a reflected plane wave. We see, then, that whatever phase mixing occurs in the $\vec{k}_f$ is due to contributions from other planes parallel to P. If we define a coordinate x normal to P, it is clear that x alone is required to specify the problem.

The scattering vector $\Delta\vec{k}$, where

$$\Delta\vec{k} = \vec{k}_o - \vec{k}_f$$

evidently lies along x, and the phase of a scattered component along $\vec{k}_f$, relative to that from P, is easily shown to be

$$e^{i\Delta kx}$$

Consequently, one may write the total scattered amplitude as simply an integral over all these phase planes. Let us assume that $\phi = \pi/2$, i.e., that we are observing normally to $\vec{E}_o$, the incident electric vector. Then

5) $$E_s = \frac{r_e}{r} E_o \int n(x) e^{i\Delta kx} dx,$$

The density $n(x)$ is understood here to be already integrated over the other two coordinates normal to x; i.e., $n(x)$ is in units of cm^{-1}.

The intensity is obtained by convolving this electric field with its complex conjugate, giving

6) $$I_s = I_o \left(\frac{r_e}{r}\right)^2 \int\int n(x)n(x') e^{i\Delta k(x-x')} dx dx'$$

We assume now that the density $n(x)$ is nearly uniform but subject to local fluctuations. Thus,

7)
$$n(x) = n_o + \delta n(x)$$
$$n(x') = n_o + \delta n(x')$$

Then,

$$n(x)n(x') = n_o^2 + n_o |\delta n(x) + \delta n(x')|$$
$$+ \delta n(x)\delta n(x')$$

We define $\Delta x \equiv x' - x$ so that

$$dx' = d\Delta x$$

LIGHT SCATTERING

Now we substitute the time average of 8) into 6) and integrate, to obtain a time average of the scattered intensity.

The first term, containing n_o^2, integrates to zero, since only $e^{i\Delta k \Delta x}$ is left in the integral. The second term also vanishes because the fluctuations average to zero. The third term remains:

$$9) \quad I_s = I_o \left(\frac{r_e}{r}\right)^2 \int dx \int \overline{\delta n(x) \, n(x+\Delta x)} e^{-i\Delta k \Delta x} d\Delta x$$

The inner integral is the spatial wavenumber spectrum of the autocorrelation function of electron density fluctuations.

The above treatment can be generalized to include the time; the result is

$$10) \quad I_s(\Delta k, \omega) = I_o \left(\frac{r_e}{r}\right)^2 \int dx \int \delta n(x,t) \delta n(x+\Delta x, t+\tau) e^{i(\Delta k \Delta x - \omega \tau)} d\tau d\Delta x$$

Thus, the scattering depends upon Δk according to the spatial wavenumber spectrum of fluctuations, and upon ω according to the frequency spectrum of fluctuations.

Stated in perhaps more physical terms, we find that if the "granular" plasma has a granule size approximating $1/\Delta k$, the scattering will be strong. Furthermore, any frequency at which the fluctuations tend characteristically to oscillate will produce a modulation of the scattered light at that frequency. Equivalently, one can look for <u>sidebands</u> in the scattered frequency spectrum separated from the incident frequency by the modulation frequency.

An important corrolary implicit in the above is that a perfectly uniform plasma will not scatter light, except directly forward, in which case one is merely discussing the classical refractive index in an equivalent way. (The perfect uniformity referred to here, is, of course, perfect uniformity over distances comparable to a wavelength of light.)

Before reviewing the formal work which has been done on computation of plasma fluctuations, and hence, of the scattered spectrum, it will be worthwhile, on simple physical grounds, to speculate upon the scattering one might expect under various conditions.

First, in the low density limit, it is already clear that the scattered intensity can be predicted from the Thompson cross-section alone. The spectrum, however, will not be monochromatic since electron motions will doppler-shift the emitted frequency.

A measurement of the spectrum should, accordingly, allow an inference of the electron temperature.

For the higher densities, we may guess that the characteristic length and frequency associated with a plasma, i.e., the Debye length λ_D, and the plasma frequency ω_p, will appear as important parameters. One may recall that λ_D is the approximate upper limit to the size of fluctuations in a plasma; hence, if $\lambda_D \ll 1/\Delta k$, we may expect scattering to be very small, and indeed this is true.

Furthermore, the frequency which tends to dominate all others in plasma fluctuations is ω_p itself; we could then reasonably expect the scattered spectrum to contain strong sidebands separated from ω_o, the light frequency by an amount $\pm\omega_p$. For high densities, these sidebands are observed to be extremely prominent.

c. <u>Results of Fluctuation Theory</u>

The calculation of electron density fluctuations has been done through application of plasma kinetic theory, and its details are beyond the intended scope of this paper. We will describe the results of this work, employing notation due to Saltpeter[7].

To delimit the low and high density regimes of the problem, we define a dimensionless parameter, :

$$\alpha \equiv \frac{1}{\Delta k \lambda_D}$$

where, we recall

$$\vec{\Delta k} = \vec{k}_o - \vec{k}_f$$

so that

$$\Delta k = 2k_o \sin \frac{\theta}{2}$$

and

$$\lambda_D = \left(\frac{kT}{4\pi m_e e^2} \right)^{1/2}$$

Roughly then, α is a measure of the ratio of the light wavelength to the Debye length. For $\alpha \ll 1$, Thompson scattering predominates, and for $\alpha > 1$ collective effects are important.

The results of Saltpeter's calculations are given as a "form factor"

$$S(\Delta k, \omega)$$

LIGHT SCATTERING

which is the double integral appearing on the right side of equation 10, divided by the total number of scattering electrons. That is,

11) $$I_s = I_o \left(\frac{r_e}{4}\right)^2 S(\Delta k, \omega) \cdot N$$

The Δk dependence is only the dependence upon Θ, since the scattered wavelength is very near to λ_o; ω here means the <u>difference</u> frequency between that observed and the incident frequency.

Subject to the usually attainable condition that ion and electron temperatures are not so different as actually to make their velocities comparable, Saltpeter finds that $S(\Delta k, \omega)$ can be decomposed into two separate terms, one due alone to electrons, and one to ions; hence, these are dubbed the "electron feature" and the "ion feature":

$$S(\Delta k, \omega) d\omega = \Gamma_\alpha(x_e) dx_e + Z \left(\frac{\alpha^2}{1+\alpha^2}\right) \Gamma_\beta(x_i) dx_i$$

where

$$x_e \equiv \frac{\Delta \omega}{\Delta k v_e}$$

with

$$v_e = \sqrt{\frac{2kT_e}{m_e}}$$

x_e and the corresponding x_i are frequency shifts normalized to the thermal doppler shift. Z is the mean ion charge.

Without elaborating further upon the functional form of Γ_α and Γ_β which are roughly similar, we may note that the ion feature will tend to be very closely localized at the laser frequency (the x's scale with thermal speeds) where the electron feature occupies a much broader spectrum.

Figure 2, taken from Saltpeter, shows the scattered spectrum $\Gamma_\alpha(x)$ for various values of α. At $\alpha = 0$, we see our previous reasoning confirmed, i.e., that the spectrum is a simple Gaussian whose width is determined by electron doppler shifts. As α increases, however, the spectrum becomes more complicated, and finally, at large α, all that is left is a very strong "satellite" peak; this is the sideband, removed by $\sim\omega_{pe}$ from the laser frequency. (This figure has, of course, omitted the ion feature, which for large α becomes large itself.)

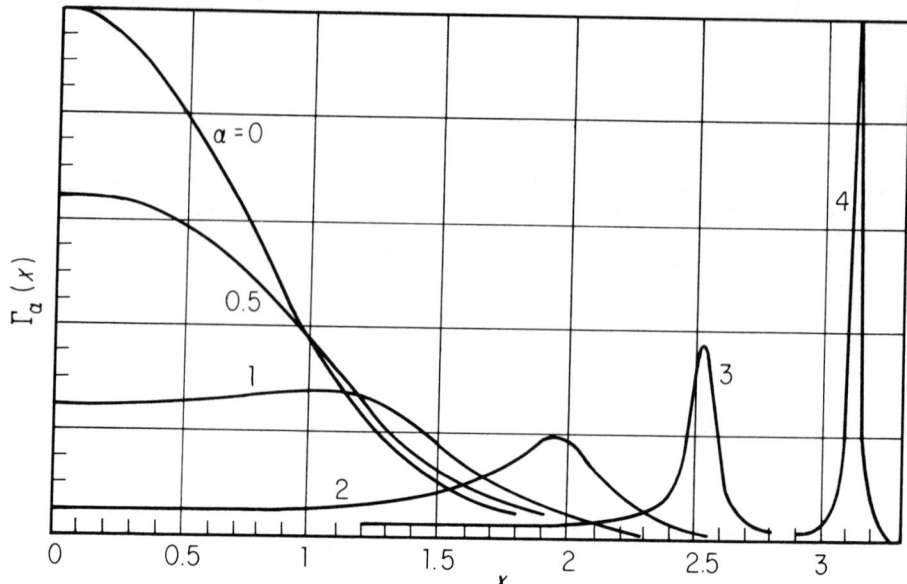

Fig. 2. The spectral distribution of scattered radiation for several values of α.

It is very important at this point to realize the dependences of these spectra in the limits of α. For α = 0, the spectral shape is determined by electron temperature alone; it is a pure Gaussian, descriptive of the electron velocity distribution. At the high α limit, the spectral shape is determined by density alone, since ω_p is only density dependent:

$$\omega_p^2 = \frac{4\pi n_e e^2}{m_e}$$

The spectral shape for the transition region, i.e., for 0.5 < α < 2.0, is determined by density and temperature together, however, and it is in precisely this α ≈ 1 region where scattering becomes a uniquely powerful tool. It turns out that for a given (or experimentally measured) electron spectrum there is one unique combination of T_e and n_e which will give a best fit. Thus, no absolute intensity measurement or calibration is necessary in order to get n_e as well as T_e. The spectral shape is all that is needed.

LIGHT SCATTERING

When $S(\Delta k,\omega)$ is integrated over frequency, we obtain the total scattering cross section in terms of the Thomson cross section. For example, at $\phi = \pi/2$,

$$\frac{d\sigma}{d\Omega} = r_e^2 \, S(\Delta k)$$

where

$$S(\Delta k) = \int S(\Delta k,\omega) d\omega$$

In the Saltpeter approximation,

$$S(\Delta k) = S_e(\Delta k) + S_i(\Delta k)$$

where

$$S_e(\Delta k) = \frac{1}{1+\alpha^2}$$

and

$$S_i(\Delta k) = \frac{Z\alpha^4}{(1+\alpha^2)(1+\alpha^2+Z\frac{T_e}{T_i}\alpha^2)}$$

These are interesting to examine in the limits of α. For example, as $\alpha \to 0$,

$$S_e \to 1$$

and

$$S_i \to 0$$

which again demonstrates that at low density, the total scattering cross section per electron is just the Thompson value, and that ions do not contribute.

For $\alpha \gg 1$, however, the situation is reversed. Here,

$$S_e \to 0,$$

$$S_i \to \frac{Z}{1+Z\frac{T_e}{T_i}}$$

We see that the total scattered energy from electrons, now relegated to the two satellite peaks at about $\omega_o \pm \omega_p$, becomes negligible compared to energy contained in the ion feature. The physical nature of the ion feature scattering is probably best elucidated by the "dressed test particle" model of Rosenbluth and Rostocker[9]. They show that statistically, ions can be said to be shielded by clouds of electrons which tend to follow their motions. These shielding clouds impose fluctuations upon the mean background and thus scatter radiation; however, since such a fluctuation moves with its parent ion, the doppler-shift of the scattering corresponds to the ion velocity, and not to the electron thermal speed.

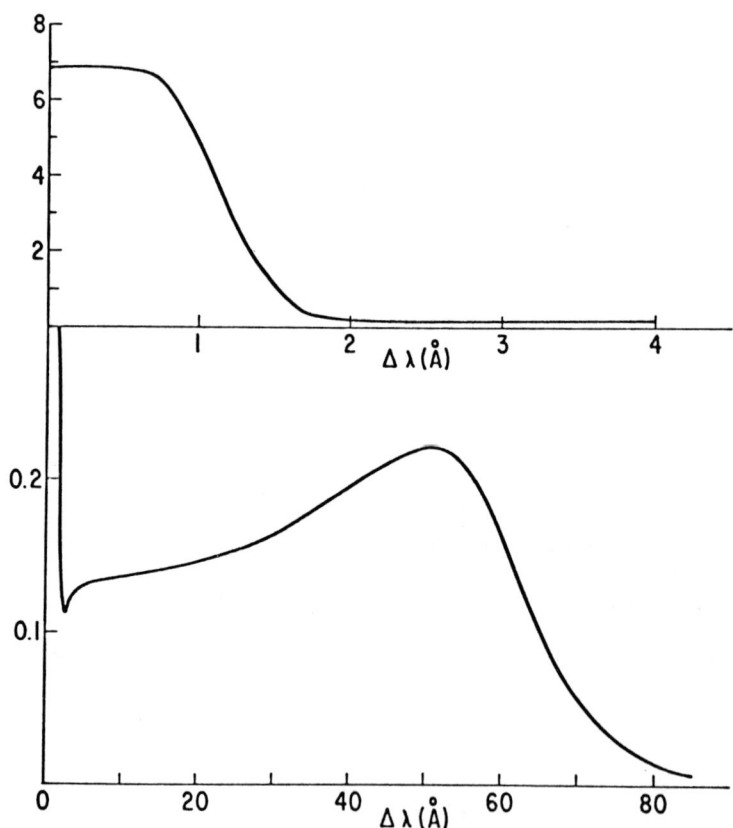

Fig. 3. The spectral distribution of scattered radiation (at $90°$) for $\lambda_o = 6943$ $\text{A}°$, $n_e = 5 \times 10^{16}$ cm^{-3}, $T_e = T_i = 4 \times 10^4$ $°K$. The upper graph is expanded in λ and compressed in amplitude to better depict the ion feature.

A sufficiently careful measurement of the usually narrow ion feature spectrum can yield the ion temperature and density in a way exactly analogous to that described earlier for electrons.

Figure 3 is an example of the spectrum from an equilibrium hydrogen plasma having $n_e = 5 \times 10^{16}$ cm^{-3}, $T_e = 4 \times 10^4$ °K; $\Theta = 90°$. Two different wavelength and intensity scales show the electron and ion features.

III. EXPERIMENTAL PROCEDURES

We will discuss here the principal problems associated with light scattering experiments. This cannot possibly be an exhaustive treatment; for many details, we refer the reader again to Evans and Katzenstein, who have nearly succeeded, in a single paper, in being exhaustive.

a. Sources

It should be clear from the discussion in section I that the source of light for a scattering experiment must be a high power pulsed laser. The object of the experimental design must be to collect during the duration of the experiment, a sufficiently large number of scattered photons to override significantly the intrinsic detector noise as well as whatever light is sent into the acceptance angle and bandwidth of the detector from the luminous plasma.

In the event (nearly universally so) that the detector output is observed as a function of time on an oscilloscope, a significant difference between the duration of the laser pulse and the plasma pulse--if indeed it is a pulsed system--can provide excellent discrimination between scattered and emitted radiation from the plasma. For example, one will typically observe a very short but distinct signal pulse corresponding in time to the actual laser output, this pulse being superimposed upon a longer, though perhaps stronger component from the plasma itself.

Since a majority of plasma systems typically investigated by light scattering are themselves impulsive in character, there is a very distinct advantage in employing as a source a laser having its output compressed into the shortest possible time. The Q-switched ruby laser fills this requirement almost ideally and has been used in nearly all experiments to this date. Readily available lasers produce power levels of tens to hundreds of megawatts in pulses lasting a few tens of nanoseconds.

b. Detectors

Without exception, photomultipliers are used as detectors of scattered radiation. No other available device can meet the requirement for nearly single-photon sensitivity and risetime of a few nanoseconds. Red-sensitive cathodes, e.g., S-20, are always used.

c. Scattering System Geometry

The requirements for a light scattering system are in many ways similar to those for a nuclear scattering experiment. For example, the incident beam must be well collimated, and one should employ a system of baffles around the incident beam in order to reduce as much as possible scattered incoming light from entrance windows or other optics. It is usual in laser scattering to bring the incident beam to a focus at or near the scattering volume. Since the angle of convergence of the beam is usually not large, the focus is not, in general, fine enough to produce breakdown or other nonlinear effects; however, the scattering region can typically be made as small as one or two millimeters in diameter by this means, and thus, excellent spatial resolution transverse to the beam can be achieved.

The scattering volume is usually imaged by a collecting lens upon a detector input aperture. The length of beam image actually inside this aperture then defines the length of the scattering volume.

In the case of very low-angle scattering experiments, it is usual to collect the scattered light in an annular lens surrounding the incident beam. A quite large effective solid angle of detection can be achieved in this way.

d. Beam Dumping

A most characteristic and severe requirement of laser scattering experiments is that the unscattered beam which has passed through the test chamber must be absorbed with such great efficiency that any part finding its way back into the system and thence to the detector must be negligible compared to the real signal, which is only 10^{-12} or so of the incident power.

Great ingenuity has gone into the design of efficient beam dumps, and a survey of this field would itself constitute a major article. However, one rather simple device has proven to be remarkably effective, and is probably used today more than any other. It is simply a polished sheet of blue glass, set at the

LIGHT SCATTERING

Brewster angle with respect to the beam, and rotated in azimuth so that the incident beam is polarized in a plane normal to the glass. A cascaded set of two such glasses, together with a well-baffled entry to the chamber they occupy, seems to be adequate for most experiments.

Since "equipment scattered" light is harmful only to the extent that it actually enters the detector, it is also common to provide a light dump into which the detector itself looks.

e. <u>Dispersing Instruments</u>

Inference of plasma parameters from the spectral distribution of scattered light requires that one have at hand some means of dispersing or filtering the input to the photomultiplier. This may be accomplished by means of tunable monochromators or by multi-layer dielectric bandpass filters which can be effectively tuned by tilting with respect to the incident radiation. The design of the dispersing system is critical in the sense that it must admit as much as possible of the scattered light to the detector consistent with wavelength resolution adequate for determining the shape of the scattered spectral distribution. For example, one will usually find that a high resolution spectrograph (or monochromator) of high f-number, i.e., low optical speed, is totally unsuitable for this work. Coarse resolution, typically several angstroms or so, is usually adequate to get a good spectral profile; an amount of light approximately proportional to width of the spectral sample is transmitted to the detector, so that effective sensitivity tends to vary inversely with resolution.

Tunable Fabry-Perot etalons and interference filters have been most frequently favored because of their high optical speeds; however, small monochromators of modest resolution and low f-number (e.g., the Jarrell-Ash model 82-410) which are well-matched to this application have become available and are being successfully used in scattering work.

In the event that a single filter or monochromator is used to analyze the scattered spectrum, it is clear that many discharges of the system must be employed to accumulate a complete spectrum. This procedure requires, first of all, that the plasma under study be highly reproducible from shot to shot, and further, that the laser pulse reproduce well. It is usual to provide some kind of monitor for the laser beam power (displayed on an oscilloscope together with the scattered signal) so that possible shot-to-shot laser fluctuations can normalized out of the scattering data.

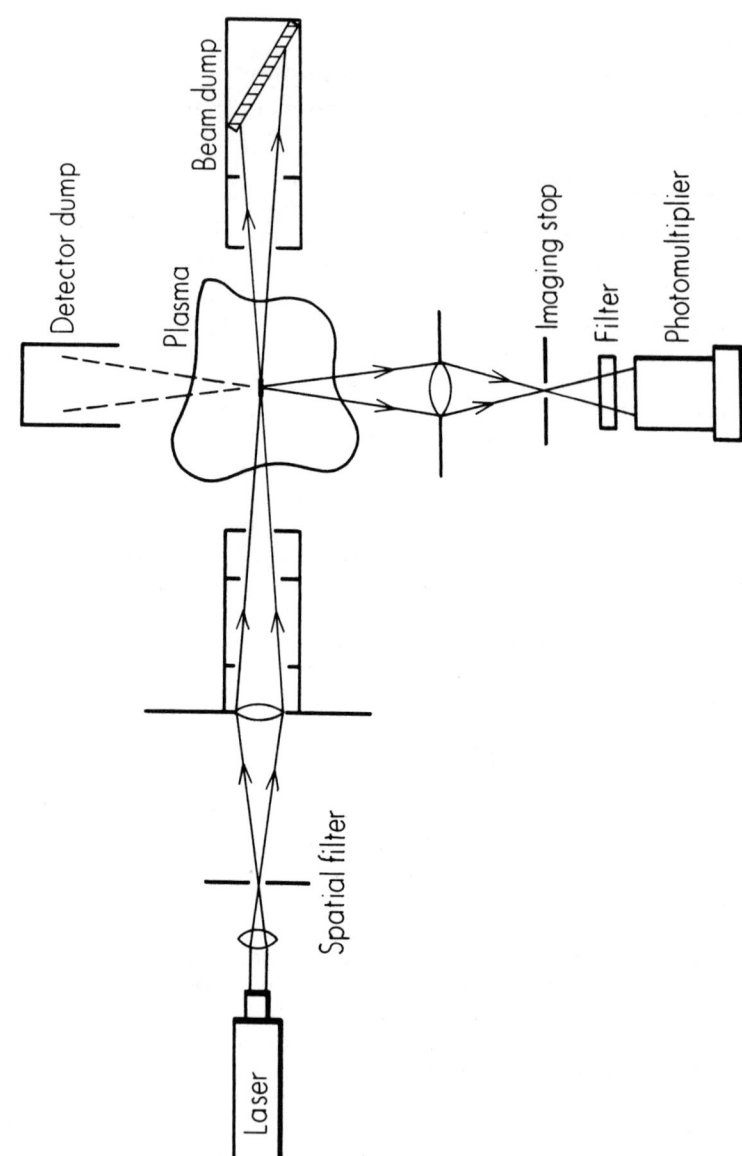

Fig. 4. Schematic layout of a typical scattering experiment.

Alternatively, one may provide a multi-channel dispersing system, e.g., one in which a number of detectors are placed in the image plane of a spectrograph, so that the entire spectrum is determined at a single shot. Various systems of this sort are described in reference 7.

In Figure 4, we present a highly schematic diagram of a scattering experiment. The only component in this figure not discussed in the text is the "spatial filter" just following the laser. This is simply a "pinhole" at which the main laser beam is brought to a focus. Spurious side-lobes of the laser beam, which occur very commonly, are focused in this plane as separated spots, and do not pass through the aperture; thus, the collimation of the input beam is greatly improved.

IV. A SCATTERING EXPERIMENT

We will use the results of a single recent experiment, that of R. E. Siemon[8], in order to illustrate the use of the technique in an investigation, where very high spatial and temporal resolution were required, and where the plasma density and temperature profiles were combined with magnetic field data to obtain a rather complete picture of the dynamics of a complicated plasma system.

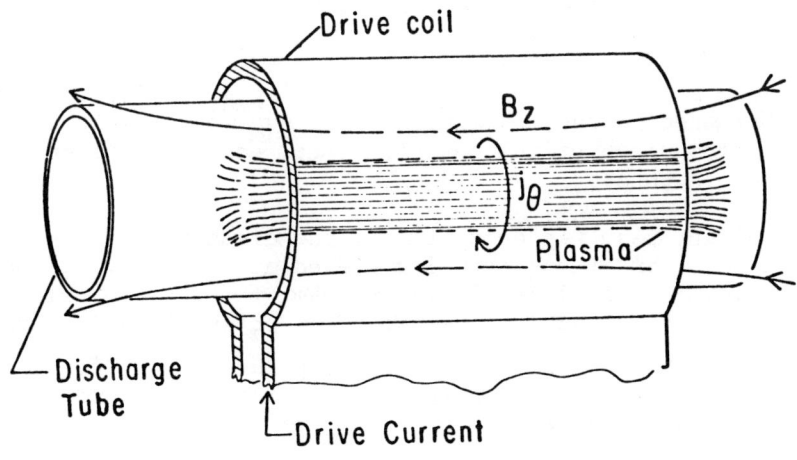

Fig. 5. Plasma fields, and currents in a theta-pinch. These data are taken near the peak density region of the plasma shell.

The overall problem in this experiment was to study the "snowplow" process in an imploding theta-pinch. The latter term designates an induced cylindrical current flow in a plasma, where the interaction between the external induction coil, a one-turn solenoid, and the plasma current itself is to drive the cylindrical current sheet rapidly toward the system axis (Figure 5). In contracting radially the current sheet overtakes, ionizes, and entrains within itself nearly all of the neutral gas within the discharge tube. The azimuthal current flowing in the plasma is confined to a shell whose thickness is much less than the discharge tube radius; hence, one may view the current-plasma distribution as locally planar.

Parameters characterizing this experiment are the following:

Discharge tube diameter	10 cm
Discharge tube length	30 cm
Maximum plasma current	2×10^5 a.
Implosion Time	10^{-6} s.
Test gas fill	50 mtorr N_2

Light scattering was employed to determine the distribution within the imploding current sheet of both n_e and T_e. (In this particular experiment the ion feature was neglected). Since the spatial distribution is itself in rapid motion radially, it was necessary in this work to have a very short laser pulse and exceedingly good shot-to-shot plasma reproducibility, since a timing jitter of 0.1 μsec of the laser pulse relative to the main current flow would introduce a radial position error equal to the entire current sheet thickness.

The incident beam was brought into the discharge tube parallel to the tube axis, and the scattering viewed at $\theta = 90°$. In order to allow a radial scan of the incident beam, the optical relay system of Figure 6 was used. The two relay prisms P_1 and P_2 serve a double function: They rotate the laser polarization from vertical to horizontal so that one may have $\phi = \pi/2$ for the upward scattering; also, the assembly consisting of P_2, the focussing lens L_2, and the input collimator can be moved up and down to effect the necessary radial beam scan.

The scattered radiation is relayed by prism P_3 through L_3 which images scattering volume upon the entrance slit of an 0.25 meter, f:3.3 Jarrell-Ash monochromator.

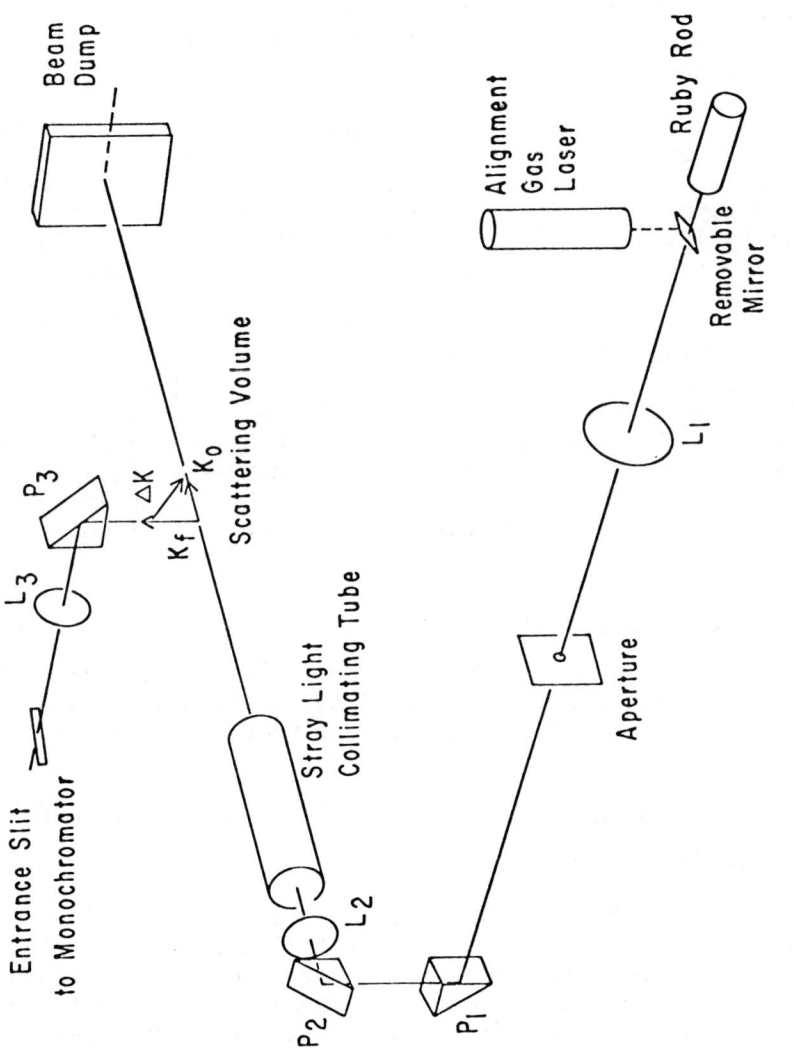

Fig. 6. Optical layout of the Siemon experiment.

A blue glass Brewster-angle beam dump is located beyond the discharge region. The actual sequence of operations is as follows:

First, the pump lamps in the ruby laser are turned on. 600 μsec later, the main spark gap switch in the induction coil circuit is closed. A small pickup coil, coupling flux from the main coil, produces a rapidly rising voltage pulse which is sent through a variable delay line to trigger the Pockels Cell Q-switch in the laser. Thus, the laser pulse may be set to occur at any time in the history of the discharge.

Figures 7 and 8 are two typical spectra obtained for the same instant of time in the discharge history, but at two slightly different radii. The current sheet here was at about half the tube diameter. Figure 7 is typical of the $\alpha = 1$ region, while Figure 6 is typical of low α. Note that the uncertainty in plasma density for the low α case has become very large.

Each point shown on these figures represents data from one shot of the machine. The solid curves are best-fit Saltpeter curves which have been determined by a computer program which treats temperature, density, and a normalizing factor as adjustable parameters. The uncertainty figures are also given by this program. It is particularly interesting to note that in spite of the substantial scatter in the data of Figure 7, the uncertainties in n_e and T_e obtained through this fitting procedure are only about 5%.

In Figure 9, the radial distribution of n_e, T_e, and the axial magnetic field are given for the same instant at which Figure 7 and 8 were taken. In the peak density region, α is near unity, and both n_e and T_e carry small errors; however near the inside of the shell n_e carries a particularly large relative error and the uncertainty in T_e is also growing. For smaller radii, the scattered signal was overcome by noise.

It is interesting to note that it was possible in this experiment to take advantage of a very beneficial effect associated with electron-feature experiments. It is that even if equipment scattering of the incident beam is relatively serious (inadequate dump, etc.), this scattering is all at ω_o, the central laser frequency. Spectra extending out to as much as 100 A° on each side of center are essentially uncontaminated. In this experiment, for example, it was possible to look directly through the rather dirty tube wall, to despense altogether with a farside detector dump, and still to obtain excellent spectra. Such liberties could never be taken were one interested in the ion feature which lies extremely close to the incident laser frequency.

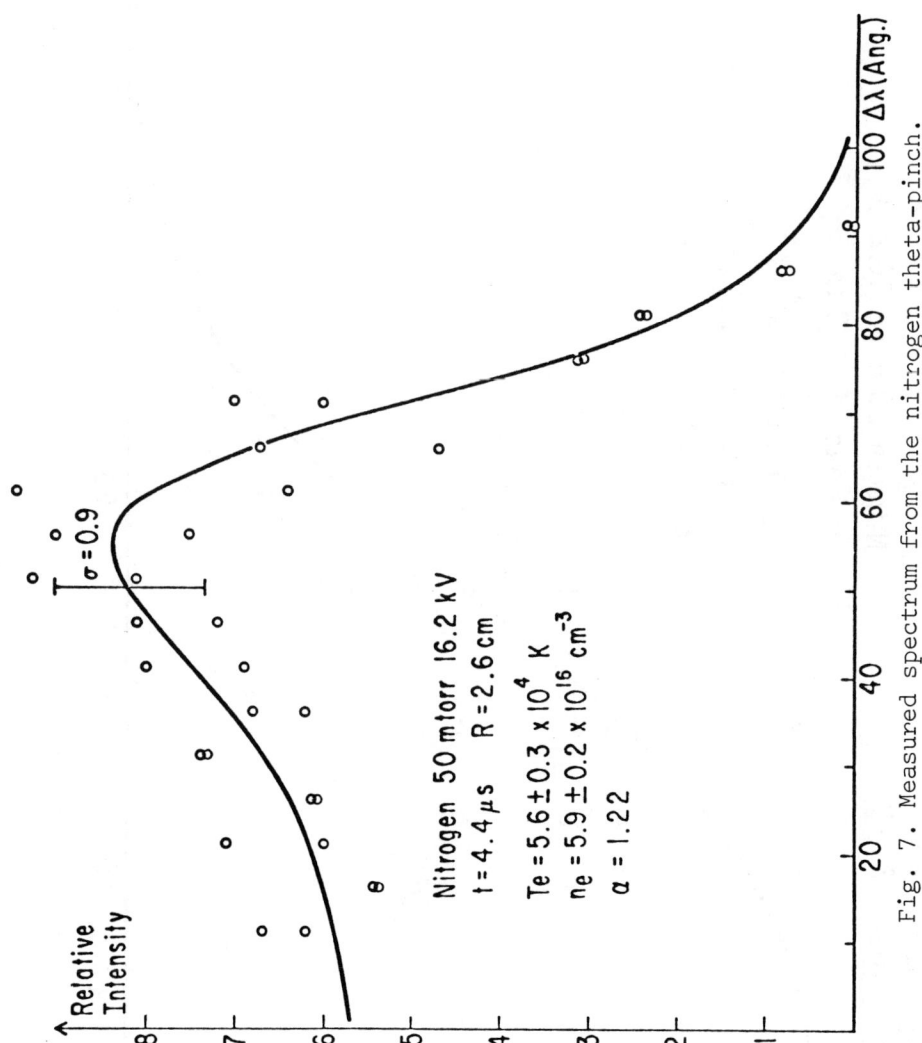

Fig. 7. Measured spectrum from the nitrogen theta-pinch.

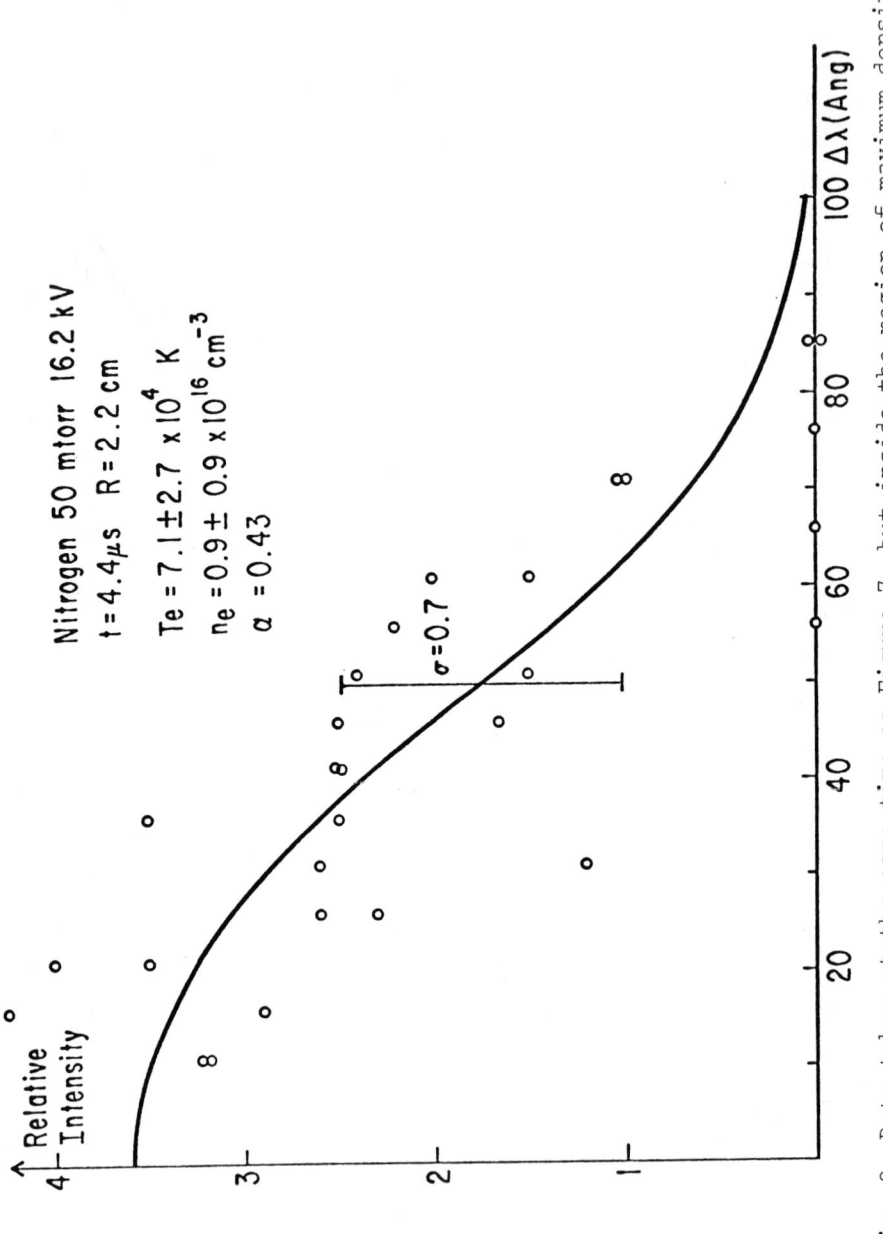

Fig. 8. Data taken at the same time as Figure 7, but inside the region of maximum density.

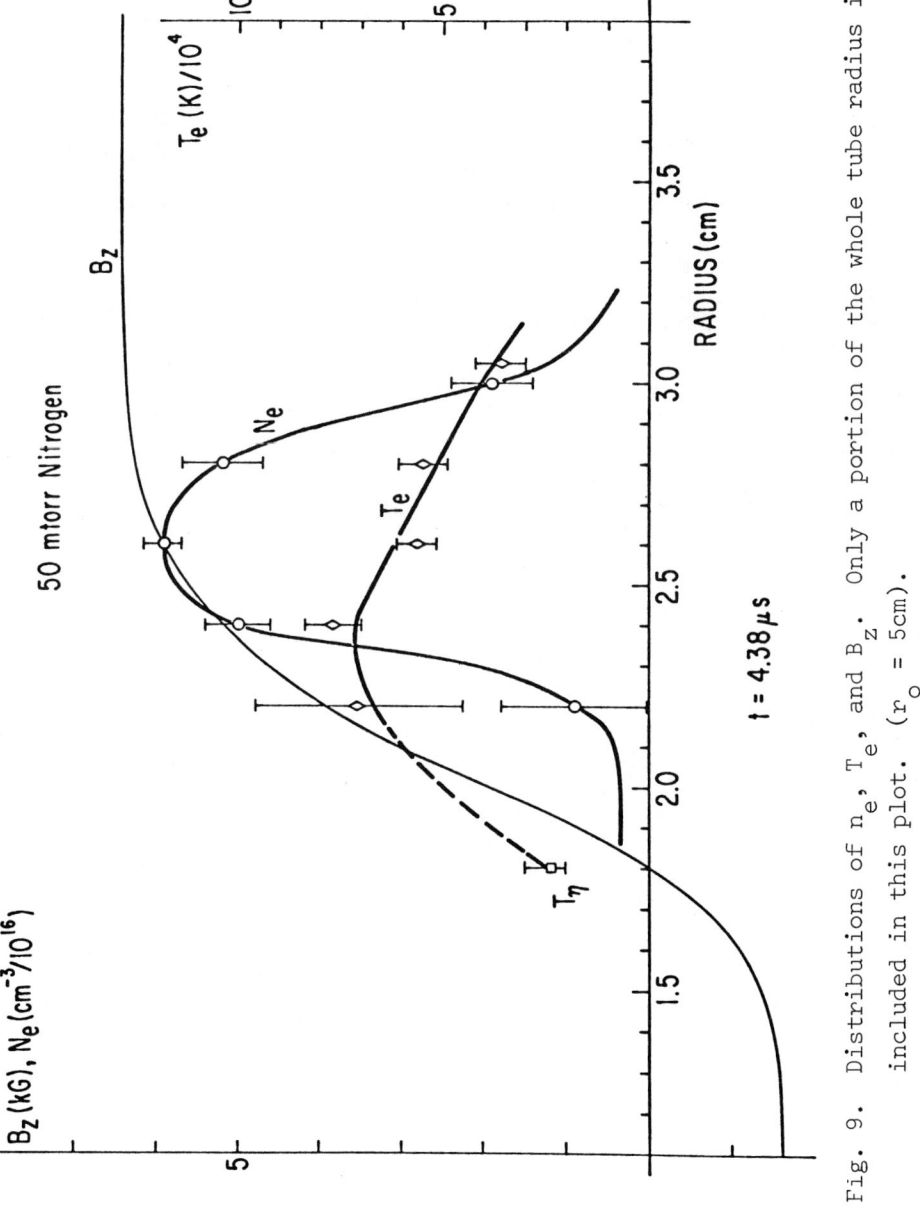

Fig. 9. Distributions of n_e, T_e, and B_z. Only a portion of the whole tube radius is included in this plot. (r_o = 5cm).

References

1. E. E. Saltpeter, *Phys. Rev. 120*, 1528 (1960)
2. J. A. Fejer, *Can. J. Phys. 38*, 1114 (1960)
3. J. P. Dougherty and D. T. Farley, *Proc. Roy. Soc. A259*, 79 (1960)
4. K. L. Bowles, *Adv. Electron Phys. 19*, 55 (1964)
5. I. B. Bernstein, S. K. Trehan, and M. P. Weenink, *Nucl. Fusion 4*, 61 (1964)
6. H. J. Kunze, "The Laser as a Tool for Plasma Diagnostics" in *Plasma Diagnostic Techniques*
7. D. E. Evans and J. Katzenstein, *Rep. Prog. Phys. 32*, 207 (1969)
8. R. E. Siemon, Ph.D. Thesis, Univ. of California, San Diego (1960)
9. M. N. Rosenbluth and N. Rostocker, *Phys. Fluids 5*, 776 (1962)

LASER - GENERATION OF RAREFIED PLASMA FLOWS*

David W. Koopman

Institute for Fluid Dynamics and Applied Mathematics

University of Maryland, College Park, Maryland 20742

ABSTRACT

A focused pulse from a Q-switched laser incident on a solid target is used to generate a plasma which expands into a vacuum o low density background gas. In a typical situation, conditions measured 33 cm from a aluminum target are: flow velocity $U = 10^7$ cm/sec, $n_e = \Sigma n_i Z_i = 5 \times 10^{11}/cm^3$, $T_e = 3$ eV, $U^2/2 = 2$ keV, $T_i \leqslant 10$ eV, $\lambda_{ii} \sim 100$ cm, and Debye distance $\lambda_D \sim 10^{-3}$ cm. These properties have been measured with Langmuir probes and microwave interferometers incorporating Lecher - wire elements. Applications of this technique of plasma flow generation to investigations of probe response in flowing plasmas and to studies of collisionless counterstreaming plasma instabilities will be described.

---

*Work supported by the U.S. Atomic Energy Commission.

INTRODUCTION

Gas dynamics research has, in the past thirty years, advanced from the study of sub-sonic, continuum fluids to investigations in which basic physical interactions and properties of the constituent molecules, ions, and electrons play an increasingly important part. With the advent of satellites, sounding rockets, and space vehicles, gas dynamics has merged with plasma physics in attempting to understand situations in which Coulomb forces between charged particles, and particle interactions with electric and magnetic fields, become the dominant mechanisms governing the behavior of an ionized medium. In order to investigate such phenomena, it is useful to have a laboratory source of ionized flows; if we wish to separate the strictly electrical effects from those associated with the presence of a magnetic field, a facility which requires no gross electrical current or magnetic field is desirable. We also wish to isolate the <u>collisionless phenomena</u>, which are the results of the collective behavior of large numbers of ions and electrons, for investigation. This paper will describe an experimental program using laser-produced ionized flows that is being developed to investigate collisionless electrostatic plasma phenomena. Data obtained on Langmuir probe response in flowing plasmas and on counterstreaming instabilities will be presented as examples of the type of research that can be performed.

As previous investigations have shown, a Q-switched laser pulse focused on a solid target produces an initial high density, high temperature plasma, collisionally dominated. If such a plasma is generated in a vacuum or near-vacuum environment, it is free to expand radially, the initial thermal energy being rapidly converted to ordered flow kinetic energy.[1-4] Calculations for typical cases indicate that by the time an initially dense plasma has expanded to dimensions on the order of .01 to .10 cm, the temperature has fallen to a point where the bulk of the energy is in ordered flow.[5-6] In this simple model, a subsequent outward expansion at constant velocity for each plasma element is expected, with the plasma velocity, U, following a similarity model of the form

$$U(R) = (R/R_{max})U_{max}$$

where R_{max} is the radius of the leading edge of the expansion, and U_{max} is the constant velocity at which the leading edge expands. Let us examine some <u>advantages</u> of the laser-technique of plasma production that make it an attractive method for investigating certain classes of ionospheric aerodynamic problems. No external magnetic field or electric current is associated with the plasma production or acceleration, so that electrostatic effects alone can be examined. Additional magnetic fields, of arbitrary direction and

magnitude, can be added as desired. After the initial effects
associated with heating and collisionally dominated radial acceleration, a flowing, expanding plasma exists in the experimental chamber.
For each firing of the laser, a wide range of flow velocities and
densities can be observed by choosing the distance from the target
at which some diagnostic element is located and the time at which
the observation is made. Under typical experimental conditions,
ion mean free path, λ_{ii}, is large, so that the plasma is collisionless.

In the flowing plasma, the charge neutrality condition,
$n_e = \Sigma n_i z_i$ requires that the mean flow velocity of electrons and
ions be equal. Therefore, the ordered kinetic energy of the plasma
is largely in the ions due to their greater mass, and it is evident
that electrostatic phenomena involving ion interactions must be
important in any collisionless process that significantly affects
the plasma flow. If the laser plasma is allowed to expand into
a low density plasma environment, the relative velocity between the
streaming and the background plasma can lead to various classes of
streaming instabilities and provides an experimental situation
suitable for investigating the collisionless electrostatic shock
waves predicted theoretically and possibly observed in earlier
laser plasma investigations.[9-12]

Moreover, if a solid body is placed in the plasma flow at some
arbitrary potential with respect to the plasma, the electrostatic
field from the combined space-charge and body voltage influences
the particle dynamics in the neighborhood of the body. The particle
flux, flow pattern, and potential and density distribution in the
neighborhood of a probe immersed in a flowing plasma, or near a
satellite or space vehicle moving through a plasma, can therefore
be studied in a laser-produced plasma flow.[13]

Other possible applications for laser-produced plasma flows
include simulation of exhaust from hypothetical thermonuclear
reactors, where anomalous ion heating due to relative streaming may
influence concepts for exhaust dynamics, direct electrostatic energy
conversion, or "fusion torch" material processing.[14]

APPARATUS

The apparatus for the investigation described in this report
is centered around a cylindrical vacuum chamber 21.5 cm in diameter
and 76 cm long, constructed of 304 stainless steel with internally
welded seams. Sixteen access ports to the chamber allow connection
to the vacuum pumps, pressure gauges, and gas supply valves, provide
for adjustment of target material, allow the laser beam to be focused
on the target, and provide stations at which diagnostic instruments
can be located. All ports are sealed by Viton O-rings.

The laser used in this program is a Korad Model K2-Q, rated for a maximum output power of 500 Megawatts at 6943 Å, and typically operated at a measured 30 nanosecond pulse half width at six joules energy. The ruby rod is 5/8" diameter and eight inches long and produces a beam divergence of 2.1 milliradians 1/2 angle for 1/2 power. The system includes a temperature controlled water cooler for maximum pulse reproducibility. The laser output is monitored with a fast photodiode, calibrated against a ballistic thermopile.

The transient electrical signals from the plasma diagnostic devices described in later sections of this chapter are recorded photographically from a bank of oscilloscopes. A schematic diagram of the basic apparatus is shown in Fig. 1.

The target holder consists of a wheel approximately 10 cm in diameter, driven by a hand-rotated vacuum feedthrough, gears, and a Viton belt, and located so that the laser focus occurs near the outer edge of the wheel surface, the wheel axis being parallel with, but about 4.5 cm above, the chamber axis. The wheel may be constructed of the target material itself (e.g., copper, titanium, or aluminum) or may be a holder for thin foil or special material targets (e.g., gold, aluminum foil, or polyethylene sheet). With this arrangement, it is possible to make well over one hundred laser

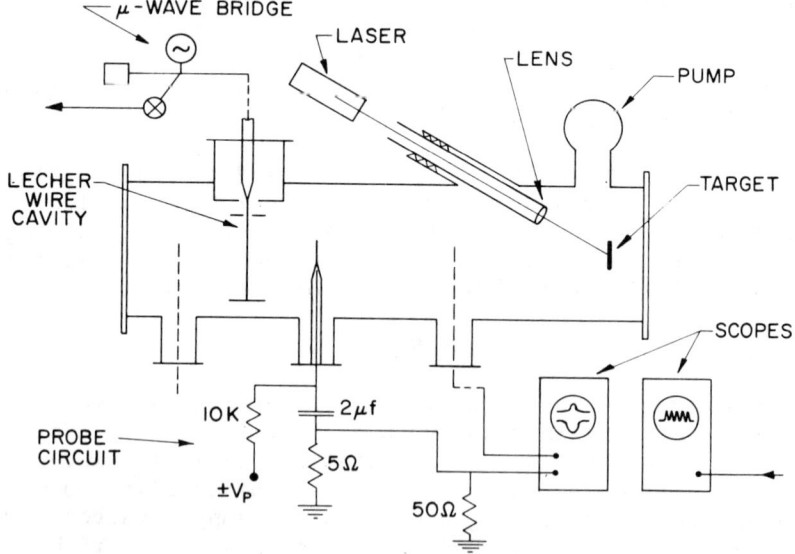

Fig. 1. Diagram of experimental system, showing typical locations for diagnostic devices.

firings at successive positions on the target and to change target materials as desired, without breaking the vacuum.

Diagnostic measurements are made with Langmuir probes and with microwave interferometers. The electrical circuitry of the probes was chosen for fast time response and for well defined probe potentials relative to the grounded metal vacuum vessel even when the probe is collecting plasma species. The current to the probes, as a function of time, is displayed on oscilloscope channels. By using multiple probes and successive laser firings at various probe voltages, curves of current vs. voltage can be constructed for specific time after the passage of the plasma front. Note that the experiment is also fitted with a Lecher wire probe coupled to a microwave interferometer.[15] The use of an open twin-wire guiding section as the flow-probing element has the advantage that the spatial resolution to density variations in the plane of the probe is approximately equal to the wire spacing (1 cm); moreover the wires do not greatly hinder the plasma flow. Reflecting plates define a resonant cavity as the probing element, so that the increase of wavelength as the plasma density increases yields a reestablishment of boundary conditions and bridge balance at successively lower mode numbers. Electron densities in the range of 10^{10}/cc to 1.2×10^{12}/cc can be measured in the rarefied flow generated by the laser pulse, with X-band (10 Gc, 3 cm) microwave equipment.

MEASUREMENTS ON LASER-PRODUCED FLOWS

The properties of the rarefied plasma flow generated by the expansion of a laser-produced plasma in a vacuum have been measured by the Langmuir probe and microwave interferometer diagnostics for various target materials. The microwave density measurements have been valuable in verifying the probe measurements, since the behavior of Langmuir probes in streaming ionized gases has received less attention and is less well understood than the stationary plasma case. Much of the early theoretical work has been based on the assumption that there is no interaction between the positive and negative species, an assumption not valid in many cases of practical laboratory interest; on the other hand, much of the experimental study has been done under conditions that failed to satisfy the criteria assumed in the interpretation of the data. Therefore, the same studies which have served to measure the flow variables have also allowed a critical investigation of probe response in the type of flowing environments produced by laser ionization.

In the present paper we will direct our attention to the case of transverse, cylindrical probes of length ℓ and radius r in metallic plasmas generated by pulsed laser heating of solid targets in a vacuum chamber. We assume, and can later document, that the ions stream out with high kinetic energy (up to several kev) and little thermal motion. They all leave the target within a short time interval, so that their velocity can be calculated by the time-of-flight to the probes, located 10-40 cm away. The ions carry with them a swarm of electrons; if the electrons have a temperature of more than a fraction of an eV, $c_e = \sqrt{8kT_e/\pi m_e} \gg U$, and the streaming velocity, U, can be neglected with respect to the electron thermal velocity, c_e, in describing the electron distribution. Also, the plasma is assumed to be collisionless and the Debye radius is shorter than the probe radius.

Consider how the probes function. Because the directed ion energy is so high, over a wide range of voltage, the ion current will remain constant, and be given by the product

$$I_i = e \cdot 2r\ell \cdot \Sigma n_i z_i \cdot U$$

as shown in Fig. 2. The electron current will be given by the thermal flux to the upstream side of the probe (where the ion density is constant) when the probe is at the plasma potential V_o; for more negative probe voltages, an exponential decrease is expected. The net probe current will therefore fall to zero at some $V = V_f$ (floating potential) and thereafter to the saturated positive currents I_i. We expect the electron current to the rear of the probe to be very small, because in the ion shadow, the electron density will also be low to maintain plasma nuetrality.

Note that the measure of the quality $\Sigma n_i z_i$, the positive charge density, can be obtained from the saturated ion current and from the time-of-flight calculation of $U \cdot T_e$ can be measured, if feasible, from a semi-log plot of I_e vs V; if our expectations are correct, the point $V = V_o$ and the value of I_{eo} should be identifiable on the graph. From T_e and I_{eo}, the value of n_{eo} can be calculated

$$I_{eo} = -\pi r \ell e\, n_{eo}\, c_e/4 \ .$$

Because

$$n_{eo} = \Sigma n_i z_i \ ,$$

in a dense plasma, the two measured values of these quantities should agree.

LASER PLASMA FLOWS

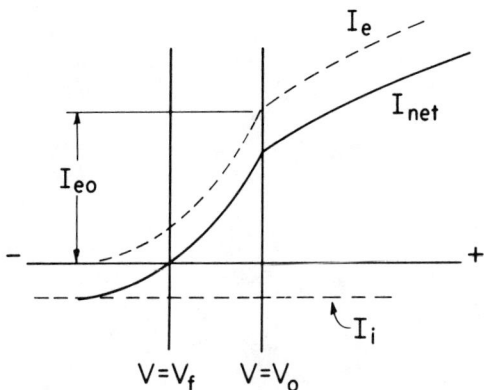

Fig. 2. Representation of net Langmuir probe current in a flowing plasma as the sum of a constant ion current and a variable electron current.

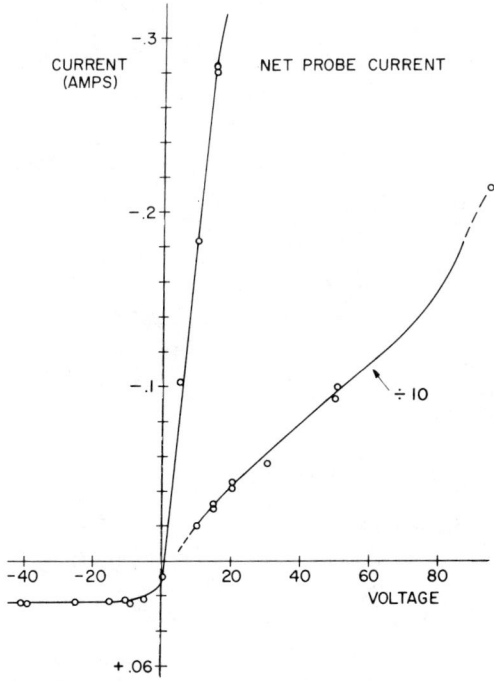

Fig. 3. Measured probe current as a function of probe voltage.

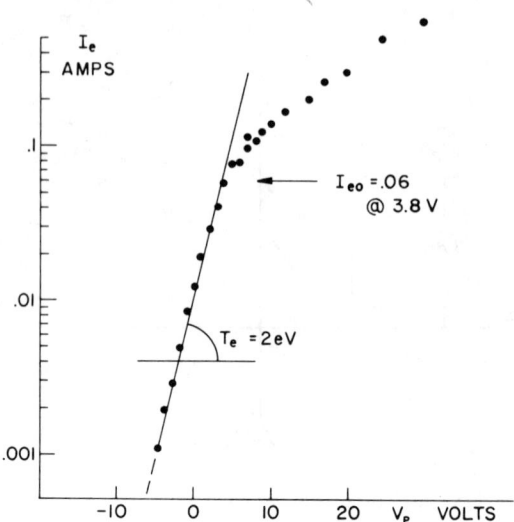

Fig. 4. Semi-log plot of electron current as a function of probe voltage.

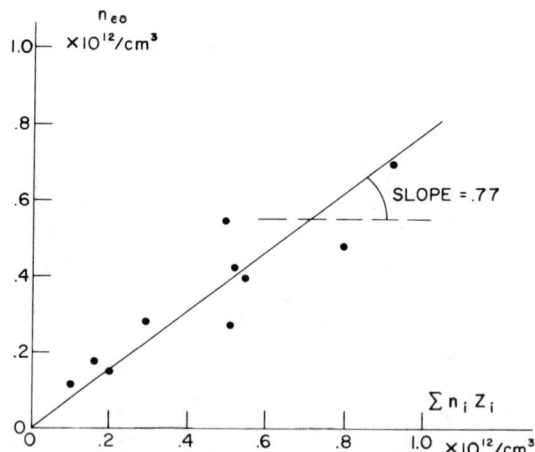

Fig. 5. Comparison of electron density as calculated from electron current slope and break with ion charge density measured by ion saturation currents in flowing plasmas.

LASER PLASMA FLOWS

Let us now turn to actual data obtained from a cylindrical probe (2.5×10^{-2} cm. diam., 1 cm long) located 32.6 cm from a copper target giving a plasma front velocity of 1.1×10^7 cm/sec when illuminated by a 30 ns laser pulse of 5 joules energy focused into a .03 cm^2 area. The current-voltage relation shown in Fig. 3 is obtained one microsecond after the passage of the plasma front, a time at which the plasma density at the probe is near its peak. Note the ion saturation region for negative voltages < 10 eV, followed by a sharp rise of electron current which continues to rise approximately linearly even at large positive potentials. The floating potential is low; $V_f = 1$ volt, a result not surprising, since the plasma is also in contact with the grounded metal vacuum wall and the net current to the wall must be zero to maintain plasma neutrality.

When the constant ion component of the net current is identified and substracted, the resulting value of the electron current, when plotted, shows the characteristics given in Fig. 4. Note the linear region from which T_e may be determined, a calculation which invariably yields T_e such that

$$c_e \gg U \quad \text{and} \quad MU^2/2 \gg kT_e$$

which were the necessary conditions for the applicability of our treatment.

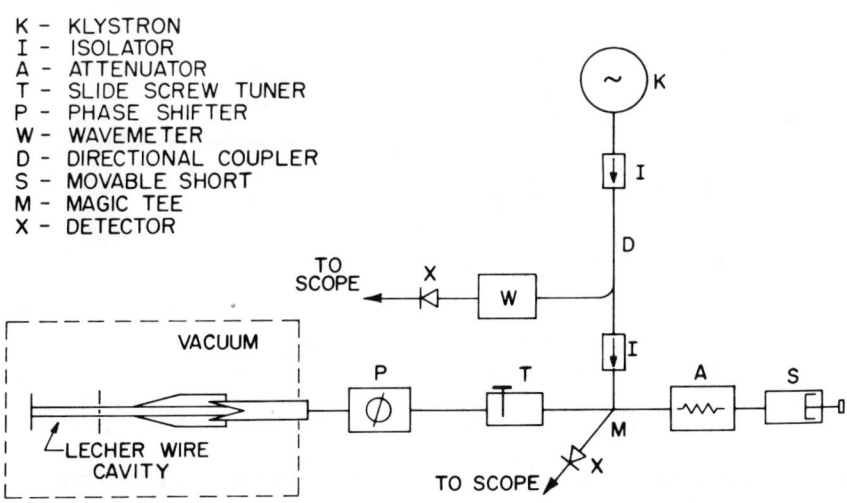

Fig. 6. Block diagram of X-band microwave interferometer coupled to Lecher wire resonant probe.

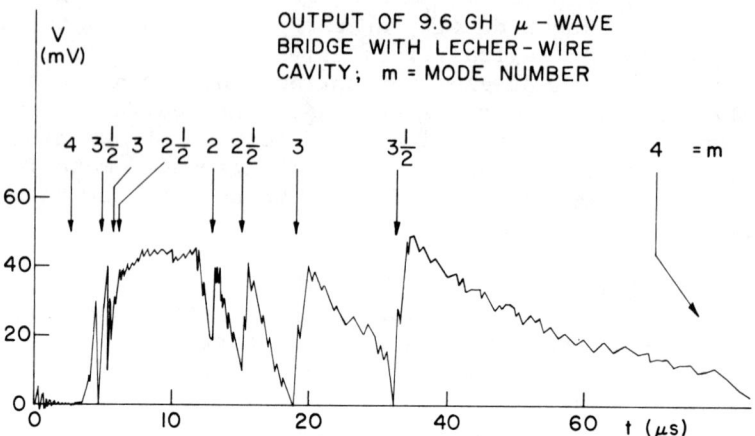

Fig. 7. Signal obtained from microwave probe in flowing plasma, showing dips corresponding to cavity modes.

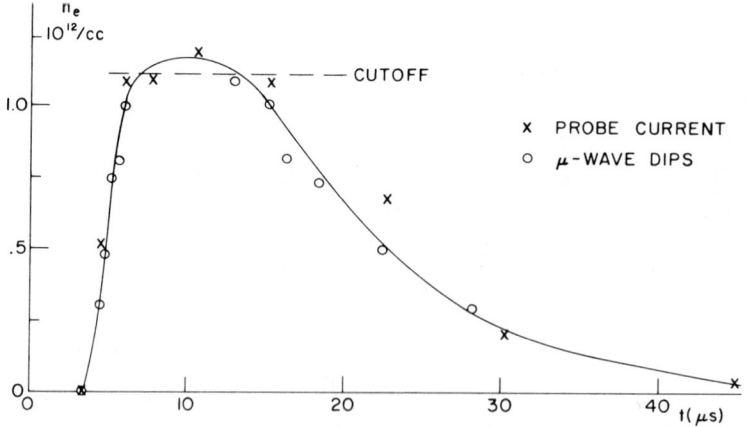

Fig. 8. Comparison of electron densities measured by Langmuir and by microwave probes.

A comparison between the values of n_{eo} and $\Sigma n_i z_i$, measured from the electron current break and the ion current saturation, respectively, for a series of experiments, is shown in Fig. 5. Note that the the agreement can be considered satisfactory, in view of the simple model chosen and the possible sources of error. This self-consistency is further evidence that the interpretation of the probe signals is based on a good approximation to the behavior of probes in such a plasma.

The probe measurements have been augmented by microwave determinations of electron density. Using the balanced microwave interferometer with waveguide coupled to the twin-line or Lecher wire element in a resonant cavity, as shown in Fig. 6, it has been possible to obtain sufficient spatial and time response to resolve features in the passage of the plasma. In Fig. 7, the output of the microwave bridge is shown as a function of time after the laser firing. Note the dips occurring when the plasma density attains values such that the dielectric constant allows the initial boundary conditions to be reestablished at lower mode numbers, corresponding to successive half-integer wavelengths in the cavity. By computing the dielectric constant appropriate for each dip, the electron density may be found.

Fig. 8 presents the quantity $\Sigma n_i z_i$ measured from ion saturation probe signals and the quantity n_{eo} measured from the microwave interferometer resonance dips, both as a function of time during the flow duration of the plasma generated by a laser firing. Note the density determined from our probe interpretation is accurately and independently confirmed by the microwave measurement.

Finally, consider the phenomenon that the probe currents do not saturate in the electron collecting region, even though $\lambda_D \sim 1/6 \, r_p$. For flowing plasmas of the type studied in our experiments, the high ion energy means that the voltages applied to the probe do not affect the ion density. Accordingly, an ion-free sheath, through which space-charged limited electron current flows and which limits the range of the probe potentials, does not form. Preliminary calculations treating the charge density and resulting potential distribution in flowing plasmas near probes indicate that the influence of positive potentials ($V > V_o$) will extend for distances many times larger than λ_D and that the current will continue to rise with increasing potential, in qualitative agreement with the experimental results.

The behavior of the probes in the present experiment should be contrasted with the reports of other investigators in which Langmuir probe type interpretations were given to currents collected from laser plasmas.[16] In the earlier work, often done in insulating or intermittently grounded chambers, the "probes" were often large enough to markedly affect the plasma, and were not designed as low impedance elements. Although it is difficult to determine exactly how some of the previous measurements were made, it is clear that high electron temperature, nonlinear effects, and density estimates derived from studies where probe potentials relative to any reference are uncertain, and where self-consistency checks and comparisons to independent microwave density measurements have not been made, must be viewed with caution.

In the present work, numerous series of measurements have been made with targets of copper, titanium, aluminum, carbon, and polyethylene. We will summarize with some general observations.

(a) The electron temperature is found to vary between 0.7 and 12 eV, the lower temperatures generally being associated with the leading edge of heavy ion (tungsten, copper) plasmas, and the higher value more characteristic of light ion (carbon, hydrogen) plasmas and of later times in the plasma flow.

(b) The velocity of the leading edge is constant as the plasma streams along the axis.

(c) Floating potentials range between 0 and $.1 T_e$ (volts).

(d) With a sharply focused laser spot ($10^{-3} cm^2$) on the target, the plasma flow tends to be non-reproducible (40% fluctuations) and characterized by small scale fluctuations. Density-vs-time profiles at a fixed position are peaked at earlier times and decay rapidly. With a broader laser focus ($10^{-2} cm^2$) the expansion velocity is reduced by $\sim$ 10-20% and the density profiles in time at a probe is relatively constant after the peak, indicating that the plasma profile is spatially sharply peaked toward the center. With the broader focus, shot-to-shot reproducibility is good.

(e) Use of orthogonal transverse and target-facing plane probe pairs has shown negligible transverse velocity components in the ion motion. The transverse currents observed are typically 3-5% of the current due to streaming, a result in the range expected from slight probe misalignment, and indicating that T_{ion} perpendicular to flow direction is at most on the order of 1-10 eV. Substantial transverse currents are observed only for low energy streaming light ions at late times in the flow, when

$$\tfrac{1}{2} M_i u^2 \approx - e(V - \phi_{plasma})$$

and $\lambda_D \geq r_p$, conditions under which ion saturation is observed to no longer occur, and when attraction of ions, as in a nearly-stationary plasma, is expected. The constant speed for the leading edge demonstrates that T_{ion} parallel to the flow is also small, comparable to the perpendicular value.

Table I

Comparison of Streaming Plasma Properties Measured 32.6 cm from Target with 5 Joule, 10^{10} Watt/cm^2 pulse

Target	Cu	Ti	Al	C	$(C_2H_4)_N$
Thickness	solid	solid	solid	.025	.0025
Front velocity (cm/sec x 10^{-6})	8.6	10.4	14.4	17.5	19.0
Front ion kinetic energy (Kev)	2.3	2.6	2.8	1.85	0.84*
Peak electron density (#/cc x 10^{-11})	2.8	3.9	6.8	10.5	14.0
Peak velocity (cm/sec x 10^{-6})	4.9	6.5	9.0	13.6	14.0
Density at twice t_{peak} (#/cc x 10^{-11})	1.8	2.8	2.2	1.8	1.4
T_e (eV)	2	4	6.5	9	12

* M_i = 4.7 assumed.

Table I presents a comparison of features measured for a series of targets, using a probe located 32.6 cm from the target, at sharp laser focus. The power density was approximately 10^{10} watts/cm^2, with 5 Joule pulse energy.

STUDY OF COUNTER-STREAMING IONIZED FLOWS

It has been found that when a laser-produced plasma is generated in a low density background, e.g., 2×10^{-3} mm H_2, the photon flux from the initial high density-high temperature target material at the focal point is sufficient to produce an appreciable ionization of the background gas. The degree of ionization is $\sim 100\%$ within 1 cm of the focal point, falling to $\sim .1\%$ at 30 cm from the target. Thus it is possible to generate experimentally a counter-streaming of the laser plasma into a partially ionized background.

The mean-free-path for collisional interactions can be expressed as $\lambda = 1/n\sigma$, where σ is the cross section for a collisional process, and n is the number density of the scattering species. For typical atomic processes, e.g., charge exchange of the expanding laser-plasma ions with the background gas, $\sigma \sim 10^{-16}-10^{-15}$ cm^2.[17] For multiple-encounter Coulomb collisions,

$$\sigma \simeq 2\pi e^4 z^2 \ln \Lambda / W$$

is an effective 90° elastic multiple scattering cross-section for ion-ion or electron-electron interactions. In this expression, $\ln \Lambda \simeq 15$, and Z is the charge on a test-particle characterized by energy W with respect to the scattering elements. Substitution of numerical values appropriate to a laser-produced plasma streaming into $\sim 3 \times 10^{-3}$ mm of background gas indicates an attenuation by neutralization of the laser-plasma on a 10-100 distance scale; the multi-coulomb scattering mfp for ion, λ_{ii}, would be on the order of 10^2-10^3 cm. Thus, little collisional heating and randomization of the distribution function representing the counter-streaming ion species is expected. If, however, the counter-streaming of the laser-plasma through the background is unstable, when collisionless plasma phenomena are considered, then a broadening of the ion distributions functions would be expected until random velocities were on the order of U, the relative drift velocity. If the distance scale over which such energy exchange and momentum exchange occurred were small compared to experimental dimensions, the laser-plasma would be expected to act as a piston, sweeping background plasma ahead of it. The leading edge of the disturbance would then be a boundary between undisturbed, cold, background plasma and hot, compressed background mixed with laser plasma material. At this point, the appropriate description would be that of a sharp discontinuity across which, for a single material, the collisionless counter-streaming instability mechanisms would have to lead to the compression and heating of the background to satisfy the Hugoniot relations.

The criteria for stability between two cold ion beams in a plasma can be derived from a dispersion relation of the form:

$$\frac{\omega_{p1}^2}{(\omega/k)^2} + \frac{\omega_{p2}^2}{(\omega/k-U)^2} + \frac{\omega_e^2}{(\omega/k)^2-c_e^2} = k^2$$

where ω_{p1}, ω_{p2} and ω_e are the two ion plasma frequencies and the electron plasma frequency, c_e is the electron thermal velocity, and U is the relative drift velocity.[18-20] In the range where $c_i \ll U \ll c_e$, as in our studies, the condition for stability becomes

$$U^2 > c_e^2(\omega_{p2}^2/\omega_e^2)\{1 + (\omega_{p1}/\omega_{p2})^{2/3}\}^3 .$$

For typical cases, the stable region can be defined as

$$U^2 > 3 \times 10^{12} \frac{T_e}{n_e} [(\frac{n_1 Z_1}{M_1})^{1/3} + (\frac{n_2 Z_2}{M_2})^{1/3}]^3$$

where T_e is given in eV, M in amu, and Z in charge per ion. In the present laser-plasma study, the stable region for ion-ion electrostatic counter-streaming has an upper boundary at $U \simeq 2 - 7 \times 10^6$ cm/sec. Hence this mechanism is not expected in the range of $U = 1 - 2 \times 10^7$ cm/sec, although it was perhaps observed in previous laser-plasma counter-streaming investigations.[11]

A second class of instability which may be present in counter-streaming electrostatically coupled plasma is the ion-electron interaction which may be unstable in the range $\sqrt{kT_e/M_i} < U < \sqrt{kT_e/m_e}$, when $T_e \gg T_i$, the exact range of instability being given by a detailed study of the dispersion equation for electrostatic waves in a plasma of species j, which has the form

$$k^2 + \sum_j \frac{1}{\lambda_j^2} \cdot I(L_j) = 0$$

where $\lambda_j = c_j/\Omega_j$, $c_j = \sqrt{kT_j/M_j}$, $\Omega_j = (4\pi n_j e_j^2/M_j)^{1/2}$, $L_j = (\omega/k-U_j)/\sqrt{2}c_j$, and $I(A) = 1 + i\pi^{1/2}A \cdot W(A)$, where W is the error function of complex argument.[20] Maxwellian distributions are assumed for each species. It is this class of instability which would be expected to give rise to interactions between the streaming and background plasma in the present experiment. Typical growth rates might be in the range

$$\gamma = 10^{-1} - 10^{-3}\Omega_i$$

so that $\gamma \sim 10^6 \text{sec}^{-1}$ would be appropriate to a background density in the range of $10^{11}/\text{cm}^3$. Thus a growth rate comparable to the experimental time scale is expected.

Under such circumstances, a sharp discontinuity between laser-plasma and background would not occur, and snow-plow pickup of background would not be expected. A slow heating of the background, until $\sqrt{kT_i/M_i} \sim U$, might be expected to be observable. The experimental data does not show this effect clearly, and continued investigations of this type of instability are being made.

In the absence of collisionless interactions which can lead to heating and acceleration of the background by the laser-plasma, it is clear why well-formed collisionless shocks of the type proposed by Tidman have not been observed in the present study.[9] Such shocks require $T_e > T_i$, a supersonic flow:

$$U > \sqrt{kT_e/M_i},$$

and an instability criteria

$$1 \gg .16 \left(\frac{U}{c_i}\right)^2 \left[\frac{3m_e T_i^3}{2M_i T_e^3} \ln \frac{M_i T_e^3}{m_e T_i^3}\right]^{1/2}$$

all to be satisfied. In these expressions, all quantities refer to conditions in the undisturbed plasma. In our studies, these conditions are fulfilled, but it appears to be the absence of an effective piston action by the expanding laser plasma that prevents the electrostatic shock mechanism vrom being tested.

Recent theoretical and numerical studies on the behavior of counter-streaming plasmas in a magnetic field have shown that relatively modest fields are able to influence electrostatic collisionless interactions between the plasma ion beams. In particular, it has been shown that the addition of a B-field transverse to the flow can render the counter-streaming unstable, even though the situation is stable in the absence of a B-field and though the field is too weak to appreciably affect the ion motion.[21,22]

The analysis indicates that in the limit $\omega_{ce}/c_e \gg k$, the dispersion relation for a cold plasma can be written

$$\frac{\omega_{p1}^2}{(\omega-kU/2)^2} \frac{\omega_{p2}^2}{(\omega+kU/2)^2} = 2\{1+(\frac{\omega_{pe}}{\omega_{pe}})^2\}$$

where ω_{p1} and ω_{p2} are the plasma frequencies for the two ion components streaming at relative velocities $\pm U/2$ with respect to the electrons, and where ω_{pe} is the electron plasma frequency and ω_{ce} is the electron cyclotron frequency. For this case, a growth rate for an instability which randomizes the ion motion at the expense of the streaming energy is given by

$$\gamma = \omega_{pi}/2\{1 + \frac{\omega_{pe}}{\omega_{ce}}\} \sim \{\omega_{pi}\frac{\omega_{ce}}{\omega_{pe}}\} \ .$$

The width of the turbulent region can be expected to be on the order of

$$W = U/\gamma \simeq U \ \omega_{pe}/\omega_{pi} \ \omega_{ce} \simeq 50 \ U\sqrt{M}/10^7 B$$

where M is in amu, U is cm/sec, and B is given in gauss. Thus we see that for a hydrogen background, magnetic fields in the 100 gauss range can greatly enhance the instability growth rate above that expected in the absence of a field, expected transition regions on the order of one cm. One dimensional computer calculations have supported the validity of this treatment in some idealized cases.

In our preliminary laboratory studies, a number of large ceramic magnetics were placed along the bottom of the experimental chamber, in order to give a transverse B-field in the range of 200 gauss to 10 gauss, depending on the irregular placement of the magnets. Cylindrical probe and microwave data of the type already described were taken for a range of gas pressures.

Fig. 9 presents a comparison of the signals obtained with and without the field when a copper target was expanded into a 2 micron hydrogen background. Note in both cases the photoionization background ($n_e \sim 5 \times 10^{10}/cm^3$) formed in the hydrogen. The laser-plasma front arrives at the probe and microwave location, 33 cm from the target, at approximately 2.8 microseconds, as expected. There is however, a precursor signal when the B-field is present; also the form of the electron collection is modified immediately after the arrival of the plasma front.

The reason for the observed probe signal profiles is better understood on Fig. 10, where characteristic curves are displayed at different times after the laser firing.[23] Consider the floating potential, i.e., the point at which the net probe current is zero. This is seen to occur at approximately at zero volts at 0.5 microseconds (before the precursor) and after 7 microseconds. For intermediate times, particularly when the laser-plasma front reaches

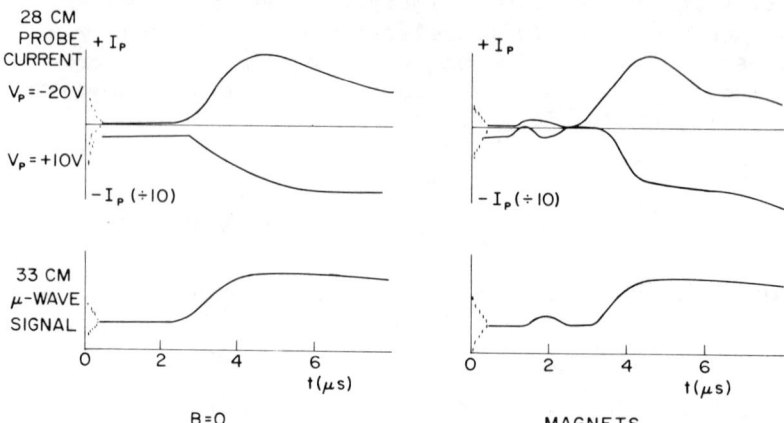

Fig. 9. A comparison of cylindrical probe and microwave signals from a copper plasma streaming into a 2.0μ hydrogen background without and with a magnetic field in the experimental volume. Note the precursor signal in the ionized background at ~1.5μs after the laser firing when a B-field is present.

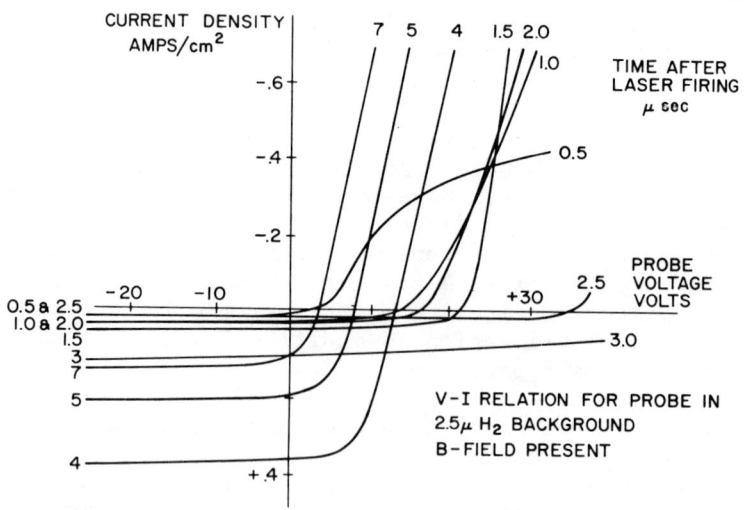

Fig. 10. Voltage-current relationships obtained at various times as a copper plasma streams through a preionized hydrogen background with a B-field present.

the probes at 2.5-3.5 microseconds, the floating potential becomes high, greater than 40 volts at the three microsecond point when the ranges of the probe voltage were inadequate to complete the data curve.

Note also the gradual slope of the 2.5 and 3.0 microsecond curves, which may represent either a high electron temperature or modest local ion energy. Transverse and target facing probe pairs in this region have an ion current ratio of $\sim .5$ indicating an appreciable flow superimposed on background ion collection. Thus it appears that the addition of a magnetic field enhances the interaction between the streaming laser-plasma and the stationary ionized component of the background environment. A uniform, transverse field of ~ 200 gauss will be added to the experimental facility to investigate these magnetic enhancements of counterstreaming interactions more completely.

REFERENCES

[1] H.F. Haught and D.H. Polk, Phys. Fluid 9, 2047 (1966).
[2] D.W. Gregg and S.J. Thomas, J. Appl. Phys. 38, 1729 (1967).
[3] D.W. Koopman, Bull. Am. Phys. Soc. 11, 464 (1966); Phys. Fluids 10, 2091 (1967).
[4] E.W. Sucov, J.L. Pack, A.V. Phelps, and A.G. Englehardt, Phys. Fluids 10, 2035 (1967).
[5] J.M. Dawson, Phys. Fluids 7, 481 (1964).
[6] N.G. Basov and O.N. Krokhin, Sov. Phys.-JETP 19, 123 (1964).
[7] W.J. Fader, Phys. Fluids 11, 2200 (1968).
[8] R.G. Rehm, Phys. Fluids 13, 921 (1970).
[9] D.A. Tidman, Phys. Fluids 10, 547 (1967).
[10] D.A. Tidman, J. Geophys. Res. 72, 1799 (1967).
[11] D.W. Koopman and D.A. Tidman, Phys. Rev. Letters 18, 533 (1967).
[12] M. Lubin, Bull. Am. Phys. Soc. 13, 320 (1968).
[13] D.W. Koopman and S. Segall, Bull. Am. Phys. Soc. 14, 1011 (1969).
[14] B.J. Eastlund and W.C. Gough, The Fusion Torch-Closing the Cycle from Use to Re-use, USAEC Division of Research Publication WASH-1132 (U.S. Gov't Printing Office, Wash., D.C., May, 1969).
[15] R.C. Ajmera, D.W. Koopman, and H. Lashinsky, Bull. Am. Phys. Soc. 14, 1011 (1969).
[16] S. Namba, P.H. Kim, T. Itoh, T. Arai, and H. Schwartz, Papers Inst. of Phys. and Chem. Res. (Tokyo) 60, 101 (1966).
[17] H.J. Zwally and D.W. Koopman, Proc. VI International Conf. on the Physics of Electronic and Atomic Collisions, MIT Press (1969), p. 1025.
[18] A.A. Vedenov, E.P. Velikhov and R.Z. Sagdeev, Sov. Phys.-Uspekhi 4, 332 (1961).
[19] M.D. Gabovich and G.S. Kirichenko, Sov. Phys.-JETP 23, 785 (1966).

[20] T.E. Stringer, J. Nuclear Energy C6, 267 (1964).
[21] I. Haber, D.A. Hammer, R.C. Davidson, J. Dawson, N.A. Krall, K. Papadopoulos and R. Shanny, Bull. Am. Phys. Soc. 15, 641 (1970).
[22] K. Papadopoulos, private communication.
[23] D.W. Koopman, Bull. Am. Phys. Soc. 15, 643 (1970).

"A SUMMARY OF RECENT RESEARCH ON CONTINUOUS-WAVE CHEMICAL LASERS"

Terrill A. Cool

Department of Thermal Engineering

Cornell University - Ithaca, New York 14850

I. INTRODUCTION

It has been a little over 5 years since the first pulsed chemical laser was developed by Kasper and Pimentel (1). During this time, important advances have been made contributing to our understanding of the kinetic mechanisms by which population inversions can be created in chemically reacting gases. Much of this information comes to us from the pulsed chemical laser experiments of Airey, Gross, Moore, Pimentel, and others (2-10), and from the pioneering experimental and theoretical studies of infrared chemiluminescence by Polanyi and co-workers (11-13).

Largely because of this growing background of knowledge the chemical kinetic aspects of laser operation have not been, at least at this early stage, a limiting factor in cw chemical laser development. Progress with cw chemical lasers has been quite rapid since the initial demonstrations of continuous operation were performed over a year ago (14-16). Output power levels of several hundred watts have been achieved (17, 18); the continuous conversion of chemical energy into laser output without an external source of energy to initiate or sustain laser action has been demonstrated (19); cw chemical laser operation at a variety of wavelengths ranging from about 2.5 to 11 microns is now possible (17, 20-25).

II. FOUR BASIC PROCESSES INFLUENCE CW CHEMICAL LASER OPERATION

A fundamental problem in the development of present cw chemical lasers has been the establishment of suitable operating conditions within the limitations imposed by the relationships between the rates of four basic processes. The cw chemical lasers developed to date are pumped by the vibrational energy release of simple exothermic atom exchange reactions of the type

$$A + BC \rightarrow AB^* + C, \qquad (1)$$

extensively studied by Polanyi, et al (11-13). Other types of chemical reaction have produced chemical laser action on a pulsed basis (16), and no doubt will form the basis for eventual cw laser operation. It has been demonstrated that many atom-exchange reactions of the type of eq. 1 can provide an efficient conversion of the energy release of reaction into selective vibrational excitation of specific states of the reaction product molecule AB (8, 9, 11-13). It has further been shown that such reactions can lead to total population inversions of product molecules under conditions of laser interest because the reaction rates can exceed the rates for the collisional redistribution of vibrational energy among product molecules by the vibration $\rightarrow$ vibration (V $\rightarrow$ V) energy transfer processes (8, 9, 27, 28)

$$AB(v=n) + AB(v=m) \rightarrow AB(v=n-1) + AB(v=m+1) . \qquad (2)$$

This fact, and the additional relationship that the rates for the collisional transfer of the vibrational energy of a product molecule into the translational motion of an arbitrary molecule or atom by the V $\rightarrow$ T processes

$$AB(v=n) + M \rightarrow AB(v=n-1) + M \qquad (3)$$

tend to be small compared to the rates of processes 1 and 2, form the kinetic basis upon which pulsed chemical lasers of substantial gain operate under conditions of total population inversion.

The fourth basic process which must be considered in understanding the operation of present cw chemical lasers has a decisive influence upon the operating characteristics and efficiencies of such devices. This process is the continuous mixing of gaseous reactants,

$$\boxed{A} \; \boxed{\genfrac{}{}{0pt}{}{BC}{BC}} \; \boxed{A} \longrightarrow \boxed{A + BC} \qquad (4)$$

which occurs by means of an injector suitably designed to break the flow into unmixed portions of small dimensions to enhance the rates of diffusive mixing and chemical reaction. At pressures of a few torr or so, the rates at which reactants can be mixed by practical means tends to limit the otherwise rapid rates of chemical reaction to values which may fall below the rates of the redistribution of vibrational energy by V $\rightarrow$ V transfer among the reaction products by processes 2. Thus, for a given reaction mechanism and a given characteristic dimension for unmixed portions of the reacting flow (determined by the gas mixing injector design), a definite upper limit on operating pressure exists such that a total vibrational population inversion can no longer be sustained in competition with the collisional redistribution of vibrational energy. The ultimate rates of mixing that can be achieved set severe limits on the approaches to the achievement of cw chemical laser operation.

In broad terms, there are three approaches that are presently being followed to achieve cw chemical laser action, despite this mixing rate limitation, which are discussed in the following sections. These three approaches differ from one another in the degree to which laser operating characteristics are influenced by mixing rate limitations; the second and third approaches are essentially attempts to achieve efficient laser operation by lessening this influence.

III. APPROACH ONE: LASER ACTION DIRECTLY FROM PRODUCT MOLECULES IN FLUID MIXING CHEMICAL LASERS

In the first approach an attempt is made to achieve laser action directly from the product molecules of reactions of the type of eq. 1, under conditions of either partial (27) or total inversion among vibration-rotation states. A total inversion will only be possible if the mixing time, τ_m, associated with process 4 is less than or about equal to the time, τ_r, associated with chemical reaction. This is necessary to permit the characteristic time, τ_{mr}, of mixing and reaction to be less than the characteristic time, τ_{vv}, associated with the vibration-vibration (V $\rightarrow$ V) energy transfer processes 2. On the other hand, even if $\tau_m \gg \tau_r$ and $\tau_m > \tau_{vv}$ a partial inversion may result; provided, of course, that $\tau_m > \tau_{vt}$, where τ_{vt} is the time associated with processes 3.

There are severe practical difficulties in obtaining sufficiently fast mixing rates to insure the inequalities necessary for total population inversion; this tends to limit the operating pressures at which such a condition is possible.

Perhaps the earliest serious attempt to achieve cw chemical laser action by total population inversion was that reported by Anlauf, et al (28). They showed that total vibrational population inversions in HCl and HBr could be sustained continuously and observed intermittent spikes of laser radiation, but only at pressures too low to provide gains sufficient for cw operation.

Continuous-wave chemical laser operation in product molecules under conditions of partial inversion was demonstrated by Airey and McKay, of the Avco Corporation, in a shock-induced flow shown schematically in Fig. 1 (16). These authors reported laser emission

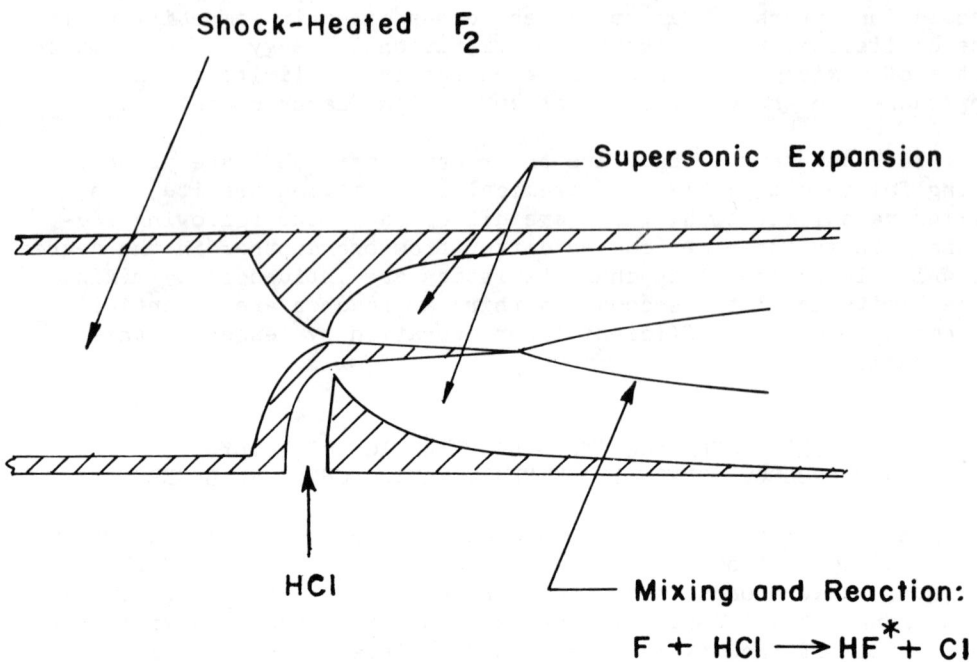

Figure 1: Schematic Diagram of Shock-Induced Apparatus Used to Demonstrate CW Chemical Laser Operation (ref. 16).

from vibrationally excited HF at a wavelength of about 2.8 microns. The HF was formed by the reaction of F-atoms (formed by thermal dissociation of F_2 behind the shock wave used to produce the supersonic flow of Fig. 1) and HCl, i.e.,

$$F + HCl \rightarrow HF + Cl, \quad \Delta H = -32 \text{ kcal/mole} \qquad (5)$$

Laser action could be sustained for the duration of the flow which lasted for about 2 milliseconds.

At about the same time as the Avco work, Spencer, Jacobs, Mirels, and Gross (14) of the Aerospace Corporation observed cw laser emission from HF at about 2.8 microns with a continuous supersonic flow system shown schematically in Fig. 2. An electric arc heater was used to thermally dissociate SF_6 to provide F-atoms. The heated flow was then expanded supersonically to cool it to near room temperature and was then rapidly mixed with H_2. (See Fig. 2.) The chemical reaction forming HF in this case was

$$F + H_2 \rightarrow HF + H, \quad \Delta H = -32 \text{ kcal/mole} \tag{6a}$$

Both of these successful early experiments operated at relatively low powers and were probably only partially inverted, owing to the mixing rate limitations present in these devices. Subsequently, however, the latter authors have obtained total vibrational population inversions in HF and DF by use of an improved mixing injector which gave very rapid mixing rates (17).

They have reported cw power outputs near 500 watts by reaction 6. The efficiency of conversion of the 32 kcal/mole of the chemical energy released by reaction 6 into laser output exceeds 12% for the HF laser case under conditions of maximum power output (17).

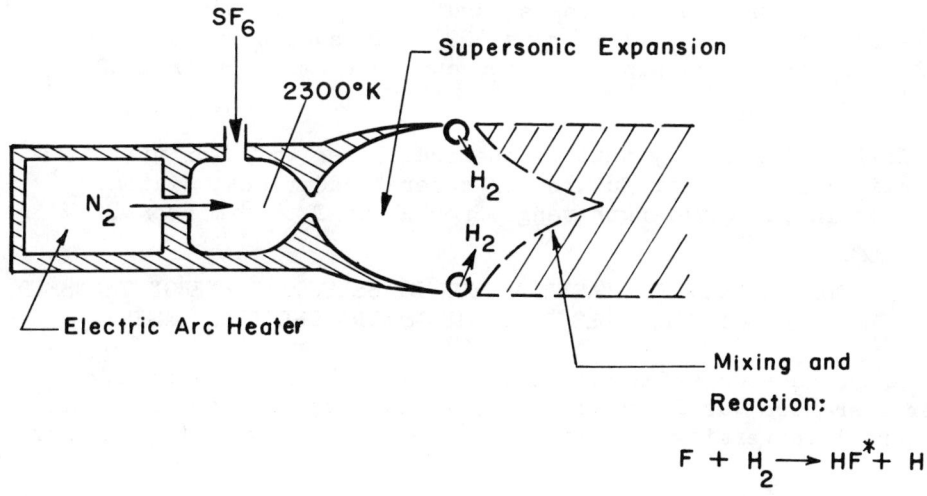

Figure 2: Schematic Diagram of HF (or DF) CW Chemical Laser (ref. 14).

More recent work on devices of this first category for which either total or partial population inversion is produced directly within product molecules is directed toward operation at higher powers and efficiencies for the HF and DF chemical lasers and to the achievement of cw operation in other molecular systems.

Continuous laser operation near 5.2 microns from vibrationally excited CO formed by the reaction

$$O + CS \rightarrow CO + S \tag{6b}$$

has been reported by Wittig, Hassler and Coleman (24) and subsequently by Suart, Kimbell, and Arnold (25).

Partially inverted HCl formed from the reactions

$$H + Cl_2 \rightarrow HCl + Cl \tag{7}$$

and

$$Cl + HI \rightarrow HCl + I \tag{8}$$

has been found by Cool, Stephens, and Shirley to provide cw chemical laser output near 3.7 microns (22). In subsequent studies, Naegli and Ultee (23) also observed cw laser emission from HCl by reaction 8

It is quite likely that cw chemical laser emission will also be realized in the near future for several additional diatomic molecules in the wavelength range from about 2 to 8 microns.

IV. A SECOND APPROACH: LASER ACTION BY SELECTIVE ENERGY TRANSFER FROM PRODUCT MOLECULES IN FLUID MIXING CHEMICAL LASERS

A second successful approach to the problem of cw chemical laser operation was demonstrated by Cool, Stephens, and Falk (15) at Cornell University at about the same time the work (14, 16) of the Avco and Aerospace groups was performed. This approach was based upon the achievement of total population inversion in CO_2 by selective intermolecular vibrational energy transfer from vibrationally excited hydrogen- or deuterium-halide reaction products. Examples of such processes are the pumping of CO_2 by vibrationally excited HCl and DF first observed in the respective pulsed chemical laser experiments by Moore, et al, (5) and Gross (4), e.g.

$$HCl(v=1) + CO_2(00°0) \rightarrow HCl(v=0) + CO_2(00°1), \Delta E = 537 \text{ cm}^{-1} \quad (9)$$

or

$$DF(v=1) + CO_2(00°0) \rightarrow DF(v=0) + CO_2(00°1), \Delta E = 558 \text{ cm}^{-1} \quad (10)$$

where ΔE is the energy separation between centers of the relevant vibrational bands.

Fig. 3 gives a qualitative representation of the energy levels involved in the energy transfer mechanisms presumed to be responsible for chemical pumping of the upper CO_2 laser level (00°1) in the $HF-CO_2$, $DF-CO_2$, $HCl-CO_2$ and $HBr-CO_2$ chemical lasers (4, 5, 19-21). The left-hand side of Fig. 3 shows the (00°0), (00°1), (10°1) and (02°1) levels of CO_2. The energies corresponding to the various possible P and R branch transitions for the $1 \rightarrow 0$, $2 \rightarrow 1$, $3 \rightarrow 2$, and $4 \rightarrow 3$ vibrational bands of HF and DF are shown on the right-hand side (the energy levels of HCl are quite similar to those of DF). Fig. 3 qualitatively indicates the relative numbers of the hydrogen- or deuterium-halide molecules available for each of the various vibration-rotation transitions for an assumed vibrational temperature of 10,000°K, a rotational temperature of 500°K, and a translational temperature of 400°K.

Fig. 3 illustrates several important features:
(A) The over-all resonance defect measured as the difference between the center of a given vibrational band in DF or HCl and the center of the $CO_2(00°1) \rightarrow CO_2(00°0)$ band is large compared to the average translational energy of approach kT_t, except for the higher vibrational bands. These higher bands, however, would be expected to account for relatively fewer molecular transitions because of the reduced occupation of the higher levels. A similar situation exists in HF with respect to the transfer of energy to the (10°1) and (02°1) levels of CO_2.
(B) Because of the large rotational spacings of the energy levels in HF, DF, and HCl a significant number of the higher J

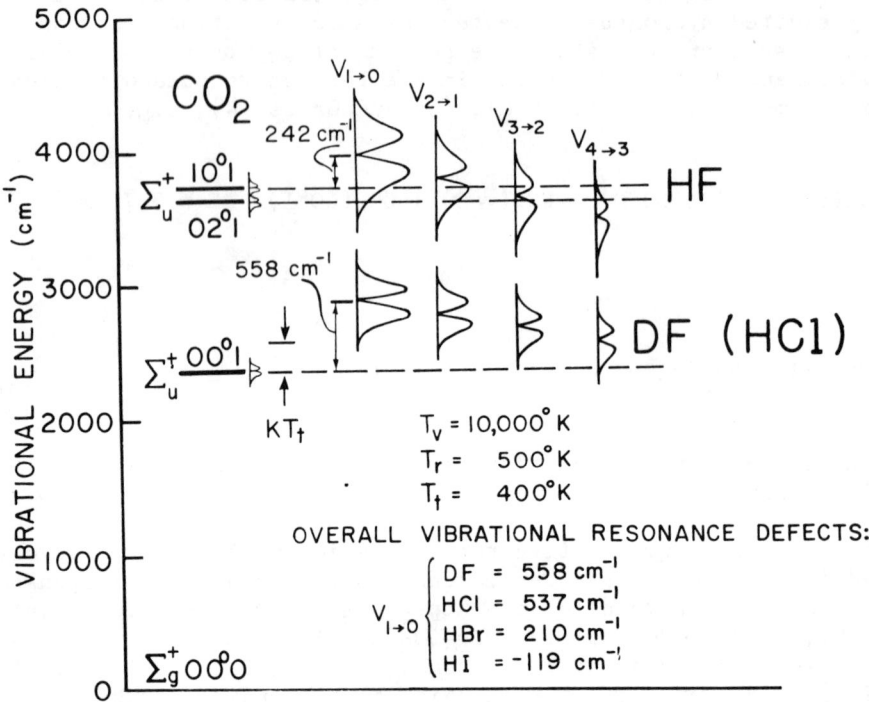

Figure 3: Vibrational Energy Level Diagram for the HF-CO_2, DF-CO_2 and HCl-CO_2 Chemical Lasers.

states of the P-branches of given vibration-rotation bands lie within kT_t of the center of the $CO_2(00°1) \rightarrow CO_2(00°0)$ band in the cases of the DF-CO_2 and HCl-CO_2 lasers and of the $CO_2(10°1) \rightarrow CO_2(00°0)$ and $CO_2(02°1) \rightarrow CO_2(00°0)$ bands for the HF-CO_2 case.

(C) Because of (B) above, the probability for energy transfer by the processes of eqns. 3 and 11 may be large for P-branch transitions involving states of high J in HF, DF or HCl. The rapid rotational thermalization of molecules in a given vibrational level could perhaps provide the means for energy to be coupled efficiently from HF, DF or HCl to CO_2 by means of these few transitions of relatively high probability.

A satisfactory theoretical description of these intermolecular $V \rightarrow V$ transfer processes, e.g.,

$$AB(v=n) + CO_2(00°0) \rightarrow AB(v=n-1) + CO_2(00°1) \quad (11)$$

is not yet available. The measured rate of transfer (5) for the case where AB = HCl and n = 1 (eq. 9) is fairly fast; a similar result is expected to be found for the DF, HBr and HI cases based upon the effective coupling which must be assumed to explain the cw laser operation observed for the HCl-CO_2, DF-CO_2 and HBr-CO_2 systems (20, 21). One of the challenges to be met by an adequate theoretical description of processes 11 will be a proper explanation of the apparently large rates for such processes despite the fairly substantial energy differences that exist between the initial and final states (e.g., for DF with n = 1, an overall vibrational resonance defect of 558 cm^{-1} exists, see Fig. 3).

Another challenge is an accurate description of the vibrational energy transfer processes in the HF-CO_2 chemical laser. The author has proposed that a two-step mechanism may be responsible for the observed pumping of CO_2 (12), i.e.,

$$HF(v=n) + CO_2(00°0) \rightarrow HF(v=n-1) \qquad (12a)$$
$$+[CO_2(10°1) \text{ or } CO_2(02°1)]$$

followed by,

$$[CO_2(10°1) \text{ or } CO_2(02°1)] + CO_2(00°0) \rightarrow \qquad (12b)$$
$$[CO_2(10°0) \text{ or } CO_2(02°0)] + CO_2(00°1).$$

It is clear that much experimental work is yet needed to help suggest the proper theoretical approach to be followed in the kinetic description of processes 11 and 12.

The principal advantage of this second technique for achievement of cw chemical laser operation is the fact that the detailed distribution of vibrational energy among the product molecules is unimportant in contrast to the case of laser action directly from the product molecules themselves. Total population inversion is provided by the selectivity of processes 11. Most of the vibrational energy of the product molecules distributed over a number of vibrational levels is made available to produce total inversion in CO_2. Therefore, efficient operation is possible with this technique under conditions for which total or partial inversion in the reaction products themselves would not be maintained. That is, the requisite relationships between the rates of the 4 basic processes of chemical reaction, V → V and V → T energy transfer, and fluid mixing are much less restrictive than in approach one. The mixing and reaction rates need not be particularly fast, but only sufficient to insure that the intermolecular coupling by processes 11 occurs on a time scale short compared to the overall vibrational relaxation time of the CO_2 $(00°1)$ state and to the time scale associated with the processes 3. The inequalities $\tau_m < \tau_{vt}$ and τ_{vv} (for processes 11) $< \tau_{vt}$ must be met for efficient operation; but these are much more easily satisfied than the previously mentioned inequalities requisite for total inversion in the reaction products themselves.

Fig. 4 shows the experimental apparatus we used at Cornell University for these early studies (15, 20, 21). The design is a fairly direct adaptation of the designs for similar flow systems we had used previously in experiments with He-Ne and N_2-CO_2 fluid mixing lasers (29, 30). It features an optical axis aligned along the flow direction which provides a versatile arrangement for small-scale exploratory studies, but is not conveniently scaled to high power outputs. Table I summarizes the results achieved with

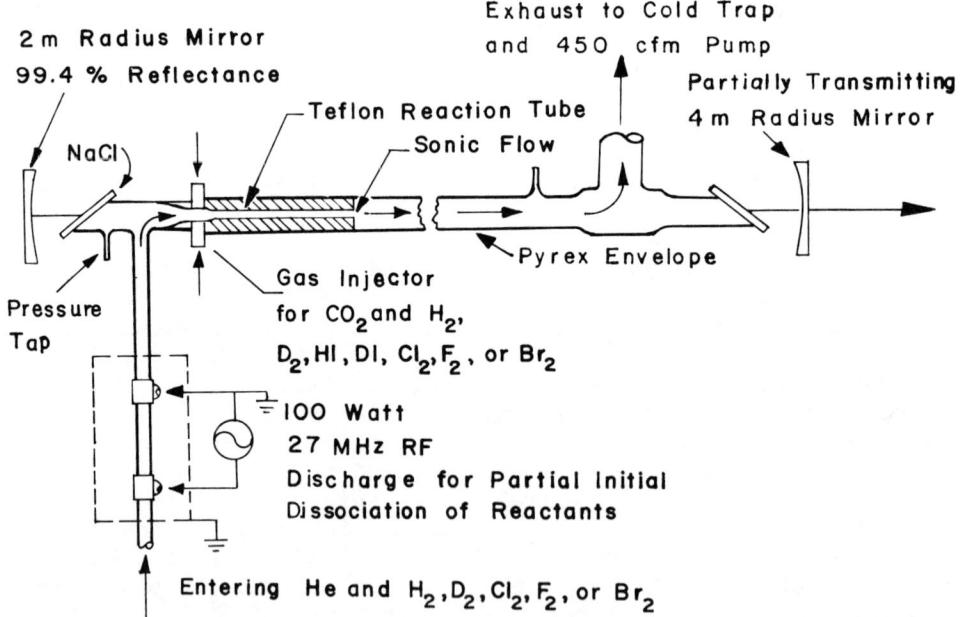

Figure 4: Schematic Diagram of Subsonic Flow CW Chemical Laser for the HF-CO_2, DF-CO_2, HCl-CO_2, and HBr-CO_2 Systems (refs. 15, 20, 21).

TABLE I: CW CHEMICAL LASER PERFORMANCE AT 10.6 MICRONS

Case	Initiating Reaction	Flow Rates, (10^{-3} moles/sec)	Pressure (torr)	Laser Output (watts)
A	$F + D_2 \rightarrow DF^* + D$	He: 3.15, CO_2: 0.60, D_2: 0.25, F_2: 0.16	12.3	0.74
B	$D + F_2 \rightarrow DF^* + F$	He: 21.8, CO_2: 2.10, F_2: 0.70, D_2: 0.73	50	2.9
C	$F + DI \rightarrow DF^* + I$	He: 2.91, CO_2: 0.68, DI: 0.046, F_2: 0.064	9.6	0.084
D	$F + H_2 \rightarrow HF^* + H$	He: 8.74, CO_2: 1.27, H_2: 0.20, F_2: 0.28	23	0.079
E	$H + F_2 \rightarrow HF^* + F$	He: 11.2, CO_2: 1.79, H_2: 0.26, H_2: 0.16	26	0.050
F	$F + HI \rightarrow HF^* + I$	(no output was detected)	--	--
G	$H + Cl_2 \rightarrow HCl^* + Cl$	(only intermittent output spikes were detected)	--	--
H	$Cl + HI \rightarrow HCl^* + I$	He: 3.15, CO_2: 0.88, HI: 0.10, Cl_2: 0.099	11.1	0.020

this system for a variety of atom-exchange reactions providing pumping of the CO_2 ($00°1$) state with varying degrees of success. This initial work established that reactions B and C (or E and F) of Table I occurred together as the chain propagation steps

$$F + D_2 \rightarrow DF^* + D \tag{13a}$$

$$D + F_2 \rightarrow DF^* + F \tag{13b}$$

under conditions of laser interest. This observation led to the development by Cool and Stephens (19) of the first purely chemical laser, shown schematically in Fig. 5. The kinetic mechanisms of importance in the $DF-CO_2$ purely chemical laser are given in Table II. In contrast to all of the other chemical lasers discussed above, this system requires no external energy source to initiate or sustain chemical laser excitation. It operates solely by the

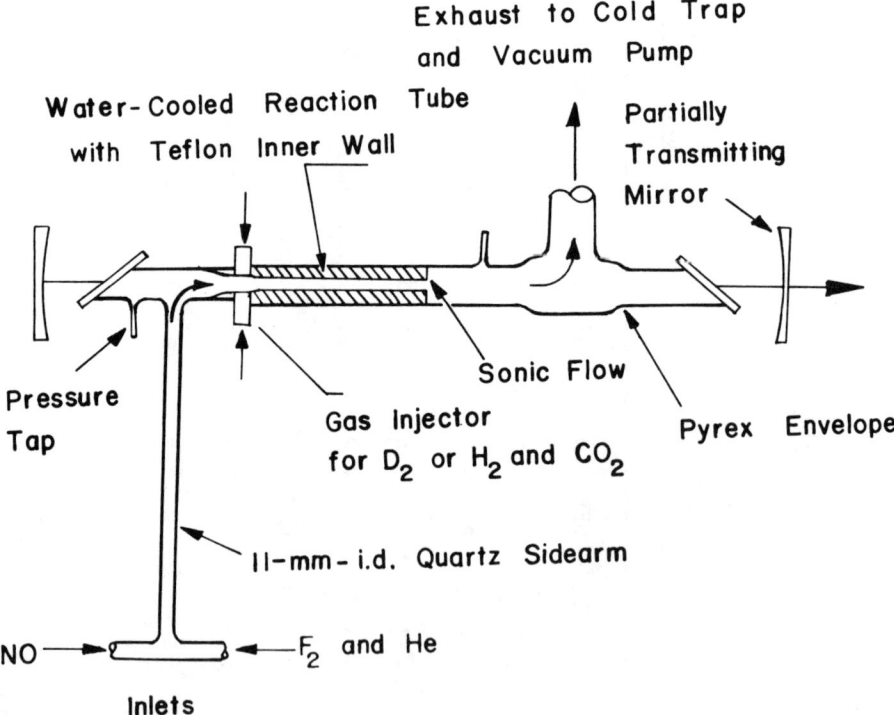

Figure 5: Schematic Diagram of the First Purely Chemical Laser (ref. 19).

TABLE II: A PURELY CHEMICAL LASER

(A) $F_2 + NO \rightarrow ONF + F$

(B) $F + NO \rightarrow ONF$

(C) $F + D_2 \rightarrow DF^* + D$

(D) $D + F_2 \rightarrow DF^* + F$

(E) $DF(v = n) + DF(v = m) \rightarrow DF(v = n - 1) + DF(v = m + 1)$

(F) $DF(v = n) + CO_2(00°0) \rightarrow DF(v = n - 1) + CO_2(00°1)$

(G) $CO_2(00°1) + M \rightarrow CO_2(10°0) + M$

(H) $DF(v = n) + M \rightarrow DF(v = n - 1) + M$

(I) $CO_2(01^10) + He \rightarrow CO_2(00°0) + He$

simple subsonic mixing of commercially available bottled gases. Fig. 5 shows the apparatus used in the original experiments (19). A steady-state F-atom concentration was reached by means of the reactions A and B of Table II within the sidearm tube leading to the main flow laser tube. This steady-state ratio of F-atom concentration to F_2 molecule concentration is given by the ratio of the rate constant of reaction A to that of reaction B (Table II). Subsequent mixing with D_2 and CO_2 within the main flow tube of Fig. 5 permitted the chain propagation reactions (13a) and (13b) to produce chemical pumping of the CO_2 by processes 11. The reactions were driven rapidly to completion with the subsequent ultimate decay of vibrational energy (in the absence of stimulated emission) governed by the processes G and H of Table II.

The laser of Fig. 5 operated at a maximum power output of 8.1 watts giving a continuous conversion of 4.1% of the total energy release of the overall reaction $D_2 + F_2 \rightarrow 2DF$ into laser output at 10.6 microns (19).

Since this initial work was reported, a system with a 5 path optical axis, folded transverse to the flow direction, has been built at Cornell, as shown schematically in Fig. 6. This device

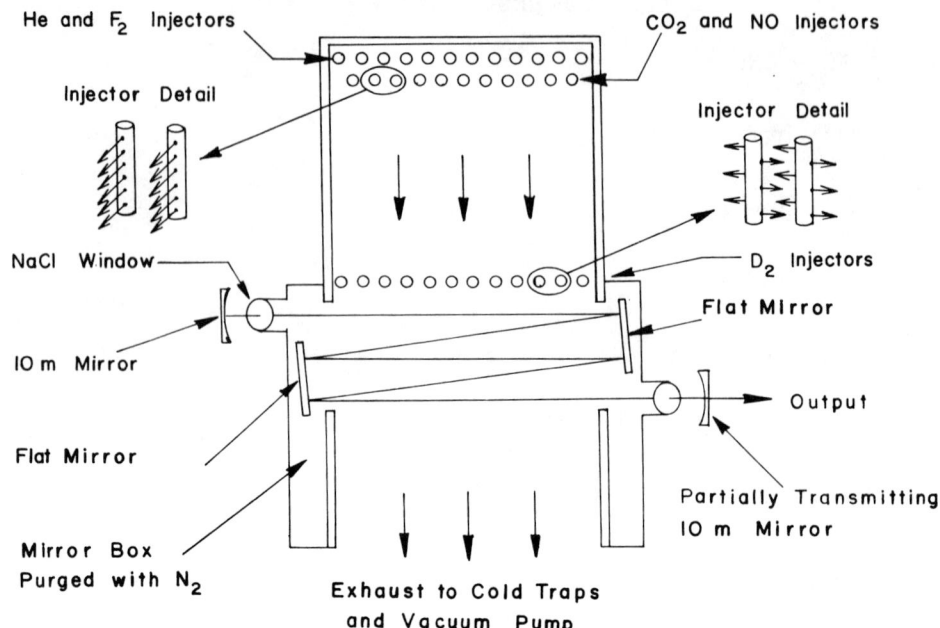

Figure 6: Schematic Diagram of a Transverse Flow DF-CO_2 Multimode Chemical Laser (ref. 19).

has operated at 162 watts with a chemical efficiency based on the overall reaction $F_2 + D_2 \rightarrow 2DF$ of 4.6% (18).

A Comparison of Approaches One and Two

A series of experiments was recently conducted (22) to assess the difference in performance between lasers operating by means of partial inversions in the reaction products and lasers that are chemically pumped by selective energy transfer to produce total population inversions in CO_2. The apparatus was the same as that of Fig. 4 except that the mirrors and Brewster angles had to be suitably modified to accommodate the relevant wavelengths. Table III summarizes the results obtained under conditions giving a maximum power output for mirrors of the same fixed value of 0.5% transmission as was employed for the data of Table I. A comparison of the two tables suggests that under the conditions of these experiments ($\tau_r < \tau_m$, $\tau_{vv} < \tau_m$, $\tau_m < \tau_{vt}$) a significantly greater laser power output can be achieved with total inversions in CO_2 rather than by partial inversions in the product molecules. There are some other important differences between the two cases that must be

TABLE III: CW CHEMICAL LASER PERFORMANCE AT WAVELENGTHS FROM 2.7 TO 4.0 MICRONS

Case	Initiating Reaction	$-\Delta H^\circ_f$ (kcal/mole)	Flow Rates (10^{-3} moles/sec)			Pressure (torr)	Output Power (mW)
A	$F + H_2 \rightarrow HF^* + H$	32	He: 17.7,	H_2: 0.18,	F_2: 0.10	24	53
B	$H + F_2 \rightarrow HF^* + F$	98	He: 20.8,	H_2: 0.28,	F_2: 0.22	27	50
C	$F + HI \rightarrow HF^* + I$	64	He: 11.0,	HI: 0.062,	F_2: 0.11	17	4
D	$F + D_2 \rightarrow DF^* + D$	31	He: 15.7,	D_2: 0.18,	F_2: 0.17	21	83
E	$D + F_2 \rightarrow DF^* + F$	99	He: 24.6,	D_2: 0.37,	F_2: 0.45	32	80
F	$F + DI \rightarrow DF^* + I$	64	He: 14.4,	DI: 0.12,	F_2: 0.13	19	2
G	$H + Cl_2 \rightarrow HCl^* + Cl$	45	He: 37.3,	H_2: 2.00,	Cl_2: 0.28	45	1
H	$Cl + HI \rightarrow HCl^* + I$	32	He: 17.8,	HI: 0.045,	Cl_2: 0.044	21	12
I	$H + Br_2 \rightarrow HBr^* + Br$	36	No output observed			--	--
J	$Br + HI \rightarrow HBr^* + I$	16	No output observed			--	--

taken into account (31), but the general conclusion may be drawn that when the inequality $\tau_m < \tau_{vv}$ can not be satisfied, then selective energy transfer from the product molecules offers a superior means to achieve laser output compared to the case of operation by partial inversion within the product molecules.

Despite the inefficiencies inherent in the partially inverted systems, the wavelengths of laser oscillation achievable in such systems may be desirable for some applications. At present some very important and interesting uses for such lasers are being investigated. The laser induced fluorescence technique developed by Javan, et al (32), Moore, et al (5, 33), and Rosser, Wood and Gerry (34) has been enormously successful in measurements of vibrational energy transfer probabilities for specific processes that can not be examined with other methods. Pulsed chemical lasers or small Q-spoiled cw chemical lasers are convenient sources of

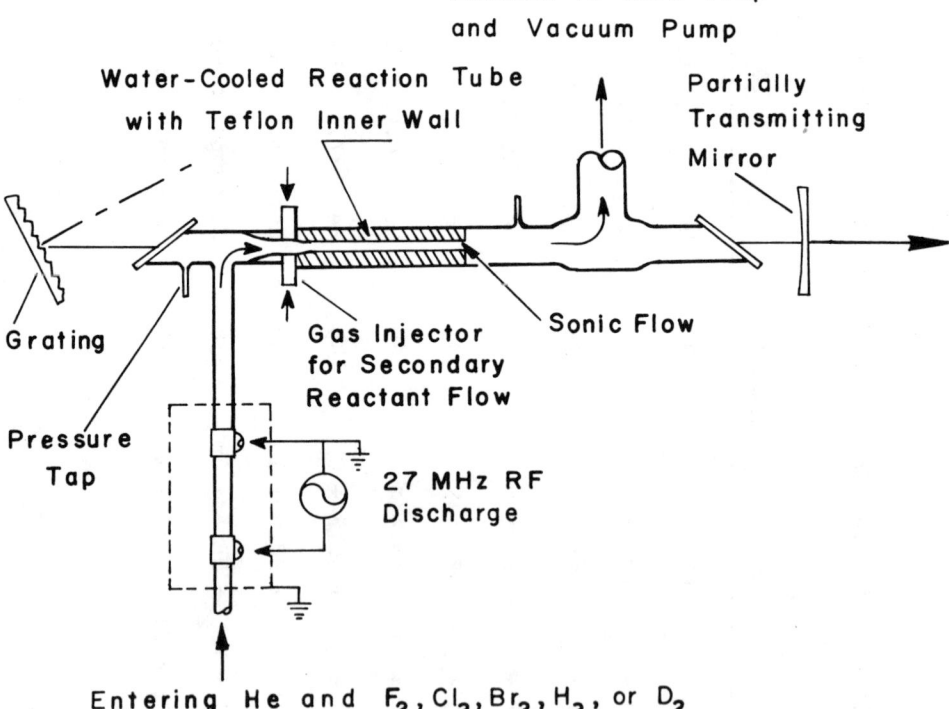

Figure 7: Apparatus for Single Mode, Single Frequency CW Chemical Laser Operation on Selected P-Branch Transitions in Partially Inverted HCl, HF, and DF (ref. 22).

radiation for such experiments provided they operate on 1 → 0 band laser transitions. Pulsed chemical lasers have been used by Moore, et al (5) for this purpose. An alternative approach would be to Q-spoil a laser of the type of Fig. 7 as a convenient source of selected laser output on any one of a number of P-branch 1 → 0 transitions in HCl, HF or DF. Lasers of the type shown in Fig. 7 have been recently operated in the author's laboratory under conditions of about 40 mw cw output on 1 → 0 band transitions in HF and DF, and 20 mw cw output for 1 → 0 transitions in HCl.

Another possible use of partially inverted cw lasers may be in the exciting new experiments being done by Javan and co-workers with the frequency mixing of infrared laser outputs (35). Very accurate direct comparison of infrared wavelengths (to 1 part in 10^6) from several lasers is now possible from about 9 to 337 microns (35). The existence of cw chemical lasers of substantial power at wavelengths near 2.5 microns provides additional comparison sources for creation of a standardized scale of frequencies ranging from the visible to the microwave spectral regions.

Total Population Inversion

If on the other hand, the inequality $\tau_m < \tau_{vv}$ can be satisfied, then highly efficient and powerful chemical laser action can be realized from totally inverted product molecules, as the results of Spencer, et al (17) have demonstrated. There is hope that total inversion can be achieved in many other chemical systems for which the above inequality can be met.

A problem of current concern with the HF and DF systems is the fact that fairly large concentrations of F-atoms (or H- and D-atoms) appear to be necessary to drive the chain propagation reactions (13) at a sufficiently rapid rate so that $\tau_r < \tau_{vv}$. The experiments of Spencer, et al (14, 17) were concerned with only the $F + H_2 \rightarrow HF + H$ step of the chain; a sufficiently rapid rate of formation of HF was provided by a very large initial concentration of F-atoms provided by dissociation of SF_6. It is hoped that the requisite F-atom concentrations can be achieved purely chemically by some means analogous to the reaction $F_2 + NO \rightarrow ONF + F$ used by Cool and Stephens for the purely chemical $HF-CO_2$ and $DF-CO_2$ lasers (19). This particular reaction is probably too slow to provide total inversions in HF, but perhaps other reactions can be utilized.

The recent experiments of Meinzer (36) are interesting in this regard. An HF flame was used to thermally dissociate F_2 which was then expanded and mixed with D_2 in a supersonic flow. The chemical efficiencies for laser action from vibrationally excited DF formed from the reaction $F + D_2 \rightarrow DF + D$ were, however, quite low, suggesting the existence of only a partial vibrational population inversion.

V. A THIRD APPROACH: A CW EXPLOSION OR DETONATION WAVE CHEMICAL LASER

An approach of considerable merit, would be to avoid the mixing problem entirely by premixing reactants under conditions such that reaction occurs very slowly, if at all. Rapid reaction (explosion) would then be abruptly initiated by an external triggering source (37, 38) or, in the case of a supersonic flow, by means of a standing detonation wave (39-41).

Several supersonic combustion studies have been performed in which reactions were initiated and stabilized in steady supersonic flows (42-44). Most of this work was done in connection with the development of supersonic combustion ram jets. In general, these flows were at somewhat higher pressures than those at which present cw chemical lasers operate; the mixing problems were exclusively those pertaining to turbulent flows. Nevertheless much of this work will be quite valuable in the development of cw detonation wave lasers.

In some cases, it may be possible to combine the processes of gas mixing and shock-initiation of chemical reaction at the same

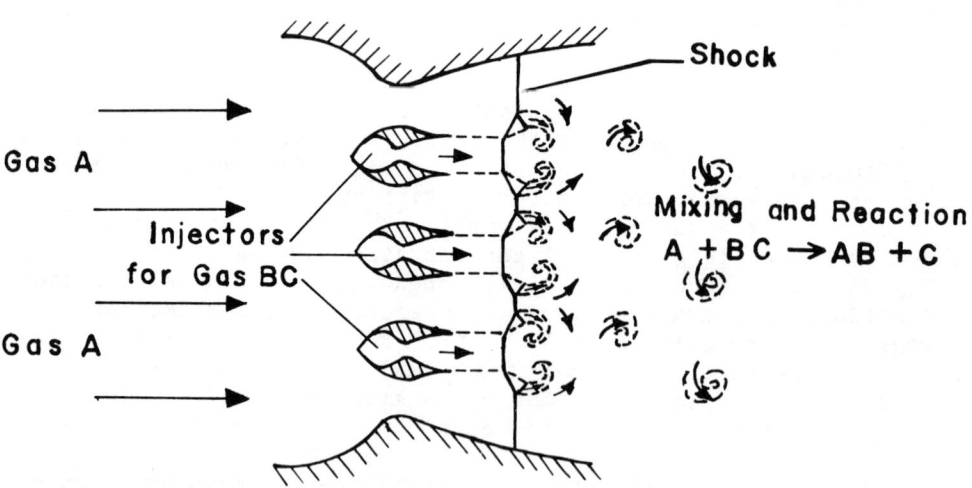

Figure 8: A Shock Mixing Technique as it Might be Applied to a Chemical Laser System (ref. 40).

flow region (20). Fig. 8 shows such a scheme. An initially stratified flow is rendered abruptly unstable as it passes through a shock wave. Large pressure and velocity differences are produced as the flow passes through the shock because of the smaller momentum of gas BC with respect to that of gas A. These differences cause flow recirculation and vorticity generation leading to rapid mixing and chemical reaction in the high temperature gases behind the shock.

At present, several groups are actively engaged in pulsed chemical laser experiments which have immediate relevance to the cw laser problem.

Pulsed emission from HF behind detonation waves in $F_2O + H_2$ mixtures was observed by Gross, Giedt, and Jacobs (41) in a shock tube experiment. They have also demonstrated similar emission from $SF_6 + H_2$ mixtures.

In contrast to most of the pulsed chemical laser work done during the past five years, several Soviet researchers (37, 38) have extensively investigated the possibility of pulsed lasers operating with chain branching mechanisms. The objective of such research is the establishment of conditions under which reactants can be stably premixed and then triggered at will either by an electrical discharge, or by means of a shock wave. In principle, the total laser energy output can far exceed the requisite trigger energy. Basov, et al (38) have recently been concerned with $H_2 + F_2$ mixtures and the chain propagation reactions of eq. 13. The chain branching mechanism

$$HF(v=4) + F_2 \rightarrow HF(v=0) + 2F \qquad (14)$$

has been proposed to explain the explosion limits observed for the $H_2 + F_2$ system (38, 45).

Quite recently Gregg, et al (46) have reported initial results of a very interesting approach to the problem of the abrupt, uniform initiation of chemical reaction in premixed gases. They have used a 1.2 Mev, 50-n sec pulsed beam of electrons to initiate intense chemical laser emission from HF product molecules of the reactions of $N_2F_4 - H_2$ and $NF_3 - H_2$ mixtures at pressures from 10 to 100 torr. This method may offer a means to achieve uniform volume dissociation leading to chemical reaction and the release of energy far exceeding that of the dissociating electron beam.

A similar goal has been sought by Jacobson and Kimbell (47) in their achievement of substantial laser pulse energies from chemical reactions in the O_2-CS_2 systems with a multiple spark discharge directed transverse to the optical axis.

Though much yet needs to be done with premixed pulsed systems along the above lines, it appears possible that cw chemical laser operation by similar mechanisms may ultimately be successful at much higher operating pressures than those at which the fluid mixing chemical lasers described in the preceding sections appear to be limited.

VI. CONCLUSION

The potential advantages cw chemical lasers should ultimately offer over electrically or thermally excited lasers include higher over-all efficiencies, a self-contained energy storage capability independent of sources of electrical power, and operation at shorter wavelengths than have heretofore been achieved in molecular laser systems.

During the past year, much progress has been achieved in efforts to demonstrate these advantages. A comparison of over-all efficiencies, for example, with efficiencies presently obtainable with more extensively developed molecular lasers is interesting: Thermally pumped (gasdynamic) lasers can be built to very large sizes, but operate at an efficiency (defined as the ratio of laser power output to the rate of chemical energy release in the combustion heating process) of less than one percent (48). Electrically excited flow lasers can also be built to large sizes with an electrical efficiency of about 20%. If a thermal efficiency of 40% is assumed in the conversion of the chemical energy of fossil fuels into the requisite electrical energy, then an over-all efficiency of about 8% is obtained. Over-all efficiencies of about 5% (17, 19) are possible with chemical lasers at this early stage of development, and ultimate values of 15% do not appear to be unreasonable extrapolations.

The energy storage capability of chemical lasers for terrestrial applications is presently limited by the need to provide flow pumping for present devices. The purely chemical DF-CO_2 laser operates independently of electrical power sources, and even at its present stage of development could be deployed in satellite applications where gases could be exhausted directly into space. Efforts are being made to obtain cw chemical laser operation at higher operating pressures to minimize pumping requirements for systems operating in the terrestrial environment.

The short wavelengths of the DF, HF, and HCl chemical lasers are important in many applications where problems in signal detection and propagation are critical. In the laboratory, these new wavelengths are already being utilized in laser-induced fluorescence studies, and should find application in other areas of research in molecular physics and infrared spectroscopy.

Intensive further study should be given to the development of improved methods for the rapid mixing of reasonably large volumes of gases. The rates at which gases can be mixed limit in a fundamental way the performance and feasibility of presently conceived cw chemical laser systems. Possible techniques for achieving more rapid mixing might include (40):

A. Secondary components can be mixed with a primary gas by initial dispersion by droplet injection, followed by rapid induced vaporization and diffusive mixing in the gas phase.

B. Rotating flows can provide extremely high accelerations to drive the mixing of gaseous components of differing densities.

C. Ultrasonic, electrostatic, and electromagnetic oscillation may be useful in inducing break up of gaseous interfaces.

D. Magnetogasdynamic interactions could aid in mixing ionized gases.

E. Stratified flows can be mixed by gasdynamic interactions behind stationary shock waves.

A second important current problem in the development of cw chemical lasers is the achievement of single mode, single frequency operation from optical cavities of high Fresnel number (similar to that of Fig. 6). The highest powers achieved so far (17, 18) have been for multimode beams of poor quality and large divergence. Present high power flowing gas electrically and thermally excited molecular laser systems have also been beset with similar difficulties, which can be apparently minimized with proper design (48). However, these problems are complicated for the HF, DF and HCl chemical laser by the relatively high optical gains inherent in these systems.

ACKNOWLEDGEMENT

The author's research program has been supported by the Advanced Research Projects Agency monitored through the Office of Naval Research under contract N00014-67-A-0077-0006 and by the National Aeronautics and Space Administration under contract NGL-33-010-064.

REFERENCES

1. J. V. V. Kasper and G. C. Pimentel, Phys. Rev. Letters, 14, 352 (1965).
2. J. Richard Airey, IEEE J. Quantum Electron, QE-3, 208 (1967).
3. J. Richard Airey, J. Chem. Phys., 52, 156 (1970).
4. Rolf W. F. Gross, J. Chem. Phys., 50, 1889 (1969).
5. H. L. Chen, J. C. Stephenson, and C. Bradley Moore, Chem. Phys. Letters, 2, 593 (1968).
6. C. Bradley Moore, IEEE J. Quantum Electron, QE-4, 52 (1968).
7. P. H. Corneil and G. C. Pimentel, J. Chem. Phys., 49, 1379 (1968).
8. K. L. Kompa, J. H. Parker, and G. C. Pimentel, J. Chem. Phys., 49, 4257 (1968).
9. J. H. Parker and G. C. Pimentel, J. Chem. Phys., 51, 91 (1969).
10. The atom exchange reaction, $O + CS \rightarrow CO + S$, is believed to be responsible for the chemical laser emission from CO observed with the O_2/CS_2 system. See, for example: G. Hancock and I. W. M. Smith, Chem. Phys. Letters, 3, 573 (1969), and C. Wittig, J. C. Hassler, and P. C. Coleman, Appl. Phys. Letters, 16, 117 (1970).
11. K. G. Anlauf, P. J. Kuntz, D. H. Maylotte, P. D. Pacey, and J. C. Polanyi, Discussions Faraday Soc., 44, 183 (1967).
12. J. C. Polanyi and W. H. Wong, J. Chem. Phys., 51, 1439 (1969).
13. M. H. Mok and J. C. Polanyi, J. Chem. Phys., 51, 1451 (1969).
14. D. J. Spencer, T. A. Jacobs, H. Mirels, and R. W. F. Gross, Int. J. Chem. Kinetics, 1, 493 (1969).
15. T. A. Cool, R. R. Stephens, and T. J. Falk, Int. J. of Chem. Kinetics, 1, 495 (1969).
16. J. R. Airey and S. F. McKay, Appl. Phys. Letters, 15, 401 (1969).
17. D. J. Spencer, H. Mirels, and T. A. Jacobs, Appl. Phys. Letters 16, 384 (1970); also M. A. Kwok, R. R. Giedt and R. W. F. Gross, Appl. Phys. Letters, 16, 386 (1970).
18. T. A. Cool, J. A. Shirley, and R. R. Stephens, "Operating Characteristics of a Transverse Flow DF-CO_2 Purely Chemical Laser" (to be published) Appl. Phys. Letters.
19. T. A. Cool and R. R. Stephens, J. Chem. Phys., 51, 5175 (1969); also, T. A. Cool and R. R. Stephens, Appl. Phys. Letters, 16, 55 (1970).
20. T. A. Cool, T. J. Falk, and R. R. Stephens, Appl. Phys. Letters, 15, 318 (1969).
21. T. A. Cool and R. R. Stephens, J. Chem. Phys., 52, 3304 (1970).
22. T. A. Cool, R. R. Stephens, and J. A. Shirley, "HCl, HF, and DF Partially Inverted cw Chemical Lasers" (to be published) J. Appl. Phys. The HCl cw Chemical Laser was also reported at the 1970 A.P.S. Meeting, see Bull. Am. Phys. Soc., 15, 355 (1970).
23. D. Naegli and C. J. Ultee, "A cw HCl Chemical Laser" (to be published) Chem. Phys. Letters.

24. C. Wittig, J. C. Hassler, and P. D. Coleman, "CW Laser Oscillation in a Carbon Monoxide Chemical Laser", Nature, 226, 845 (1970).
25. R. D. Suart, G. H. Kimbell, and S. J. Arnold, Chem. Phys. Letters, 5, 519 (1970).
26. See for example, M. J. Berry and G. C. Pimentel, J. Chem. Phys. 51, 2274 (1968).
27. J. C. Polanyi, Appl. Opt. Suppl. 2, Chemical Lasers, pp. 109-127 (1965).
28. K. G. Anlauf, D. H. Maylotte, P. D. Pacey, and J. C. Polanyi, Phys. Letters, 24A, 208 (1967).
29. T. A. Cool, Appl. Phys. Letters, 9, 418 (1966).
30. T. A. Cool and J. A. Shirley, Appl. Phys. Letters, 14, 70 (1969).
31. The $HCl-CO_2$, $HF-CO_2$, $DF-CO_2$ and $HBr-CO_2$ chemical lasers operated on the P(18) transition at 10.57 μ. The HF, DF, and HCl chemical lasers operated on a variety of P-branch transitions, see ref. 22. A large amount of helium was needed for the HF, DF, and HCl lasers (Table III) to keep the rotational temperature low (300°K).
32. C. K. Rhodes, M. J. Kelley, and A. Javan, J. Chem. Phys., 48, 5730 (1968).
33. C. Bradley Moore, Accounts of Chem. Research, 2, 103 (1969).
34. W. A. Rosser, A. D. Wood, and E. T. Gerry, J. Chem. Phys., 50, 4996 (1969).
35. V. Daneu, D. Sokoloff, A. Sanchez, and A. Javan, Appl. Phys. Letters, 15, (1969).
36. R. A. Meinzer, "A Continuous-wave Combustion Laser", to be published, Int. J. of Chem. Kinetics.
37. O. M. Batovskii, G. K. Vasil'ev, E. F. Makarov, and V. L. Tal'roze, ZhETF Pis. Red., 9, 341, (1969) [Sov. Phys. - JETP Lett., 9, 375 (1969)].
38. N. G. Basov, L. V. Kulakov, E. P. Markin, A. I. Nikitin, and A. N. Oraevskii, ZhETF Pis, Red., 9, 613 (1969) [Sov. Phys. - JETP Lett., 9, 375 (1969)].
39. J. R. Bowen and K. A. Overholser, Astronautica Acta, 14, 475 (1969).
40. T. A. Cool, "Fluid Mixing Lasers", NASA Conference on Gas Lasers, NASA Hdqts., Washington D. C. 15-16 July 1968.
41. R. W. F. Gross, R. R. Geidt, and T. A. Jacobs, J. Chem. Phys., 51, 1250 (1969).
42. R. A. Gross, and J. A. Nicholls, "Stationary Detonation Waves", Combustion and Propulsion, 169-177, Fourth AGARD Coll. London: Pergamon, (1961).
43. R. A. Gross and W. Chinitz, "A Study of Supersonic Combustion", J. Aerospace Sciences, 27, 517-524, 534, (1960).
44. J. A. Nicholls, E. K. Dabora, and R. L. Gealer, "Studies in Connection with Stabilized Gaseous Detonation Waves", Seventh Symposium on Combustion, 144-150, Butterworth, London,(1958).
45. G. A. Kapralova, E. M. Trofimova, and A. E. Shilov, Kinetika i

Kataliz, 6, 977 (1965) [Sov. Chem. Kinetics and Catalysis, 6, 884 (1965)].

46. D. W. Gregg, B. Krawetz, R. K. Pearson, B. R. Schleicher, S. J. Thomas, E. B. Huss, K. J. Pettipiece, and R. E. Niver, "Electron Beam Initiation of a Pulsed Chemical Laser" (to be published) Appl. Phys. Letters.

47. T. V. Jacobson and G. H. Kimbell, "A Transversely Spark Initiated Chemical Laser with High Pulse Energies", (to be published) Appl. Phys. Letters.

48. See, for example, Physics Today, July, (1970), p. 55.

EXCITATION OF GASES USING WAVELENGTH-TUNABLE LASERS

C. Forbes Dewey, Jr.

Department of Mechanical Engineering,

Massachusetts Institute of Technology, Cambridge, Mass.

ABSTRACT

Recent developments in liquid dye lasers, semiconductor lasers, and nonlinear optical techniques have made possible the production of high-intensity monochromatic radiation at wavelengths ranging from the near ultraviolet to the far infrared. Wavelength tuning opens an entirely new dimension in the spectroscopy and resonant excitation of gases.

This paper reviews the state-of-the-art in wavelength-tunable lasers. A number of applications are also discussed, including selective excitation spectroscopy, plasma ionization enhancement, and velocity determinations in high-speed gas flow.

I. INTRODUCTION

The optical study of gas dynamic phenomena is a very diverse topic. It is useful to distinguish between two important classes of optical experiments which are included within this field. In the first category, observable effects are restricted to measuring the optical properties of a medium which is not actively coupled to an external radiation field. Examples of such passive interactions are Rayleigh scattering, emission spectroscopy, and interferometry.

The second type of phenomena involves an interaction between the test medium and an incident radiation field. Among the techniques falling into this category are nonresonant energy exchanges (Thomson scattering and Raman scattering) and resonant energy exchanges (fluorescence and absorption).

This paper is concerned primarily with resonant energy exchanges between gaseous media and externally-applied radiation fields. The raison d'etre is the recent development of wavelength-tunable laser sources of high instantaneous power. Gaseous atoms and molecules have quantized energy states corresponding to rotational, vibrational, and electronic energies above the ground state. For most neutral and ionized species, transitions between adjacent electronic levels correspond to photon energies in the near ultraviolet and visible spectrum, whereas vibrational and rotational transitions correspond to photon energies in the infrared and far infrared, respectively. Absorption coefficients, which can be interpreted as measuring the strength of the particle-radiation interaction, depend upon: (a) the difference between the incoming photon energy and the energy between quantum levels in the test particle; and (b) the half-width of the absorption line.

The wavelength-tunable laser sources reviewed in this paper are capable of producing photons which are exactly resonant with the excitation process to be investigated. The coupling between the gas particles and the radiation field can therefore be made to dominate other processes for which a resonance does not occur. Tunable lasers, and dye lasers in particular, can be constructed to produce either broad-band radiation or spectroscopically narrow lines which are on the order of the natural linewidth of atomic radiation (a few hundredths of an Ångstrom).

The following Section reviews recent advances in fluid dye lasers based on organic molecules, and emphasizes the operating characteristics of such systems. With appropriate optical pumping, tunable dye laser radiation has been produced from 360nm to 1160nm (3600Å - 1.16μ). It is possible to reduce the intrinsic linewidth of the dye laser output from 5-10nm to less than 10^{-3}nm using intracavity dispersion elements (diffraction gratings and Fabry-Perot etalons).

In Sections III and IV, a summary is given of other important techniques capable of producing wavelength-tunable radiation. These methods employ semiconductor laser compounds, stimulated Raman scattering, and nonlinear optical sum- and difference-frequency generation.

Section V presents a discussion of several important applications which employ resonant interaction between gases and external radiation fields. These include plasma ionization enhancement, absorption spectroscopy, resonance fluorescence, and velocity measurements.

II. ORGANIC DYE LASERS[*]

In 1966, Sorokin and Lankard [B.85] and Schaefer, Schmidt, and Volze [B.51] independently discovered stimulated emission from

---

[*]Reference numbers preceded by B refer to the separate bibliography of dye laser papers included at the end of the references.

LASER EXCITATION

solutions containing fluorescent organic molecules. These results were presaged by the work of Stockman [B. 91, B.92] and others who suggested that four-level laser action could be achieved in organic molecules whose absorption and emission bands did not overlap. Lasing action was achieved in these early experiments by using Q-switched solid-state lasers to provide very rapid (10-40ns) excitation of the dye solution. Subsequently, other dyes were discovered in which lasing action could be initiated by fast flashlamps with rise times on the order of 100 ns. Several reviews of dye laser research have appeared which describe both laser-pumped and flashlamp-pumped systems [B.42, B.71, B.76].

The major features of the time-dependent gain characteristics of dye lasers have been investigated both qualitatively and quantitatively by a number of authors [B.2, B.32, B.34, B.43, B.76, B.78, B.87, B.96], and these results follow in a relatively straightforward way from earlier work on four-level laser systems [B.53]. There are two distinct characteristics of dye lasers which emerge from these analyses and deserve special mention. First, the radiative lifetime of electronically-excited fluorescent dyes is quite short, on the order of nanoseconds, and therefore the time-resolved intensity of laser emission follows the pumping intensity very closely. Second, most (if not all) known laser dyes exhibit nonradiative single-triplet transitions which result in absorption by the triplet state at the laser wavelength. Unless nonradiative de-excitation of the triplet state occurs [B.13, B.19, B.43], triplet absorption increases with time and the cavity gain is decreased to the point where lasing action ceases [B.2, B.76, B.87, B.88].

Table 1 presents a summary of approximately 90 dyes in which lasing action has been observed. The number assigned to each dye in the table indicates the approximate wavelength (in nanometers) of the stimulated emission produced when the laser cavity is formed by two broad-band reflectors. Wavelength tuning can be accomplished by changing the effective dye concentration, by replacing the broadband cavity reflectors with wavelength-selective dispersive elements, and by changing the characteristics of the dye solvent. There are slight differences in peak emission wavelengths between different experimental configurations and between laser pumping and flashlamp pumping of the same dye.

It is evident from the numbers assigned to the dyes in Table 1 that wavelength-tunable radiation can be produced from 360nm to 1160nm. All known dyes can be made to lase when pumped by very fast light sources with risetimes of 10-40ns. The most important rapid pumping sources are other lasers; specifically

Q-Switched Ruby	(694.3nm)
Q-Switched Nd 2nd Harmonic	(530.0nm)
Q-Switched Ruby 2nd Harmonic	(347.2nm)
Nitrogen 2nd Positive	(337.1nm)

TABLE 1
CHARACTERISTICS OF DYE LASER DYES

NO.	DYE	SPECTRAL COVERAGE (nm)	FLASH-LAMP PUMPED	REFERENCES	COMM'L SOURCE
345	p-terphenyl (p-diphenylbenzene)	343-356	X	30,31	33,44,P,27
348	2,5-diphenyl-1,3,4-oxadiazole (PPD)		X	31	44
363	2,5 diphenyloxazole (PPO)	358-391	X	12,31	44,P,33
365	isopropyl PBD	361,370	X	31	
374	2,5 diphenylfuran (PPF)	369-380	X	12,31	27
375	p-quaterphenyl		X	31	44
385	phenylbiphenylyloxadiazole (PBD)	377-415	X	12,30,31	44,P
397	α-naphtylphenyloxazole (α-NPO)	392-414	X	12,18,31	21,44,P
408	p,p'-diphenylstilbene		X	30,31	P
409	dibiphenyloxazole (BBO)	401-420	X	12,18,31	44,P
410	1,2-di-4-biphenylyl-ethelene	380-440		47	
413	2-hydroxy-4-methyl-7-amino-quinoline		X	90	
415	1,4-distyrylbenzene			47	
418	1,4-di-[2-(5-phenyloxazyl)]-benzene (POPOP)		X	12,18,24,28,27	21,44,P,27 33
420	4,4'-dichlor-1,4-distyrylbenzene			47	

NO.	DYE	SPECTRAL COVERAGE (nm)	FLASH-LAMP PUMPED	REFERENCES	COMM'L SOURCE
421	p,p'-bis(O-methylstyryl) benzene (bis-MSB)	419-425		18	P,21
425	2,2'-dimethoxy-1,4-distyryl-benzene			47	
426	1,2-di-(α-naphtyl)-ethylene			47	
429	1,4-di-[2-(4-methyl-5-phenylox-azolyl)] (DIMETHYL POPOP)	424-441		12,18	P,A,44,21
430	9,10-diphenylanthracene			41,57	P
432	1-styryl-4-[ω-vinyl-(n-biphenylyl)]-benzene			47	
435	ACRIDONE			87	
437	2,5-bis[tert-butylbenzoxazolyl(2)] thiopene (BBOT)		X	31	
441	3-ethylaminopyrene-5,8,10-trisulfonic acid			69	
454	7-acetoxy-4-methylcoumarin	441-486	X	38	
455	7-hydroxy-4-methylcoumarin (4-methylumbelliferone)	420-550 (pH)	X	20,24,32,75, 78,88	21,27
456	7-methylamino-4,6-dimethyl-coumarin	440-485 (pH)	X	90	
457	7-ethylamino-4,6-dimethyl-coumarin		X	90	

NO.	DYE	SPECTRAL COVERAGE (nm)	FLASH-LAMP PUMPED	REFERENCES	COMM'L SOURCE
458	9-aminoacridine hydrochloride	457-460	X	37	
460	7-hydroxycoumarin	450-470	X	75,76,77,96	
461	7-diethylamino-4-methyl-coumarin	438-482	X	12,25,76,83	21,27
462	7-acetoxy-5-allyl-4,8-dimethylcoumarin	458-515	X	38	
465	ESCULIN	450-470	X	76,78,88	
466	benzyl β methylumbelliferone	464-468	X	37	
467	4,8-dimethyl-7-hydroxycoumarin	455-505	X	38	
485	2,4,6-triphenylpyrilium-fluorobate			69	
510	acriflavin hydrochloride		X	31,42,54	33,27
539	1,2-dihydro-4-methoxybenzo-[C]xanthylium fluorobate			21	21
540	EOSIN			87	33
544	10-methoxy-12H-indono[3,2,6]-naptho[1,2e]pyrlium fluorobate			21	21
545	Na-Fluorescein	530-560	X	2,13,24,32 47,77,86, 87,88	33,21
546	fluorescein diacetate	541-571	X	38	
550	2,7-dichlorofluorescein		X	76,88	21,33,27
551	pyrlium salt		X	68	

LASER EXCITATION 227

NO.	DYE	SPECTRAL COVERAGE (nm)	FLASH-LAMP PUMPED	REFERENCES	COMM'L SOURCE
552	8-hydroxy-1,3,6-pyrenetri-sulfonic acid trisodium salt			21	21
565	acetamidopyrene-trisulfonate		X	68,64	
575	RHODAMINE C			46,47	
581	PYRONIN B	580, 615-632	X	24,37,46, 47,74	27
582	RHODAMINE S	578-595	X	37	
590	RHODAMINE 6G	570-620	X	2,13,24,32, 47,63,73,74, 76,77,79,83, 86,88,96	21,33,27
591	RHODAMINE G			67	
610	SAFRANINE T			47	33
611	XYLENE RED B	584-645	X	38	
615	ACRIDINE RED	600-630	X	2,76,77,86, 88	21,27
616	KITON RED S	589-642	X	38	
620	RHODAMINE B	605-645	X	2,9,13,24, 32,63,74,76, 77,86,88,96	21,33,27
660	3,3'-diethyl-oxadicarbo-cyanine-iodide			55,69	44,21
714	3,3'-diethyl-10-chloro-2,2'-(5,6,5',6'-dibenzo)-thiadicarbo-cyanine-iodide			58	

NO.	DYE	SPECTRAL COVERAGE (nm) FLASH-LAMP PUMPED	REFERENCES	COMM'L SOURCE
715	3,3'-diethyl-2,2'-thiadicarbo-cyanine iodide (DTDC IODIDE)	705-740	58,70	N,33,21
717	3,3'-diethyl-2,2'-(5,5'-dimethyl)-thiazolinotri-carbocyanine iodide		58	
720	3,3'-diethyl-10-chloro-2,2'-(4,5,4',5'-dibenzo)thiadicarbo-cyanine iodide	720,760	58	
725	3,3'-diethyl-2,2'-oxytricarbo-cyanine iodide	718-750	17,58	N
726	3,3'-diethyl-5,5'-dimetoxy-6,6'-bis(methylmercapto)-10-methyl-thiadicarbocyanine bromide	727-739	17	
735	3,3'-dimethyl-2,2'-oxatricarbo-cyanine iodide	720-770	58	
740	1,1'-diethyl-γ-cyano-2,2'-di-carbocyanine tetrafluoroborate		70	
745	1,1'-diethyl-4,4'-quinocarbo-cyanine iodide (CRYPTOCYANINE)		58,89	N,33,21,27
746	1,1'-dimethyl-11-bromo-2,2'-quinodicarbocyanine iodide		58	
749	1,1'-dimethyl-4,4'-quinocarbo-cyanine iodide		58	
754	1,1'-diethyl-4,4'-quinocarbo-cyanine bromide		58	
755	chloro-aluminum phthalocyanine (CAP)	750-780	85	21
770	3-ethyl-3'-methylthiathiazolino-tricarbocyanine iodide	738-801	17	

LASER EXCITATION

NO.	DYE	SPECTRAL COVERAGE (nm)	FLASH-LAMP PUMPED	REFERENCES	COMM'L SOURCE
771	1,1'-diethyl-2,2'-dicarbo-cyanine iodide			67	33,21,27
780	3,3'-diethyl-11-metoxythiatri-carbocyanine iodide	773-798		17	
797	1,1'-diethyl-γ-acetoxy-2,2'-dicarbocyanine tetrafluoro-borate			70	
800	1,3,3,1',3',3'-hexamethyl-2,2'-indotricarbocyanine iodide	780-830	X	17,31,58	
814	1,1'-diethyl-11-bromo-2,2'-quinodicarbocyanine iodide	780-810		58	
815	1,1'-diethyl-γ-nitro-4,4'-dicarbocyanine tetrafluoro-borate	790-830		8,9,70	
820	3,3'-diethylthiatricarbocyanine iodide (DTTC IODIDE)	794-860		17,58,67, 84,87	N,33,21
821	3,3'-diethylthiatricarbocyanine bromide (DTTC BROMIDE)	808-840		70	
825	1,3,3,1',3',3'-hexamethyl-4,5,4',5'-dibenzoiodotricarbo-cyanine perchlorate	816-833		17	
826	3,3'-diethyl-2,2'-selenatricarbo-cyanine iodide			58	
831	1,1'-diethyl-11-bromo-4,4'-quino-dicarbocyanine iodide			58	
835	3,3'-diethyl-6,7,6',7'-dibenzo-thiotricarbocyanine iodide	824-854		17	

NO.	DYE	SPECTRAL COVERAGE (nm)	FLASH-LAMP PUMPED	REFERENCES	COMM'L SOURCE
850	3,3'-diethyl-2,2'-(5,6,5', 6'-tetramethoxy)thiatri-carbocyanine iodide	845-875		58	
855	3,3'-diethyl-6,7,6',7'-dibenzo-11-methylthiatri-carbocyanine iodide	843-869		17	
856	3,3'-diethyl-2,2'-(4,5,4',5'-dibenzo)thiatricarbocyanine iodide	840-880		58	
920	3,3'-diethyl-12-ethylthia-tetracarbocyanine iodide	916-924		17	
930	1,1'-diethyl-2,2'-quinotri-carbocyanine iodide	890-960+		17,58	N
1020	1,1'-diethyl-4,4'-quinotri-carbocyanine iodide	980-1060		58	N

KEY TO COMMERCIAL SOURCES

21: Eastman Kodak (Ref. B.21)
27: Fischer Scientific (Ref. B.27)
33: Gallard Schlesinger Chemical (Ref. B.33)
44: Koch-Light Laboratories (Ref. B.44)
A: Arapahoe Chemical Co., Boulder, Colorado 80302
N: Nippon Kankoh-Shikiso Kenkyusho Co. Ltd., 123 Shimoishi, Okayama-Shi, Japan
P: Pilot Chemical Co., Watertown, Mass. 02172

The wavelength of the pumping radiation must fall within the molecular absorption band of the dye, and is always smaller than the dye laser output wavelength.

Successful flashlamp pumping has been achieved throughout the near ultraviolet and visible spectrum. The conversion efficiency of flashlamp-pumped systems (laser energy out / electrical energy into the flashlamps) is comparable to solid-state pulsed lasers, and is on the order of 0.05 - 0.6% [B.29,B.35,B.76]. Up to two joules output in a 500ns pulse has been achieved [B.29]. With proper dye circulation systems, it is possible to sustain repetition rates of 1-60Hz, with energy outputs on the order of 5-20 millijoules/pulse [B.72].

Rhodamine 6G (No. 590) is the most efficient of the dyes which have been found to lase with flashlamp excitation. The conversion efficiency of 0.6% mentioned in the previous paragraph refers to Rhodamine 6G. Several dyes have exhibited conversion efficiencies on the order of 50% when pumped with laser sources under optimum conditions. An upper limit to the conversion efficiency exists for each dye and output wavelength; it is the product of the quantum efficiency (photons emitted/photons absorbed) and the ratio of photon energies in the pump and output beams ($h\nu_{out}/h\nu_{pump}$). The ratio (ν_{out}/ν_{pump}) has a lower limit for each dye concentration and solvent, because the low frequency tail of the absorption curve overlaps the emission curve. This fact also serves to explain the shift of the peak laser output to longer wavelengths with increased dye concentration [B.23,B.87].

Two very recent papers have reported CW operation of dye lasers. Peterson, Tuccio, and Snavely [B.65] and Baxse et al [B.4] used the 5145 A° line of a CW argon laser to excite Rhodamine 6G. It is apparent from these preliminary results that triplet quenching can be satisfactorily achieved, and that CW operation is made difficult only by thermal schlieren effects and photo-depletion of the dye after long exposure.

Three commercial systems have been announced which allow laser pumping of dyes. Two of these use the so-called side-pumping scheme wherein the laser pump enters the dye cell transversely. This arrangement, illustrated in Fig. 1a, was historically the first successful system and was incorporated into the dye lasers marketed by Korad Corporation (~$3000) and Avco Corporation (~$5000). The advantages of this system include a simple cell geometry and the ease with which dispersive elements may be incorporated into the dye laser cavity. Soffer and McFarland [B.82] were the first to introduce a diffraction grating which allows continuous tuning of the output wavelength and also drastically reduces the spectral width of the output radiation.

Sorokin et al. [B.87] used a dichroic mirror on one end of the laser cavity. Pump radiation entered the cell through the dichroic mirror and was absorbed as it traversed the longitudinal axis of the system. This configuration is illustrated in Fig. 1b.

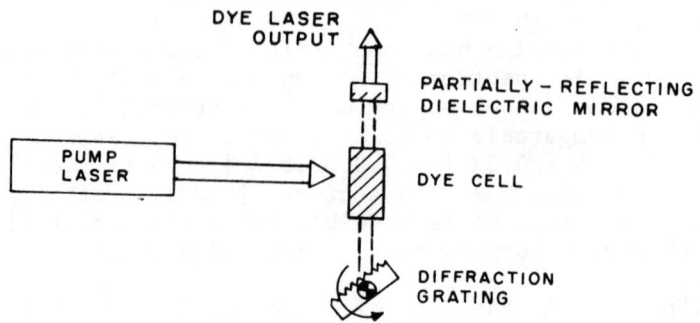

a. Transverse configuration

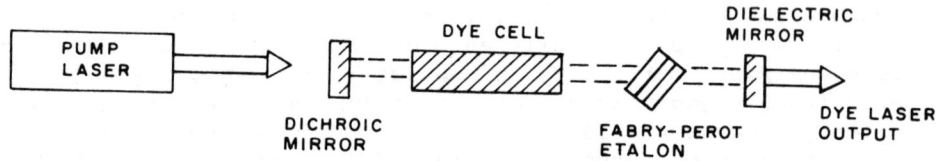

b. Longitudinal configuration

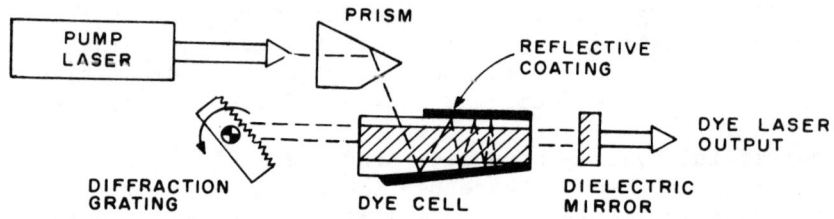

c. Multiple-pass configuration

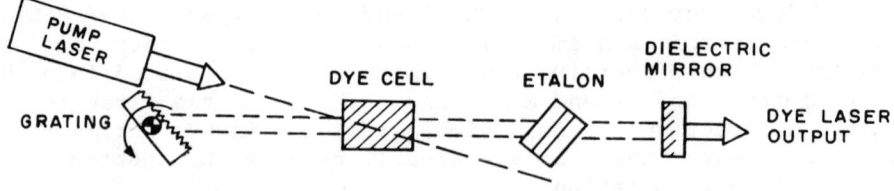

d. Angle-pumping configuration

Figure 1. Laser pumping schemes for dye lasers.

The laser-pumped dye laser produced by Xenon Corporation (~$1500) utilizes a multiple-pass pumping configuration which has not been previously reported in the literature. Pumping radiation is introduced into the dye cell at an angle and is constrained by reflective coatings on the cell to traverse the dye many times before being absorbed. The effective absorption path is approximately three times the longitudinal length of the dye cell (i.e., three times that achieved with end pumping) and twelve times the transverse dimension (i.e., twelve times the absorption length obtained with side pumping). Exponential decay of the pump beam intensity and the alternation of the pump beam direction upon successive reflections insures that each longitudinal ray along the axis of the dye laser cavity is nearly uniformly pumped. Large changes in dye concentration can be tolerated without significantly reducing the coupling between the laser pump and the dye solution.

Nonuniform lasing has been observed in several side-pumped configurations [B.8,B.12,B.84,B.87]. This arises because the dye concentration must be great enough to absorb most of the exciting radiation in one transverse pass. Absorption nearest the pumping source will therefore exceed that on the exit side of the dye cell. While this nonuniformity is eliminated by end pumping, it is then very difficult to use intracavity dispersive elements for wavelength tuning. Also, the dichroic filter required with longitudinal excitation is very difficult to obtain with high transparency at the pump wavelength and broad-band reflectivity at slightly higher wavelengths. Bradley [B.8,B.10] has alleviated both of these problems by introducing the laser pump beam at a small angle with respect to the cavity axis of the dye laser (Fig. 1c). In the two experiments in which continuous dye laser output was achieved, a longitudinal pumping scheme was employed [B.4,B.65].

Transverse pumping is extremely useful in some circumstances because of the requirement that the number of excited atoms per unit volume must reach threshold within one fluorescent decay time. (The lifetime is typically 1-10 nanoseconds for dyes of high quantum efficiency.) A N_2 laser (e.g. AVCO C950, 100 KW for 10ns pulse length) will typically produce ~3 millijoules per pulse, which is sufficient to excite only a small volume of dye. By transverse pumping a very concentrated solution (typically 10^{-3} molar of dye), only a small longitudinal filament adjacent to the dye cell wall is excited. If the dye filament is aligned to be exactly coincident with the optical axis of the laser cavity, a laser beam of very small dimension (1mm x 0.5mm) will be produced.

Flashlamp pumping was first achieved by Sorokin and Lankard [B.86] using a coaxial flashlamp [B.14] whose central bore contained a dye cell. Using a low-inductance capacitor operating at 100 joules, they were able to produce lasing action in several dyes. Many additional dyes have been found to lase when pumped with fast-risetime flashlamps [B.28,B.31,B.37,B.74,B.76]. Rapid excitation overcomes two cumulative effects which tend to destroy lasing action: triplet-state absorption and thermally-induced nonuniformities. Snavely and Schafer [B.79] found that the presence of oxygen in Rhodamine 6G solutions quenched the triplet state, but thermally-induced refractive index nonuniformities of the solution produced an effective mis-alignment of the cavity which terminated laser action after several hundred microseconds. Deutsch et al. [B.19] were the first to suggest the possibility of oxygen quenching.

Keller [43] has presented a comprehensive discussion of the effects of triplet absorption and the favorable effects to be expected from using dissolved species to quench the triplet state. Oxygen in solution, as used by Snavely and Schafer [B.79], is useful in quenching the triplet state, but it also de-excites the lasing level as well and thus reduces the effective quantum efficiency of the fluorescence. Quenching additives which selectively depopulate the triplet state would lead to higher efficiencies. A very promising quenching medium for Rhodamine 6G is cyclooctatetrene, which was recently reported by Pappalardo, Samelson, and Lempicki [B.61].

General Laser Corporation markets one of two flashlamp-pumped dye lasers which are currently available (~$5000). This unit is based on Furamoto's modifications [B.28] of the original Sorokin design. Outputs up to 200 millijoules at a repetition rate of two per minute or less are obtainable with dyes which will lase with 150ns risetime excitation. High repetition rate systems operating at lower output energies per pulse have been constructed from commercial flash photolysis systems (e.g. Xenon Corporation, ~$2000). Synergetics Corporation has recently marketed a dye laser based on the Xenon flashlamp system. (~$6000).

Because of the extremely large gain of dye lasers, cavity alignment and the optimization of flash lamp coupling are less critical than with most other laser systems. Simple close-wrap configurations using metal foil surrounding both the flashlamp and the laser tube are sufficient. A 5 inch long quartz tube with 4mm bore, a fast flashlamp system discharging 2-10 joules into a linear xenon-filled flashtube, a dye circulation system, and highly-reflecting dielectric mirrors for the cavity are sufficient to produce a simple repetitive dye laser system.

Until very recently, the paucity of dyes which would lase with flashlamp excitation inhibited development of commercial products. Among the first dyes to be used, only Rhodamine 6G exhibited high conversion efficiency, reasonably broad wavelength coverage (570-620nm), and chemical stability under repeated flashlamp excitation. Other important dyes, e.g. Dyes No. 455, 461, 540, 545, 615, and 620 were deficient in one of these three criteria.

Shank, Dienes, Trozzolo, and Myer [B.20,B.75] report very broad-band tuning of 4-methylumbelliferone and 7-hydroxycoumarin in solutions of varying pH. Although their experiments refer to rapid pumping (10ns) with a nitrogen laser, the results should be useful in developing improved dye-solvent combinations for longer-pulse systems.

Shank et al. [B.75] used 4-methylumbelliferone in ethanol and adjusted the pH of the solution by adding either sodium hydroxide or hydrochloric acid. In basic solutions, tunable lasing action was achieved from 437nm to 544nm; in a neutral solution, the range was 385-574nm; and in strongly acidic solutions, lasing occurred from 459-574nm. For a slightly acidic solution, one single dye preparation lased from 391nm to 567nm, a range never achieved before with a single dye. Using only three dyes (4-methylumbelliferone (455), Rhodamine 6G (590), and Rhodamine B (620)), the entire spectrum from 385-640nm (near ultraviolet to deep red in the visible) can be covered.

The results of Shank et al. [B.20,B.75] extend earlier experiments reported by Srinivasan [B.90] in which pH tuning from 430-490nm was achieved in aminocoumarin dyes which were pumped with a flashlamp. These papers suggest that practical systems covering wide spectral regions with a minimum number of dyes are available today.

In the near infrared spectrum (700-1100nm), ruby laser pumping remains a necessity. As is evident from Table 1, all the available dyes in this region, except chloro-aluminum phthalocyanine, are carbocyanine compounds and none have been induced to lase using a flashlamp.* Although many have exhibited temporal degradation in solution, chloro-aluminum phthalocyanine (CAP, No. 755) is extremely stable. Of the 32 dyes known to lase in the 700-1100nm range, only five (Nos. 715, 745, 755, 815, and 820) have been investigated in any detail. The most comprehensive investigations of infrared-emitting dyes were conducted by Sorokin et al. [B.872],Miyazoe and Maeda [B.57], and Derkacheva et al. [B.15]. Additional research in this important spectral region will undoubtedly produce improved dyes which are capable of spanning the near infrared with fewer individual solutions.

The simplest dye laser cavity consists of two parallel reflective surfaces, one of which transmits a fraction of the incident light at the laser wavelength. Because of the high degree of overlap between adjacent vibrational-rotational levels in the electronically-excited singlet state, organic molecules tend to produce wide spectral outputs with broad-band cavity reflectors. Typical linewidths in this configuration are 3-10nm. Soffer and McFarland [B.82] used a diffraction grating as the rear reflector, and achieved a 10:1 narrowing of the output line-width while suffering only a modest reduction in the output power. The gain of the cavity is a maximum at that wavelength which is in autocollimation with the output reflector. Typically, one can achieve a 0.5 - 1.0nm halfwidth with first-order diffraction gratings (1600-2000 lines/mm). Echelle gratings, with significantly larger dispersion, produce

---

*An exception to this statement has been reported by Furamoto and Ceccon [B.31].

linewidths of the order of 0.1nm [Bradley et al., B.8,B.10].

Figure 2 is included to illustrate the type of line profile which can be achieved with dye lasers. These data were obtained in the Fluid Mechanics Laboratory at M.I.T. using a Control Data TRG104a ruby laser pump and a Xenon 500 dye laser employing a 40% reflective dielectric output mirror. A diffraction grating (blazed at 1.6μ first order, 600 lines/mm and used in second order) was the rear cavity reflector. The laser dye was DTTC Iodide in DMSO solvent, at a concentration of roughly 5×10^{-5} molar. The laser output was spectrally resolved using a Jarrel-Ash monochrometer (dispersion 15 A°/mm) with 10μ slits, and a Tropel photodiode detector which integrated the throughput of the monochrometer over each laser pulse. Although no circulation of the dye was used and the pulse rate was two per minute, the shot-to-shot repeatability was better than ± 5%. Output power was about 20% of the input ruby pump power, yielding about 5MW/A° at the line center.

Additional spectral narrowing can be achieved by using Fabry-Perot etalons. Bradley [B.8,B.10] employed an echelle grating as a pre-dispersive element (Fig. 1d). A tilted Fabry-Perot etalon with a free spectral range of 0.09nm and a finesse of less than 10 produced a dye laser beam with a spectral width of less than 10^{-3}nm. The beam divergence was reduced to 0.5 milliradians in this configuration, the lasing action occurring in a single cavity mode.

Gibson [B.35] used a series of etalons in the cavity of a coaxial flashlamp-pumped system filled with Rhodamine 6G. The threshold for lasing action was achieved with 20 joules of electrical input to the flashlamp. The following results were obtained with 50 joules of excitation:

Free Spectral Ranges of Etalons (nm)	Output Spectral Width (nm)	Output Energy (mj)
broad-band reflector only	10	10
10	0.3	7
10, 1.5	0.05	5
10, 1.5, 0.2	0.005	3

The remarkable reduction in spectral width with only a factor of three reduction in output energy is typical of the results which have been reported with dye lasers [B.8,B.82,B.83]. The etalons used by Gibson had a finesse of less than 10, which indicates that the output linewidth is narrower than the spectral resolution of the dispersive elements by more than a factor of four.

Recently, Walther and Hall [B.95] have employed a birefringent filter in conjunction with a diffraction grating to achieve linewidths of less than 1×10^{-3}nm. The birefringence method appears to

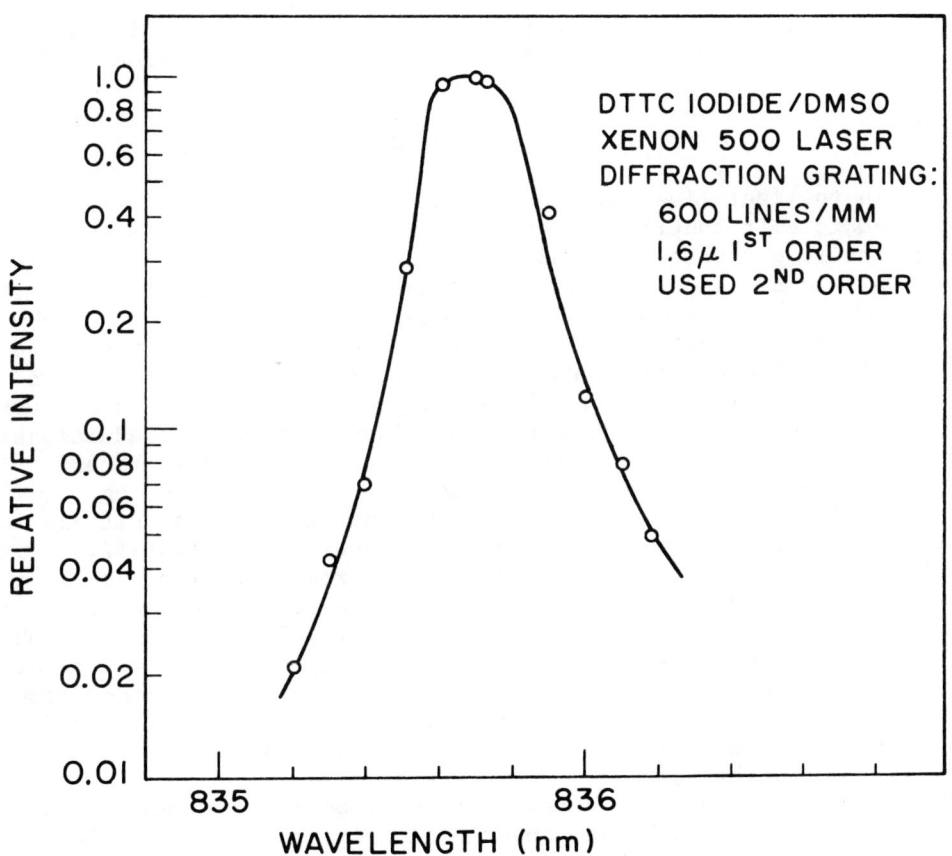

Figure 2. Typical dye laser spectral profile with moderate dispersion intracavity element.

produce a higher wavelength stability than has been achieveable with Fabry-Perot systems. In addition, the filter may be scanned over narrow spectral regions by the application of a transverse voltage to the birefringent crystals. As a demonstration of the stability of their system, Walther and Hall scanned the D_1 resonance line of sodium at 589.6nm.

The versatility and tunability of dye lasers is evidenced by the achievement of single-mode operation [B.8,B.10] and picosecond pulses from mode-locked systems [B.7-B.10, B.25,B.36,B.73,B.81]. It is interesting to note that these results were accomplished within two years of the announcement of lasing action in organic dyes.

III. SEMICONDUCTOR LASERS

In the last five years many semiconductor materials have been induced to emit coherent laser radiation efficiently at room temperature and below. In addition to the possibility of pumping via an external light source, these lasers can be excited (a) with a large reverse current through standard p-n junctions, or (b) by electron-beam injection perpendicular to homogeneous semiconductor crystals. Figure 3 illustrates typical arrangements for the two electrical pumping schemes. Ivey [1] presents a rather dated survey of semiconductor compounds exhibiting laser action, and additional citations are given by Morehead [2]. A brief, but more recent summary, is contained in the abstract by Kressel and Nelson [3].

GaAs, a favorable example of semiconductor lasers suitable for spectroscopy, has several interesting properties. First [2,4], it exhibits a high efficiency of conversion from electrical to light energy, up to 80% at 77°K. Second [5], the radiation bandwidth is relatively narrow, on the order of 0.1nm at 77°K and 0.5nm at 300°K. Third, the output wavelength is tunable by (a) temperature variation, (b) changes in impurity concentration, (c) changes in composition, and (d) the application of pressure or a magnetic field. And fourth, the diodes are extremely compact, occupying less than 1cm^3 for an average output of 10mW.

Pure GaAs can be temperature-tuned from about 800nm (4.2°K) to 905nm (300°K) although the conversion efficiency drops to 45% at room temperature [4]. Superimposed on this variation is a shift in output wavelength by $\pm$ 8.5nm with different impurity concentrations. Ternary compounds offer an additional method of wavelength variation; the combination $GaAs_xP_{1-x}$ (0.5 < x < 1) at 77°K, for example, covers the spectral range 630nm to 845nm. Fine tuning via magnetic fields is possible but has not been thoroughly investigated. Recent refinements in the junction materials used in heterogeneous diodes have improved their performance significantly [3,4], and continuous operation at room temperature has recently been achieved [6].

Significant laser power outputs from semiconductors were reported as early as 1963. Garfinkel and Engler [8] produced 1.5 watts CW in a GaAs diode array with a conversion efficiency of 30%,

LASER EXCITATION

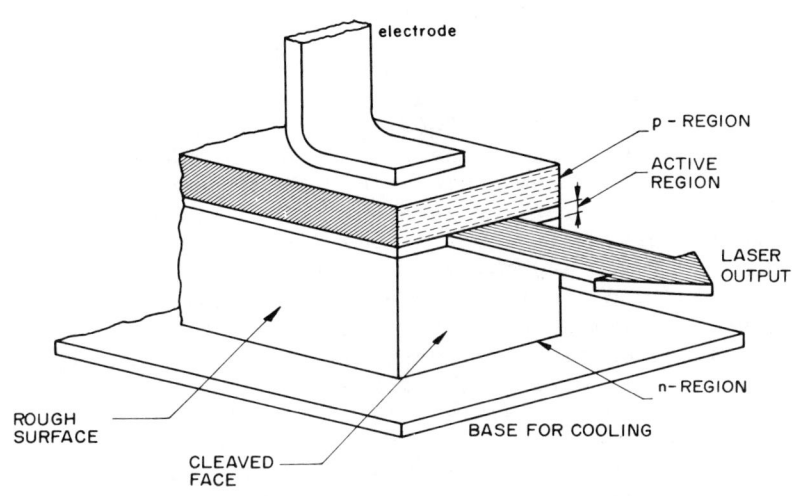

a. Current injection laser.

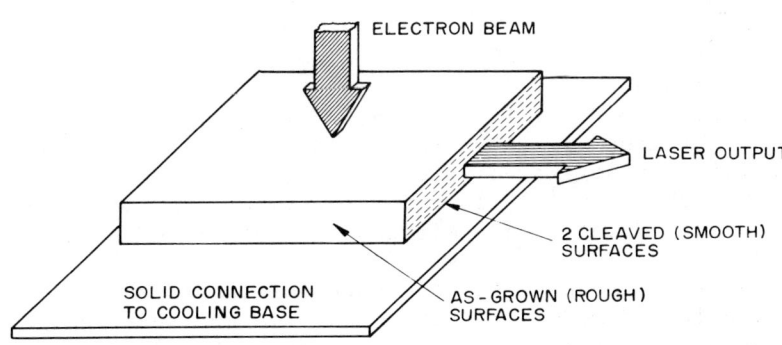

b. Electron — beam pumped laser.

Figure 3. Electrically-pumped semiconductor lasers (schematic).

and Lax [9] developed 100 watts in pulsed operation. Commercial units currently exceed this range, with laboratory prototypes producing considerably higher outputs (up to 1.5 kilowatts peak power in pulsed operation).

Figure 4 is a diagrammatic summary of the semiconductors which have been investigated for use as lasers. Compounds in which lasing action has been observed or is highly probable cover the spectrum from 325nm to 32 microns. Figure 5 gives the observations of Hurwitz [10] for the compounds Cd S_xSe_{1-x}, which covers the wavelength interval 490nm to 690nm. The data at 4.2°K and 77°K suggest the range of wavelength tuning achievable with temperature changes; for this compound, the difference is about 1.5nm with the wavelength increasing with temperature. Temperatures intermediate to these points should be achievable with proper heat sink arrangements, the use of liquid helium and controlled electrical dissipation in the diode itself, or by thermoelectric cooling.

The main limitation to pulse length is the rate at which the sample temperature rises during the excitation pulse. Hurwitz [11] and Nicol [12] have observed sample temperatures to rise as much as 80°K during 0.2 microsecond electron-beam excitation, causing shifts of about 2nm $(amp^2 - \mu s)^{-1}$ during reverse-current excitation [13]. This sweeping may be advantageous in guaranteeing coincidence between a particular spectral line and the laser line, but the effective pulse length may be reduced to a fraction of a nanosecond if the spectral line is very sharp. Wavelength shifts from heating may be minimized by employing sharp, short excitation waveforms which reach threshold rapidly. The duty cycle is limited by the average power dissipation within the laser material.

CW operation is not plagued by the spectral sweeping which accompanies pulsed operation. Hinkley and Freed [14] have achieved a frequency stability of 100kHz in $Pb_xSn_{1-x}Te$ at a wavelength of about 10.6 microns. This is equivalent to a wavelength resolution of about 22 parts per million, which is orders of magnitude higher than can be obtained with even the finest resolution spectrograph. (The comparable numbers are: spectrometer resolution, $<.05 cm^{-1}$; diode laser, $<2 \times 10^{-6} cm^{-1}$.) In this experiment, 240 microwatts of CW power were produced from a single diode at liquid helium temperatures.

Earlier experiments with semiconductor lasers showed evidence of very complicated mode structure. This is not surprising when it is recognized that the optical cavities of these small crystals were the cleaved faces of the crystals themselves. Considerable progress has been made in developing multilayer dielectric coatings for semiconductor materials which allow the small crystals to be used in external cavities, or permit high-reflectivity coatings to enhance the cavity gain of the cleaved crystals themselves. A typical spectrally-resolved output line of cleaved and uncoated ZnS [15] is given in Fig. 6; ten simultaneous output lines similar to the one resolved in this figure were observed.

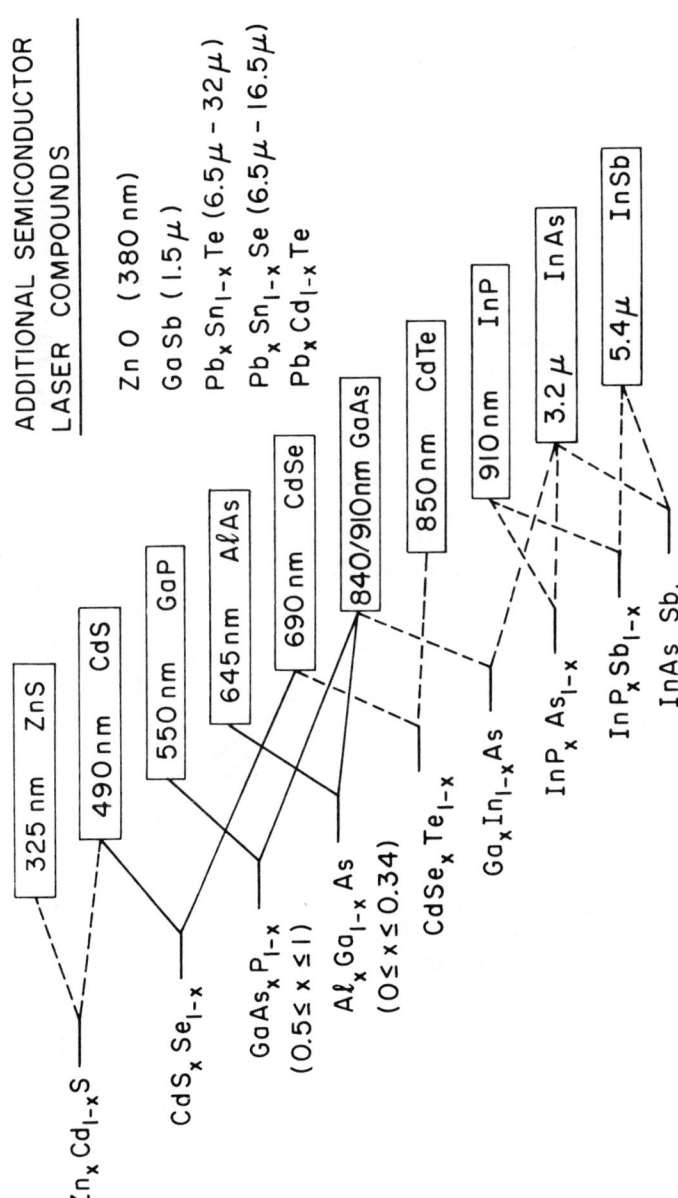

Figure 4. Semiconductor compounds exhibiting laser action. Solid lines indicate that the production of intermediate wavelengths has been observed; dashed lines indicate that a portion of the spectrum has been observed or that laser action is probable.

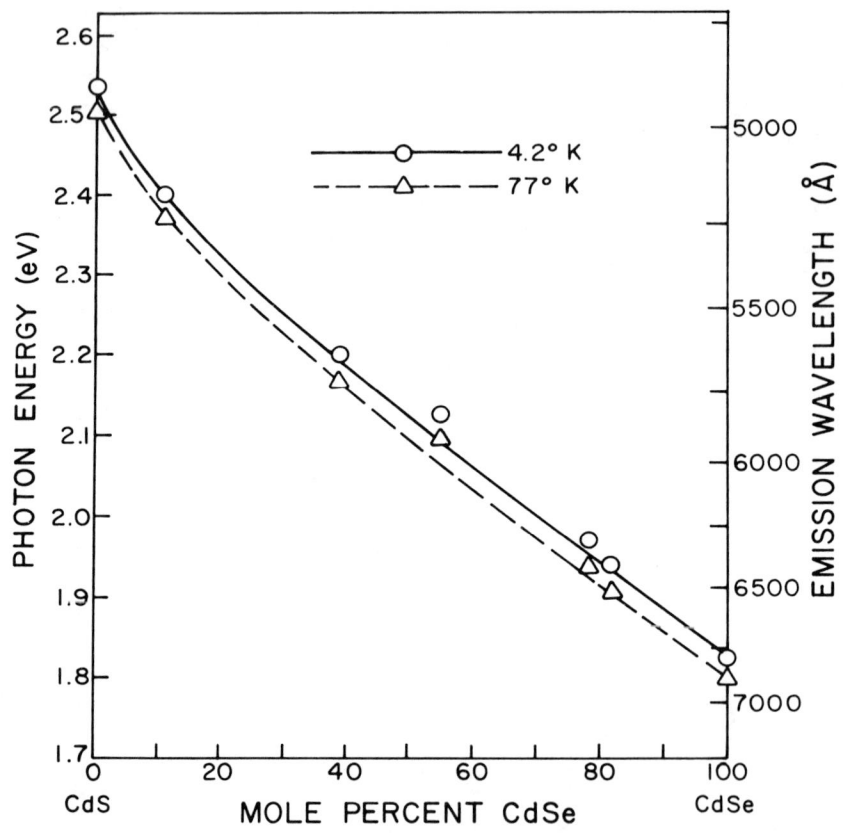

Figure 5. Wavelengths of $CdS_x Se_{1-x}$. (From C. Hurwitz, <u>Appl. Phys. Lett.</u>, <u>8</u>, 243 (1966).)

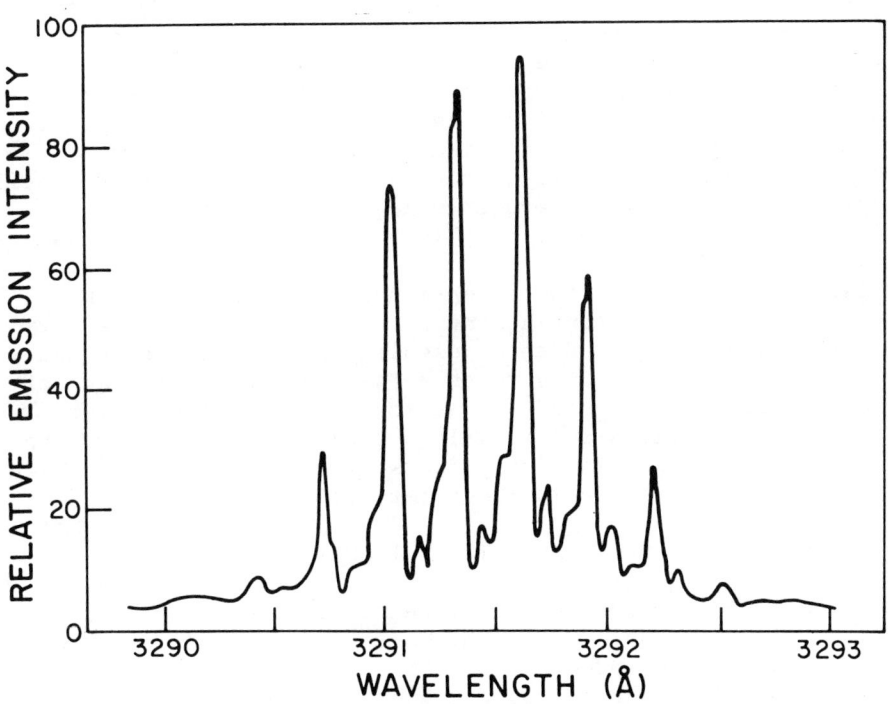

Figure 6. Resolution of a typical output line from ZnS. (From C. Hurwitz, <u>Appl. Phys. Lett.</u>, <u>9</u>, 116 (1966).)

At the present time GaAs, GaP, and $Al_xGa_{1-x}As$ are the only semiconductor lasers available commercially as off-the-shelf-items. The development of other compounds as operational spectroscopic light sources should be forthcoming as the demand develops. Probably the most significant application lies in the infrared spectrum from 2 to 30 microns, where black-body radiation sources are most frequently employed. The $Pb_xSn_{1-x}Te$ system is extremely promising for the 6.5μ - 32μ region. In As_xSb_{1-x} is a promising system for 3.2μ to 5.4μ. The need for cryogenic cooling of some semiconductor compounds represents a serious drawback in many circumstances. It is particularly cumbersome if external optics are employed.

Multimode semiconductor lasers with broad spectral outputs exhibit large beam divergences, on the order of $\pm 10°$ around the emission axis. Because of the microscopic dimensions of the emission plane, the beam divergence can be reduced to a few milliradians with suitable external optics.

Broad wavelength tuning of semiconductor lasers can be accomplished only by composition changes. For investigating specific molecular transitions, temperature tuning of a single element is sufficient to cover a range of roughly $50 cm^{-1}$. A series of elements is required to duplicate the performance of a single liquid laser dye. Figure 7 illustrates the exceptional efficiency which has been achieved with close-confinement structures [4]. The ternary system $Al_xGa_{1-x}As$ has proved to be significantly superior to $GaAs_x P_{1-x}$, both compounds covering roughly the same spectral range of 640-900nm. Nelson and Kressel [4] achieved 0.4 watts of laser output CW with a single diode roughly 0.25mm long in the direction of the cavity axis. Single-diode lasers producing 40 watts pulsed at room temperature are available commercially.

IV. NONLINEAR OPTICAL TECHNIQUES

Nonlinear interaction between intense laser sources and suitable crystals or liquids can lead to the generation of laser emission at other wavelengths. The production of second-harmonic radiation was discovered soon after the introduction of solid-state lasers. Stimulated Raman scattering has been investigated as a method of producing many relatively narrow laser lines from high-power lasers. And Parametric oscillators have been developed as sources of widely-tunable laser radiation. Recent achievements in these three areas will be reviewed with some brevity. A contemporary summary giving additional information has been published by Giordmaine [16]. Two texts on nonlinear optics are also available [17,18].

Frequency-doubling has received much attention as a means of generating new laser frequencies. Early experiments by Wang and Racette [19], Terhune et al. [20], and others achieved 10-20% conversion of Q-switched ruby and Nd laser beams to second-harmonic frequencies. Johnson [21] has achieved megawatt pulses of third-harmonic radiation from a Q-switched ruby laser by adding second-harmonic radiation from a crystal of ADP to the fundamental in a

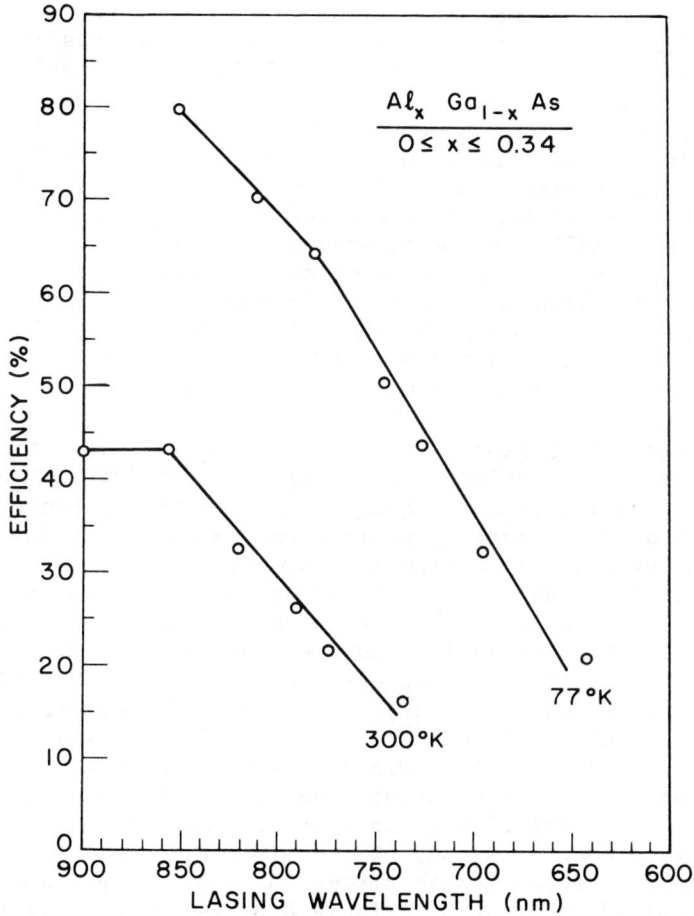

Figure 7. Efficiency of conversion of electrical energy into laser emission energy. (From H. Nelson and K. Kressel, Appl. Phys. Lett., 15, 7 (1969).)

second ADP crystal. He has also produced measurable power at the 265nm fourth-harmonic of Nd by doubling the 530nm second harmonic in a KPD crystal.

Improvements in both the coherence lengths of high-power lasers and available nonlinear optical materials has brought about dramatic improvements in the efficiency of sum-and-difference frequency generation. The new nonlinear optical materials [16, 22-25] exhibit greatly increased conversion efficiencies which arise from large nonlinear coefficients and improved resistance to optical damage. Other nonlinear materials are available for infrared wavelengths [26-28].

In 1967, Vanyukov and co-workers [24] achieved 48% second-harmonic generation efficiency in KDP using a carefully-constructed Q-switched Nd laser with a beam divergence of 0.25 milliradians. Using $BaNaNbO_3$, one of the newer ferroelectric crystals, Geusic et al. [22] produced 100% conversion of a CW Nd laser beam to green light at 530nm. Second-harmonic generation efficiencies of 50% have been achieved with KDP and ADP in converting a strong pulsed argon ion laser to its second-harmonic (0.5 watts at 257.3nm). [See Refs. 16 and 30]. Second-harmonic generation is an attractive method of producing intense laser sources in the 200-600nm region of the spectrum.

Parametric oscillation is a powerful tool for generating spectrally tunable radiation in the visible and infrared. A high-intensity pumping source of frequency ω_0 is partially converted into signal and idler waves whose frequencies ω_s and ω_i sum to ω_0.

Early experiments by Giordmaine and Miller [31,32] reported the generation of tunable coherent radiation with approximately 1% efficiency. The pump light was the second harmonic of a Q-switched Nd laser. The $LiNbO_3$ crystal which they used was constructed as a Fabry-Perot cavity with two flat parallel faces which had dielectric coatings to form a resonance for the generated frequencies. Boyd and Ashkin [33] and Giordmaine and Miller [32] give relations for computing the change in cavity dimensions with temperature. This dimension change shifts the output frequency from $1/2\ \omega_0$ to $1/2\ \omega_0 \pm \delta\omega$, where $\delta\omega$ is a calculable function of temperature. The fundamental wavelength ($2hc/\omega_0$) of the Giordmaine-Miller experiment was 1.06µ, twice the wavelength of the 530nm pump light and identical to the original Nd wavelength. Figure 8 presents their experimental results. Temperature tuning must be augmented by the application of electric fields to achieve proper phase-matching in the crystal [33].

A similar experiment was conducted by Miller and Nordland [34] in which a $LiNbO_3$ crystal was rotated within an external mirror cavity to achieve tuning. Some advantages accrue to this method of operation; for example, no electric fields need be applied to a constant-temperature crystal. The output power depends exponentially on the inverse of the cavity length and a cavity formed by external mirrors (rather than dielectric-coated crystal faces) runs the risk

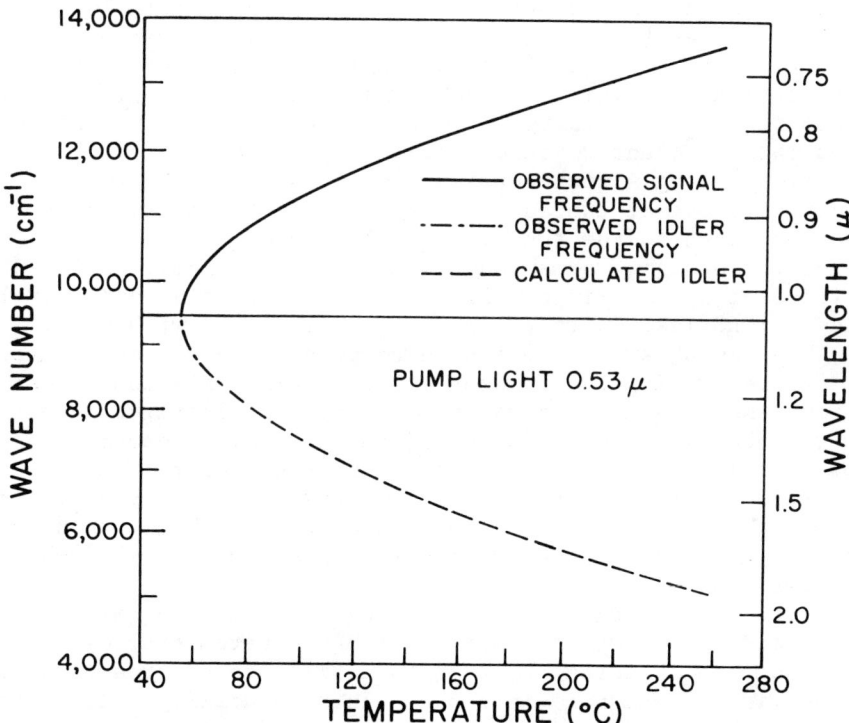

Figure 8. Parametric oscillator generation of tunable infrared radiation. (From J.A. Giordmaine and R.C. Miller, Appl. Phys. Lett., 9, 298 (1966).)

of inherently lower power outputs. The crystal must usually be anti-reflection coated to decrease losses and avoid spurious resonances.

More recent experiments on different parametric oscillator configurations have shown remarkably high efficiencies. Bjorkholm [35] reported a peak conversion of 22% between the pump input power and the tunable output signal frequency. This was accomplished by making the external cavity containing the parametric oscillator resonant to both the signal and idler frequencies. A series of Russian experiments concurrent with those of Giordmaine and Miller resulted in a 50% conversion efficiency being reported in 1968 [36].

Falk and Murray [37] used a single-cavity configuration similar to that of Bjorkholm [35], but operated the system with non-collinear phase matching. $LiNbO_3$ was used as the nonlinear medium, and it was pumped directly with a ruby laser. The external cavity was made resonant to idler wavelengths from 1.64μ to 2.05μ, and the corresponding signal wavelengths (1.20μ to 1.05μ) were coupled directly out of the cavity. As much as 45% of the pump light was converted to the tunable signal wavelength.

Amman, Yarborough, Oshman, and Montgomery [38] have achieved highly efficient parametric oscillation by operating a repetitively-Q-switched Nd:YAG laser with a $LiNbO_3$ crystal placed within the cavity. The crystal axis is oriented to produce nearly degenerate signal and idler radiation at about 2.1μ. Tuning is accomplished by adjusting the crystal temperature. An average infrared output power up to 350mW was achieved with signal and idler spectral widths of roughly $20cm^{-1}$. Generation of narrow spectral width ($\sim 0.1nm$) tunable parametric oscillations was reported by Smith [30] nearly a year earlier. He placed a crystal of $BaNaNbO_3$ within the cavity of a single-mode argon ion laser. Approximately 45-50mW of tunable radiation was produced.

The first commercial system employing a Q-switched Nd:YAG laser, frequency doubled and pumping a tunable parametric oscillator, has been announced by Chromatix. It is claimed to be broadly tunable in both the visible and near infrared. This capability will compete directly with the broad tuning capabilities of dye lasers.

Nonlinear interaction between incoming laser light and a Raman-active material results in coherent output radiation which is shifted in energy by a multiple of the Raman bond energy. The most intense shifted line corresponds to the absorption of one vibrational quantum and is referred to as the first Stokes line (S_1). Anti-Stokes shifts (AS_1, AS_2, etc.) to higher energies occur by the addition of the bond energy to the incident photon energy, and these lines are considerably less intense than the corresponding Stokes lines. As many as ten shifted lines have been observed to emanate from Raman-active materials subject to intense laser radiation, and up to 30% of the incoming photons have been observed to undergo stimulated Raman shifts.

More than 100 coherent lines have been observed from the excitation of more than 30 Raman-active materials [39]. The characteristic vibrational frequencies range from $445 cm^{-1}$ to $4160 cm^{-1}$, and many discrete laser lines can be generated by each intense laser source by simply substituting different Raman-active media.

Successful spectroscopic use of Raman-shifted lines requires coincidence between the optical transition line and the laser output wavelength. One might suppose that with some 45 Raman-active transitions and several pulsed lasers in the power range of 0.1 MW and up, a good probability would exist of overlap between one or more shifted laser lines and the wavelength of interest [39,40]. Because the half-widths of Raman-shifted lines are on the order of 2 to $40 cm^{-1}$, any coincidence within this half-width will yield acceptable power densities for many laboratory purposes.

As a test of this hypothesis, 18 atomic cesium transitions were compared with Raman-shifted wavelengths produced by ruby and ruby second-harmonic radiation. Inasmuch as the fundamental ruby wavelength changes continuously from 693.4nm at 77°K to 694.3nm at 293°K, a small amount of wavelength tuning was available for each Raman medium. Twelve of the eighteen cesium lines, ranging from 334.7nm to 876.1nm were coincident with one or more shifted ruby wavelengths to well within the Raman half-widths. A cross-check of all combinations resulting from ten strong laser lines (from 337.1nm to 1061nm) and 45 Raman-active vibrations (frequency shifts from $216 cm^{-1}$ to $4155 cm^{-1}$, S_1, and AS_1) yields a 94% coverage of the entire wavelength spectrum from 300nm to 1.6μ; for most wavelengths in the visible region, several combinations are available which would be suitable for spectroscopic purposes.

The primary advantage of Raman shifting is that the sophistication of post-laser optics is minimized. Most Raman-active substances are inexpensive liquids, e.g. benzene, acetone, carbon tetrachloride, and toluene. Some recent progress in unraveling the details of stimulated Raman emission has proved very useful in understanding this nonlinear interaction [16]. However, conversion efficiencies are generally on the order of 10%, and it appears that dye lasers are considerably more practical because of higher conversion efficiency, smaller beam divergence, wider tuning range, and spectroscopically narrower output radiation.

Wavelength-tunable infrared sources have been produced by using difference-frequency techniques. Faries, Gehring, Richards, and Shen [41] produced far-infrared radiation ($\lambda > 100\mu$) from the interaction of two ruby laser beams in $LiNbO_3$. The difference-frequency was varied by differentially cooling one of the ruby rods and shifting its output wavelength. Earlier theoretical and experimental results on difference-frequency generation may be found in Refs. 42-48.

We have recently used a difference-frequency technique to produce wavelength-tunable radiation in the infrared [49]. The experimental arrangement is given in Fig. 9. The dye laser is tunable over the spectral region 715nm to 1.1μ. Difference-

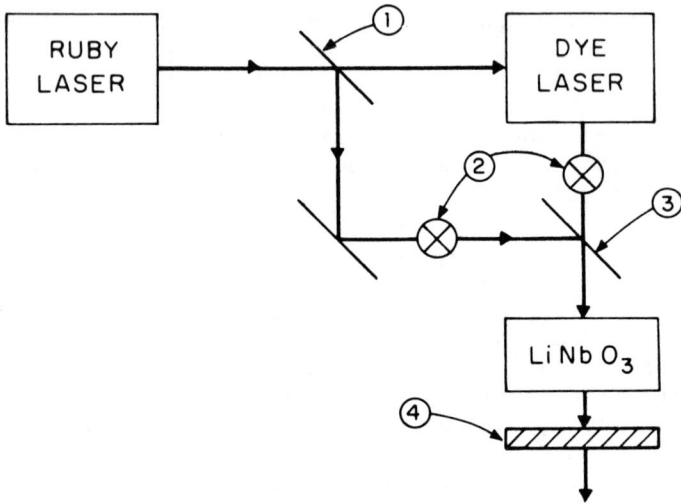

① BEAM SPLITTER
② ATTENUATORS
③ DICHROIC MIRROR
④ SILICON FILTER

Figure 9. Optical system for generating tunable infrared radiation using a dye laser. (From C.F. Dewey, Jr. and L.O. Hocker (to be published).)

frequencies generated using the 694.3nm ruby beam would be capable of covering the 2μ - 20μ region. One current difficulty is the dearth of low-absorption nonlinear materials. $LiNbO_3$ exhibits strong absorption beyond $\lambda = 4.5\mu$, and other materials such as Ag_3AsS_3 and Ag_3SbS_3 [28] begin to lose transmission around 13μ.

To illustrate the flexibility of the difference-frequency system, an absorption curve was run on polyethelene and the results compared to those obtained with a conventional spectrophotometer (Fig. 10). In this preliminary experiment, approximately 6KW of power was produced with a spectral width of less than $10cm^{-1}$. Using improved intracavity optics on the dye laser [B.95], the linewidth would be limited only by the quality of the ruby pump beam.

V. RESONANT INTERACTIONS BETWEEN LASERS AND GASES

The interest of this paper, as is evident from the preceding sections, is in spectroscopic techniques employing optical excitation of selected energy states of molecules and atoms, and subsequent interpretation of the induced effects on the gas being studied. All of the measurements to be discussed could, in principle, be accomplished with sufficiently intense conventional light sources such as flash tubes or arc sources. Laser sources, however, offer many advantages and often spell the difference between a theoretical possibility and a practical experiment.

A few comparisons between conventional spectroscopic sources and wavelength-tunable lasers are in order. Instantaneous power per unit wavelength interval is a particularly important parameter. Specifications [50] of the most powerful ablating quartz flashtube known to the author are a 1.5μsec. half-width pulse, and effective black-body temperature of 48,000°K $\pm$ 5,000°K for wavelengths longer than 200nm, effective collection angle for emitted radiation of 1/4π steradians, and a total radiating surface area of $0.25cm^2$. Total radiant energy per pulse is roughly 0.1 joule, and the peak power is approximately 10 KW. The instantaneous power spectral density of this source is 2×10^{-3} watts/cm^{-1} at 200nm. Dye lasers are capable of producing tunable radiation with power spectral densities of $10^4 - 10^7$ watts/cm^{-1} from 360nm to 1 micron. Frequency doubling the dye laser output should produce intensities of 10^3 watts/cm^{-1} at 200nm. By using difference-frequency generation with a ruby laser and ruby-pumped dye laser, tunable infrared radiation in excess of 10^4 watts/cm^{-1} should be possible from 2-20 micron in the infrared [49]. A 2000°K continuous blackbody source produces roughly 10^{-6} watts/cm^{-1} per cm^2 of emitting area at 5 microns, which is smaller by a factor of 10^{10} than the instantaneous power available from a tunable dye laser system!

In atomic-absorption spectroscopy, there has been a successful history of operation with resonance light sources [51]. These lamps are hollow-cathode arcs containing a filler gas and traces of a spectroscopic source element. By proper choice of the

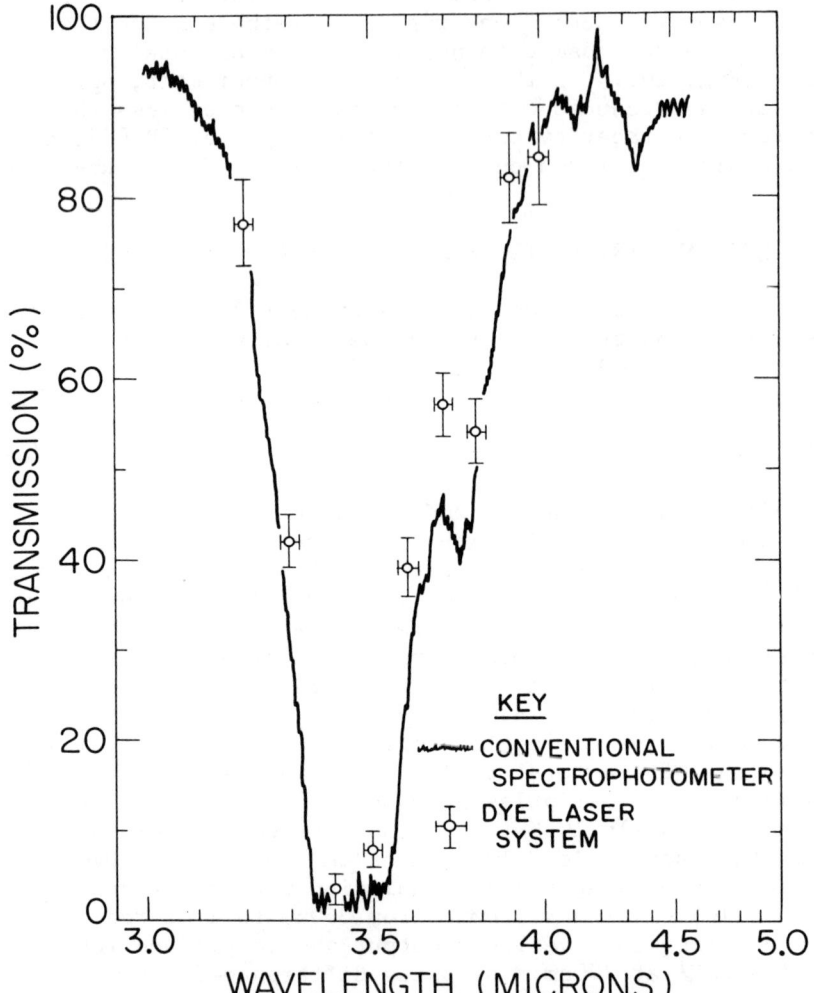

Figure 10. Absorption curve of polyethylene obtained with tunable infrared dye laser system. (From C.F. Dewey, Jr. and L.O. Hocker (to be published).)

LASER EXCITATION

operating conditions, intense atomic line radiation can be produced. The intensities are on the order of 10^{-6} watts per resonance line [52], many orders of magnitude smaller than the power spectral density available from tunable lasers. Resonance lines are extremely narrow, on the order of 10^{-3}nm, the spectral width increasing at high powers because of self-absorption and line broadening. Higher intensities in the ultraviolet, on the order of 10^{-5} watts per line, have been reported by Davis and Braun [53] in systems using microwave excitation of flowing gases; the line profiles have not been resolved, but are definitely less than 0.1nm wide.

In addition to very high power spectral densities, tunable laser sources possess a number of other advantages. The output radiation is available as a collimated beam with relatively small divergence angle and (often) some measurable coherence length. Bradley [B.6] has succeeded in producing a ruby-pumped dye laser with a beam divergence of 0.5 milliradians (about 1 minute of arc), which suggests that efficient frequency doubling is possible. The reliability of lasers being marketed currently is quite high, and they are generally much less trouble than ablating quartz flashtubes in the 30,000 + °K range. Unwanted radiation is non-existent, in contrast to black-body sources which produce a large fraction of their output in regions of the spectrum which are of no interest to the experiment. Also, low intensities and/or high levels of background radiation observed with flashlamps require high-resolution detection systems; the intense monochromatic radiation from lasers can often obviate the need for spectrographs and allow simpler detection schemes.

Resonance Scattering

The use of lasers in resonance scattering experiments can be exemplified by the following generalized procedure. Wavelength-tunable laser light is focused into the volume of gas to be studied, and photons are absorbed by one or more particular atomic or molecular excitation transitions. The population densities of the initial and final states participating in the transition are altered, thereby influencing the amount of line radiation emanating from the effected volume of gas. Observations are made of the intensities and profiles of one or more emission lines by external optics focused on the test volume.

A system of the type described above was used by Hoffman [54] and others [55-58] to determine the ion densities in barium plasmas. Hoffman's excitation source was the bright continuum of a mercury capillary arc rather than a laser. Recently, Dimock, Hinnov, and Johnson [59] reviewed the advantages of repeating this experiment using a tunable dye laser. They suggested methods of measuring ion density, ion temperature, electron temperature, and ion drift velocity by monitoring the resonance fluorescence of excited barium atoms at several wavelengths.

Measures [60] has discussed the use of resonance excitation of barium- and potassium-seeded plasmas to measure electron temperature and electron density. Because the basic features of this scheme may be useful in other situations, it is worthwhile to illustrate the process with a simple model. Suppose an upper atomic state L can be populated from two lower levels k and j by resonance absorption. The intensity of fluorescent emission induced by the excitation process k → L is therefore a measure of the state population N_k. A similar experiment pumping the transition j → L yields the number density N_j. The species temperature T would then be given by the Boltzman equation

$$N_k/N_j = (g_k/g_j)\exp - (E_k - E_j)kT$$

where $(E_k - E_j)$ is the energy difference between two states, $N_{k,j}$ are the population densities of the states, and $g_{k,j}$ are the state degeneracies. Successful interpretation of such an experiment requires that the branching ratios are known for all competing radiative de-excitation paths leading from state L to lower levels. Significant collisional de-excitation of L complicates the interpretation of the results.

A useful technique for high-temperature gases which do not possess favorable excitation transitions in the visible-ultraviolet spectrum is to use neutral alkalis or singly ionized alkaline earths as spectroscopic seed atoms. This is a common practice in the sodium line-reversal technique for measuring temperature [61,62]. The two primary difficulties are: (a) the assumption of thermodynamic equilibrium between the seed material and local parent gas; and (b) a knowledge of the de-excitation rate of the optically-excited level arising from collisional processes. The first difficulty is unimportant if one relies on excitation from the ground state and the energy of the first excited level is several kT. The second restriction must be evaluated carefully for each particular set of experimental conditions, and this requires a knowledge of the de-excitation probabilities associated with all relevant local species. Cross-sections for superelastic electron-neutral collisions are more abundant and more easily estimated than those for ion-neutral and neutral-neutral collisions [63-66].

Fluorescence methods are also important for self-luminous gases because the optical signal received by an external detector is integrated over the entire cone of view intercepting the gas. By preferentially exciting one small volume element within this viewing path with a wavelength-tunable laser, it is possible to relate the differential signal to the local properties of the excited volume.

Interpretation of Emission Line Profiles

Interpretation of the line profiles of laser-induced optical emission provides additional information on the local temperature and particle densities of a gas. The spectral line emanating from an atomic electronic transition is broadened by the natural lifetimes of the excited states, the thermal and directed motion of the atoms, and collisional interactions between the radiating particle and other atoms and charged particles. Natural broadening is normally small compared to thermal and collisional effects.

The important information contained in emission line profiles is often unavailable to the experimentor because (a) the gas is dilute and emission intensities are too low to provide accurate measurements, or (b) upper states whose emission lines are sufficiently broad to allow analysis are not populated because of low ambient temperatures. Laser enhancement of excited state populations by resonance absorption can provide large increases in the emission intensity of selected lines.

Wiese [67] has summarized current line-broadening theories which include charged particle-neutral (Stark) collisions. For dilute gases with low ionization, thermal Doppler broadening dominates the line profile and the half-width is easily related to the translational temperature of the gas. However, Doppler half-widths are quite small, on the order of 10^{-4} to 10^{-1} nm in laboratory plasmas with temperatures of a few eV or less. A most successful instrument for measuring this narrow profile is a Fabry-Perot etalon [58,61,68-70]. In cases where the gas motion has a mean velocity component parallel to the direction of observation a measurement of the Doppler shift of the line center will yield the gas velocity. However, this method of velocity determination is quite difficult in practice [61].

In plasmas, radiative transitions originating from or terminating in states far removed from the ground state are subject to significant Stark broadening from neutral-charged particle collisions. Both electrons and ions contribute to Stark broadening, but in essentially different regions of the profile [67,71]. Stark broadening often dominates the emission line profile or can be extracted from a combination of Doppler and Stark effects. Using tabulated theoretical values for Stark broadening [71] it is possible to determine the electron density from the emission line half-width. Only a rough estimate of the electron temperature is required to make this calculation. The accuracy of this plasma density measurement is no greater than the precision of the theoretical values of the Stark parameters, and hence the minimum uncertainty ranges from around 20% for some hydrogenic lines to more than a factor of two for more complicated atoms.

Molecular Excitation

Fluorescence techniques in molecular gases are considerably more difficult to interpret because of the large number of states which can be excited by the light source. The larger the spectral width of the exciting source, the greater is the number of transitions which are excited. Laser monochromaticity and high intensity have mad possible a number of molecular fluorescence studies of diverse types [72-79].

There are several common molecules which have electronic absorption bands in the near ultraviolet which are accessible to frequency-doubled dye lasers. The NO-γ band in the vicinity of 220nm has been well documented [80] and calculation schemes are available for computing the absorption coefficient as a function of gas temperature [81,82]. Typical calculations of LTE line absorption coefficients are given by Golden [83].

Provided that the natural lifetimes of the excited electronic states are short compared to the time for collisional depopulation, the ratio of incident to fluorescent light intensity represents a direct local measure of the population density in the particular NO vibrational state excited by the incident radiation. The influence of collisions on the emitted spectrum can be quite dramatic, and severe quenching of fluorescence can occur. Various collisional deactivation processes for electronically-excited molecules have been discussed by Stevens and Boudart [84].

A similar experiment appears feasible in O_2 which is highly excited in vibration. The Schumann-Runge transitions comprise a well-known system with relatively high absorption. The energy difference between the ground vibrational levels of the two participating states is 6.120eV, or slightly more than the electronic ground-state dissociation energy of 5.080eV. Other promising molecules include CO, NO_2, and O_3.

Resonance absorption of infrared radiation by vibrational-rotational bands may be useful in atmospheric pollution measurements. The large increase in absorption cross section near a resonance peak allows very dilute concentrations to be detected using fluorescent emission. Range-gated detectors can define the volume of gas along the laser path within which the scattering occurs.

Hinkley [85] has applied a single-mode tunable diode laser to high-resolution molecular absorption studies. The $Pb_xSn_{1-x}Te$ diode [14] was tuned to the 10.6μ absorption band of SF_6 and very high resolution spectra were obtained. Practical applications of this method are numerous, but are currently impeded by the difficulty of diode manufacture and the low temperatures required for CW operation. $Pb_xSn_{1-x}Te$ operates CW in a liquid helium dewar.

In molecular fluorescence experiments, there may be great difficulty in assigning accurate values to the branching ratios for competing radiative de-excitation paths. A similar problem in the electron-beam excitation of nitrogen and argon was overcome by empirical calibration [86-88] and this is a distinct possibility for laser-induced fluorescence as well. One such study on NO_2 is currently in progress [89].

Particle Tracking Measurements

Velocity measurements in flowing gases can also be accomplished using laser fluorescence techniques. At some initial time the laser is focused into the gas and a small volume is excited to upper electronic states from which it subsequently decays. Provided the characteristic spontaneous emission lifetime τ is long compared to the time required for the excited volume to move downstream a distance equal to its own diameter, the fluorescence can be tracked in time and the velocity of the gas determined. Streak photographs or an image-converter camera may be used as tracking detectors. This promising technique was discussed in an earlier publication [90], where it was shown that selective excitation provides high visibility even for self-luminous gases.

Strong Interaction Experiments

Gas-laser interactions involving strong coupling are not limited to small perturbations of the test medium. Focused laser beams of high power can produce plasmas which approach thermonuclear conditions. And strong shock waves can be generated by intense laser heating. These two examples of strong coupling do not require an energy resonance between the photons of the radiation field and the quantum states of the gas atoms or molecules.

A large class of problems exist in which resonant excitation of a gas with tunable lasers can produce signficant changes in the test medium's macroscopic properties. Oettinger and Dewey [91] have investigated the conditions under which a collision-dominated, partially-ionized gas can be excited by tunable laser radiation. Nonequilibrium population densities can be produced, and ionization enhancement by factors exceeding two is possible. A brief description of this process is given below.

In collision-dominated plasmas, recombining free electrons and ions form highly-excited neutral atoms which are subsequently de-excited by superelastic collisions with free electrons. Transitions between the upper electronic levels are frequent, because of the small energy difference between adjacent states, and the upper levels are nearly in Saha-Boltzmann equilibrium with the free electrons at the electron temperature. Because of the large number densities in the lower energy states, the first few electronic levels are in Boltzmann equilibrium with the ground state, but not necessarily with the upper levels. The details of this collisional-radiative recombination theory are well-documented in the literature [64,92-94].

The theory predicts that a collision-dominated plasma excited to nonequilibrium conditions consists of two distinct "groups" of atoms, the first containing highly-excited atoms in equilibrium with the free electrons and the second containing atoms in the lower electronic levels which are in equilibrium with the ground state. Therefore, the rate of recombination equals the net flux of atoms from the higher-energy group to the lower-energy group, according to the relation

$$\frac{dn_e}{dt} = -\alpha [n_e - (n_e)_{equilibrium}]$$

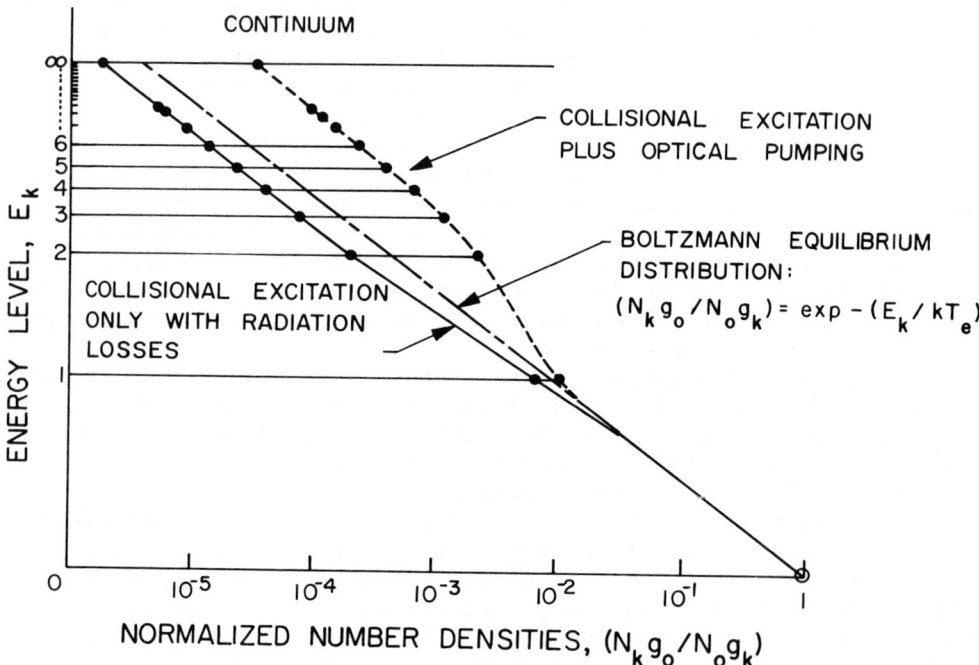

Figure 11. Schematic illustration of the effect of laser excitation on excited state populations.

where α is proportional to n_e. Quantum levels which lie between these two groups are called the recombination "bottleneck", because transitions across these levels are the rate-limiting step in recombination.

An optically-induced electronic transition which crosses the bottleneck from below represents an event which is the inverse of a recombination collision. Under continuous optical pumping, a new quasi-equilibrium is established in which the new collisional recombination rate is equal to the sum of the upward flux of atoms arising from thermal collisions plus the upward flux arising from optical excitation. A laser source tuned to a transition crossing the bottleneck will therefore produce an enhancement of the ionization level of the gas. This is illustrated in Fig. 11.

Sample calculations [91] indicate that a factor of two increase in the ionization level of several laboratory plasmas can be achieved using available wavelength-tunable laser schemes. An experiment to verify this prediction in a cesium plasma is in progress [95].

Tunable laser radiation of high intensity can be used to create nonequilibrium populations in gases. In addition to the obvious possibility of producing additional types of lasers, gas samples can be prepared which contain a large fraction of the molecules in a particular excited quantum level. This would allow the study of detailed chemical and collisional processes under conditions in which the energy levels of the participating molecules are known exactly. McIlrath [96] has conducted an important experiment of this type in gaseous calcium. The second-harmonic of a Nd laser was used to pump 3,3'-diethyloxadicarbocyanine iodide and a diffraction grating was used to tune to the 657nm transition in calcium. The existence of significant populations in upper electronic levels which were produced by laser pumping were confirmed by absorption spectra taken subsequent to the laser pulse. Collisional population of many upper levels was observed. These results hold much promise for studying astrophysical problems.

ACKNOWLEDGMENTS

This investigation was supported by the Advanced Research Projects Agency under contract N00014-67-A-0204-0040. I wish to thank G. von Gierke, J.L. Hall and R.T. Hodgson for their constructive comments.

REFERENCES

1. H.F. Ivey, IEEE J. Quantum Electronics, QE-2, 713 (1966).
2. F.F. Morehead, Jr., Scientific American, 216, 109-122 (1967).
3. H. Kressel and H. Nelson, IEEE J. Quant. Elect., QE-5, 18 (1969).
4. H. Nelson and H. Kressel, Appl. Phys. Lett., 15, 7 (1969).
5. H. Kressel and F.Z. Hawrylo, Appl. Phys. Lett., 17, 169 (1970).

6. These data are taken from specifications for RCA Developmental Type TA 2628 for a single emitting element plus corroborating information from the same laboratory appearing in the literature (see M.F. Lamorte et al., IEEE J. Quant. Elect., QE-2, 9 (1966)). These half-widths are considerably less than those reported by other investigators at equivalent temperatures. (B.A. Lengyel, Introduction to Laser Physics, Wiley, N.Y., 1966, p. 145.) Diode arrays consisting of a number of independent lasing elements in series have considerably larger half-widths, on the order of 4-5nm. Presumably proper quality control can reduce the array bandwidth to that of a single element.

7. I. Hayashi, M.B. Parish, P.W. Foy, and S. Sumski, Appl. Phys. Lett., 17, 109 (1970).

8. M. Garfinkel and W.E. Engler, IEEE Solid State Devices Research Conference, Michigan State U., 1963.

9. M. Lax, Science, 141, 1247 (1963).

10. C. Hurwitz, Appl. Phys. Lett., 8, 243 (1966).

11. C. Hurwitz, Appl. Phys. Lett., 8, 121 (1966).

12. F.H. Nicol, Appl. Phys. Lett., 9, 13 (1966).

13. T. Gonda, H. Junker, and M.F. Lamorte, IEEE J. Quant. Elect., 1, 159 (1965).

14. E.D. Hinkley and C. Freed, Phys. Rev. Lett., 23, 277 (1969).

15. C. Hurwitz, Appl. Phys. Lett., 9, 116 (1966).

16. J.A. Giordmaine, Physics Today, January, 1969, pp 39-44.

17. N. Blombergen, Nonlinear Optics, W.B. Benjamin, New York, 1964.

18. G.C. Baldwin, An Introduction to Nonlinear Optics, Plenum Press, New York, 1969.

19. C.C. Wang and C.W. Racette, J. Appl. Phys., 36, 3281 (1965).

20. R.W. Terhune, P.D. Maker and C.M. Savage, Appl. Phys. Lett., 2, 54 (1963).

21. F.M. Johnson, private communication to Dr. J. Hall, Joint Institute for Laboratory Astrophysics, Boulder, Colorado. See also F.M. Johnson, Electronics, April 18, 1966, pp 82-87; and Nature, 204, 985 (1964).

22. J.E. Geusic, H.J. Levinstein, J.J. Rubin, S. Singh and L.G. Van Uitert, Appl. Phys. Lett., 11, 269 (1967) and 12, 306 (1968).

23. S.K. Kurtz and T.T. Perry, J. Appl. Phys., 39, 3798 (1968).

24. M. Bass, D. Bua, R. Mozzi and R. Monchamp, Appl. Phys. Lett., 15, 393 (1969).

25. K.F. Hulme, O. Jones, P.H. Davies and M.V. Hobden, Appl. Phys. Lett., 10, 133 (1967).
26. D.M. Boggett and A.F. Gibson, Phys. Lett., 26A, 33 (1968).
27. J. Warner, Appl. Phys. Lett., 12, 222 (1968).
28. W.B. Gandrud, G.D. Boyd, J.H. McFee and F.H. Wehmeier, Appl. Phys. Lett., 16, 59 (1970).
29. M.P. Vanyukov, V.D. Volosov, and M.I. Rashchektaeva, Optics and Spectros., 25, 410 (1968).
30. R.G. Smith, invited paper presented at the 1969 IEEE Conference on Laser Engineering and Applications, Washington, D.C., May 26-28, 1969.
31. J.A. Giordmaine and R.C. Miller, Phys. Rev. Lett., 14, 973 (1965). See also Physics of Quantum Electronics, P.L. Kelley et al. (Eds.), McGraw-Hill, New York, 1966, pp 31-42.
32. J.A. Giordmaine and R.C. Miller, Appl. Phys. Lett., 9, 298 (1966).
33. G.D. Boyd and A. Ashkin, Phys. Rev., 146, 187 (1966).
34. R.C. Miller and W.A. Nordland, Appl. Phys. Lett., 10, 53 (1967).
35. J.D. Bjorkholm, Appl. Phys. Lett., 13, 53 (1968).
36. A.J. Kovrigen, P.V. Nikles, A.G. Piskarskas and A.J. Kholodnykh, Moscow State University, IV USSR Symposium on Nonlinear Optics, Kiev, Oct. 25-31, 1968.
37. J. Falk and J.E. Murray, Stanford Microwave Laboratory Report No. 1726, February 1969. See also J.E. Murray and S.E. Harris, J. Appl. Phys., 41, 609 (1970).
38. E.O. Ammann, J.M. Yarborough, M.K. Oshman, and P.C. Montgomery, Appl. Phys. Lett., 16, 309 (1970).
39. G. Eckhardt, IEEE J. Quantum Elect., QE-2, 1 (1966).
40. S. Yoshikawa and Y. Matsumura, Appl. Phys. Lett., 8, 27 (1966).
41. D.W. Faries, K.A. Gehring, P.L. Richards and Y.R. Shen, Phys. Rev., 180, 363 (1969).
42. N. Van Tran and C.K.N. Patel, Phys. Rev. Lett., 22, 463 (1969).
43. F. Brown, IEEE J. Quant. Elect., QE-5, 586 (1969).
44. D.C. Laine, Nature, 191, 795 (1961).
45. F. Zernike and P.R. Berman, Phys. Rev. Lett., 15, 999 (1965).
46. F. Zernike, Phys. Rev. Lett., 22, 931 (1969).
47. T. Yajima and K. Inoue, Phys. Lett., 26A, 281 (1968).
48. N. Van Tran and C.K.N. Patel, Phys. Rev. Lett., 22, 463 (1969).

49. C.F. Dewey, Jr. and L.O. Hocker (to be published).
50. Dr. F.N. Mastrup, TRW Systems Inc., private communication. See also: R. Goldstein and F.N. Mastrup, J. Opt. Soc. Am., 56, 765 (1966), and specifications of TRW Model 27A light source.
51. J. Ramirez-Muñoz, Atomic-Absorption Spectroscopy, Elsevier, New York, 1968, Ch. 7 and Ch. 9, Sect. 3.
52. F.A. Morse and F. Kaufman, J. Chem. Phys., 42, 1785 (1965).
53. D. Davis and W. Braun, Applied Optics, 7, 2071 (1968).
54. F.W. Hoffman, Phys. Fluids, 7, 532 (1964).
55. N. Rynn, E. Hinnov and L.C. Johnson, Phys. Fluids, 8, 1368 (1965).
56. M. Hashmi, A.J. Vander Houven van Oordt and J.G. Wegrove, in Proc. Conf. on Physics of Quiescent Plasmas, Vol. II, Laboratori Gas Ionizzati, Frascati, 1967, pp 523-530.
57. R.O. Motz, I.C. Rogers, and A.D. Bates, in Proc. Conf. on Physics of Quiescent Plasmas, Vol. II, Laboratori Gas Ionizzati, Frascati, 1967, pp 531-542.
58. E. Hinnov et al., Phys. Fluids, 6, 1779 (1963).
59. D. Dimock, E. Hinnov, and L.C. Johnson, Phys. Fluids 12, 1730 (1969).
60. R.M. Measures, J. Appl. Phys., 39, 5232 (1968).
61. F.P. Bundy, H.M. Strong and A.B. Gregg, J. Appl. Phys., 22, 1069 (1951).
62. S.S. Penner, in Temperature, Its Measurement and Control in Science and Industry, Vol. 3, Part I, Reinhold Publ. Corp., New York, 1962, pp 565-567.
63. M.J. Seaton, in Atomic and Molecular Processes (D.R. Bates, Ed.), Academic Press, New York, 1962, Ch. 11.
64. E.W. McDaniel, Collision Phenomena in Ionized Gases, Wiley, New York, 1964.
65. S.C. Brown, Basic Data of Plasma Physics, 2nd Ed., M.I.T. Press, Cambridge, 1966.
66. A.C.G. Mitchell and M.W. Zemansky, Resonance Radiation and Excited Atoms, 2nd Ed., Cambridge U. Press, New York, 1961.
67. W.L. Wiese, in Plasma Diagnostic Techniques, R.L. Huddlestone and S.L. Leonard (Eds.), Academic Press, New York, 1965, pp 265-317.
68. E.P. Muntz, Phys. Fluids, 11, 64 (1968).
69. J.R. Grieg and J. Cooper, Applied Optics, 7, 2166 (1968).

70. R.L. Huddlestone and S.L. Leonard (Eds.), Plasma Diagnostic Techniques, Academic Press, New York, 1965. See esp. Chs. 5,6,8,9, and 10.

71. H.R. Greim, Plasma Spectroscopy, McGraw-Hill, New York, 1964. For Stark broadening parameters, see Table 5.

72. W.B. Tiffany, H.W. Moss, and A.L. Schawlow, Science, 157, 40, (1967).

73. J.R. Novak and M.W. Windsor, J. Chem. Phys., 47, 3075 (1967).

74. K.G.P. Sulzmann, F. Bien and S.S. Penner, J. Quant. Spectros. and Rad. Trans., 7, 969 (1967).

75. A.M. Ronn, J. Chem. Phys., 48, 511 (1968).

76. J.T. Yardley and C.B. Moore, J. Chem. Phys., 48, 14 (1968).

77. W.J. Tango, J.K. Link and R.N. Zare, J. Chem. Phys., 49, 4264 (1968).

78. D.T. Phillips, Bull. Am. Phys. Soc., 13, 1684 (1968).

79. S. Ezekiel and R. Weiss, Phys. Rev. Lett., 20, 91 (1968).

80. M. Jeunehomme, J. Chem. Phys., 45, 4433 (1966).

81. J.C. Keck, R.A. Allen and R.L. Taylor, J. Quant. Spectros. Rad. Trans., 3, 335 (1963).

82. R.A. Allen, AVCO Research Report 236, April, 1966.

83. S.A. Golden, J. Quant. Spectros. Rad. Trans., 7, 225 (1967).

84. B. Stevens and M. Boudart, Ann. N.Y. Acad. Sci., 67, 570 (1957).

85. E.D. Hinkley, Appl. Phys. Lett., 16, 351 (1970).

86. E.P. Muntz, Phys. Fluids, 5, 80 (1962).

87. S.L. Petrie, The Ohio State University, ARL Report 65-122, 1965.

88. F. Robben and L. Talbot, Institute of Engineering Research Report AS-65-4, University of California, Berkeley, 1965.

89. J.R. Golin and C.F. Dewey, Jr. (to be published).

90. C.F. Dewey, Jr., Air Force Flight Dynamics Laboratory Report AFFDL-TR-68-170 (1968).

91. P.E. Oettinger and C.F. Dewey, Jr., AIAA Journal, 8, 880 (1970).

92. S. Byron, R.C. Stabler and P.I. Bortz, Phys. Rev. Lett., 8, 376 (1962).

93. D.R. Bates, A.E. Kingston and R.P. McWhirter, Proc. Royal Soc. (London), A267, 297 (1962).

94. Yu. M. Aleskovskii, Soviet Physics - JETP, 17, 570 (1963).

95. L.J. Kelly, J.R. Carpenter and C.F. Dewey, Jr. (to be published).

96. T.J. McIlrath, Appl. Phys. Lett., 15, 41 (1969).

BIBLIOGRAPHY OF DYE LASER PAPERS

1. Abakumov, G.A., A.P. Simonov, V.V. Fadeer, L.A. Kharitonov, and R.V. Khokhlov, "Ultraviolet lasers using organic-scintillator molecules", JETP Letters, 9, 9 (1969).

2. Bass, M., T.F. Deutsch, and M.J. Weber, "Frequency and time-dependent gain characteristics of laser and flashlamp-pumped dye solution lasers", Appl. Phys. Lett., 13, 120 (1968).

3. Bass, M., and J.I. Steinfeld, "Wavelength dependent time development of the intensity of dye solution lasers", IEEE J. Quantum Electronics, QE-4, 53 (1968).

4. Baxse, K., et al., "Continuous operation of a dye laser", Proc. Int. Quantum Elect. Conf., Japan, Sept., 1970 (post-deadline paper).

5. Bonch-Bruyevich, A.M., N.N. Kostin, and V.A. Khodovoi, "Selection and adjustment of generated frequencies in dye solutions", Opt. Spectry., 24, 547 (1968).

6. Bowman, M.R., A.J. Gibson, and M.C.W. Sandford, "Atmospheric Sodium measured by a tuned laser radar", Nature, 221, 456 (1969).

7. Bradley, D.J., "Generation of ultra-short laser pulses", Laboratory Practice 18, 538 (1969).

8. Bradley, D.J., A.J.F. Durant, G.M. Gale, M. Moore, and P.D. Smith, "Characteristics of organic dye lasers as tunable frequency sources for nanosecond spectroscopy", IEEE J. Quant. Elect., QE-4, 707 (1968).

9. Bradley, D.J., A.J.F. Durant, and F. O'Neill, "Generation and application of ultra-short pulses from dye lasers", IEEE J. Quant. Elect., QE-5, 16 (1969).

10. Bradley, D.J., G.M. Gale, M. Moore, and P.D. Smith, "Longitudinally pumped, narrow-band continuously tunable dye laser", Phys. Lett., 26A, 378 (1968).

11. Brock, E.G., P. Czavinsky, E. Hormats, H.C. Nedderman, D. Stirpe, and F. Unterleitner, "Coherent stimulated emission from molecular crystals", J. Chem Phys., 35, 759 (1961).

12. Broida, H.P., and S.C. Haydon, "Ultraviolet laser emission of organic liquid scintillators using a pulsed nitrogen laser", Appl. Phys. Lett., 16, 142 (1970).

13. Buettner, A.V., B.B. Snavely, and O.G. Peterson, "Triplet state quenching of stimulated emission from organic dye solutions", *Proc. Internatl. Conf. on Molecular Luminescence*. New York: Benjamin, 1969, pp 403-422.

14. Claesson, S., and L. Lindqvist, "A fast photolysis flashlamp for very high light intensities", *Arkiv Kemi*, 12, 1 (1958).

15. Derkacheva, L.L., and A.I. Krymova, "Stimulated emission of solutions of cyanine dyes", *Sov. Phys. Dokl.*, 13, 53 (1968).

16. Derkacheva, L.V., A.I. Krymova, V.I. Malyshov, and A.S. Martin, "Mode locking in polynethene dye lasers", *Opt. Spectry.*, 26, 572 (1969).

17. Derkacheva, L.L., A.I. Krymova, A.F. Vompe, and I.I. Leukeov, "Stimulated luminescence of dyes in the 720-920 nm region", *Opt. Spectry.*, 25, 404 (1968).

18. Deutsch, T.F., and M. Bass, "Laser-pumped dye lasers near 4000Å", *IEEE J. Quant. Elect.*, QE-4, 260 (1969).

19. Deutsch, T.F., M. Bass, P. Meyer, and S. Protopapa, "Emission spectrum of Rhodamine B dye lasers", *Appl. Phys. Lett.*, 11, 379 (1967).

20. Dienes, A., C.V. Shank, and A.M. Trozzolo, "Evidence for excitiplex laser action in commarin dyes by measurements of stimulated fluorescence", *Appl. Phys. Lett.*, 17, 189 (1970).

21. Eastman Kodak Company, Organic Chem. Div., 343 State Street, Rochester, New York, 14650.

22. Farmer, G.I., B.G. Huth, L.M. Taylor, and M.R. Kagan, "Time resolved stimulated emission spectra of an organic dye laser", *Appl. Phys. Lett.*, 12, 136 (1968).

23. Farmer, G.I., B.G. Huth, L.M. Taylor, and M.R. Kagan, "Concentration and dye length dependence of organic dye laser spectra", *Appl. Opt.*, 8, 363 (1969).

24. Fawcett, B.C., "Experimental comparison of tuned organic lasers", *IEEE J. Quant. Elect.*, QE-6, 473 (1970).

25. Ferrar, C.M., "Mode-locked flashlamp-pumped coumarin dye laser at 4600Å", *IEEE J. Quant. Elect.*, QE-5, 550 (1969).

26. Ferrar, C.M., "Wavelength variations of a flashlamp-pumped sodium fluorescein dye laser", *IEEE J. Quant. Elect.*, QE-5, 621 (1969).

27. Fischer Scientific Company, 461 Riverside Ave., Medford, Mass. 02155.

28. Furamoto, H.W., and H.L. Ceccon, "Flashlamp excited ultraviolet organic liquid lasers", *Bull. A.P.S.*, 15, 507 (1970).

29. Furamoto, H.W., and H.L. Ceccon, "A high-performance flashlamp for organic dye lasers", IEEE J. Quant. Elect., QE-5, 17 (1969).

30. Furamoto, H.W., and H.L. Ceccon, "Time-dependent inhomogeneities in coaxial lamp dye lasers", Bull. A.P.S., 15, 505 (1970).

31. Furamoto, H.W., and H.L. Ceccon, "Ultraviolet organic liquid lasers", Bull. A.P.S., 15, 505 (1970).

32. Furamoto, H.W., and H.L. Ceccon, "Time-dependent spectroscopy of flashlamp pumped dye lasers", Appl. Phys. Lett., 13, 335 (1968).

33. Gallard-Schlesinger, Chemical Company, 584 Mineola Ave., Carle Pl., Long Island, N.Y. 11514.

34. Gibbs, W.E.K., and H.A. Kellock, "Time-resolved spectroscopy of organic dye lasers", IEEE J. Quantum Elect., QE-4, 293 (1968).

35. Gibson, A.J., "A flashlamp-pumped dye laser for resonance scattering studies of the upper atmosphere", J. Sci. Inst. (J. Phys. E), 2, (1969).

36. Glenn, W.H., M.J. Brienza, and A.J. de Maria, "Mode locking of an organic dye laser", Appl. Phys. Lett., 12, 54 (1968).

37. Gregg, D.W., and S.J. Thomas, "New lasing organic dyes flash-lamp-pumped", IEEE J. Quant. Elect., QE-5, 302 (1969).

38. Gregg, D.W., et al., "Wavelength tunability of new flashlamp-pumped laser dyes", IEEE J. Quant. Elect., QE-6, 270 (1970).

39. Heller, A., "Liquid laser-design of neodymium-based inorganic ionic systems", J. Mol. Spectry., 28, 101 (1968).

40. Heller, A., "Liquid lasers-fluorescence, absorption and energy transfer of rare earth ion solutions in selenium oxychloride", J. Mol. Spectry., 28, 208 (1968).

41. Huth, B.G., and G.I. Farmer, "Laser action in 9,10 diphenyl-anthracene", IEEE J. Quant. Elect., QE-4, 427 (1968).

42. Kagan, M.R., G.I. Farmer, and B.G. Huth, "Organic dye lasers", Laser Focus, September, 1968, pp 26-33.

43. Keller, R.A., "Effect of quenching of molecular triplet states in organic dye lasers", IEEE J. Quant. Elect., QE-6, 411 (1970).

44. Koch-Light Laboratories, Ltd., Colnbrook Buckinghamshire, England.

45. Kogelnik, H., C.V. Shank, T.P. Sosnowski, and A. Dienes, "Hologram wavelength selector for dye lasers", Appl. Phys. Lett., 16, 499 (1970).

46. Kotsubanov, V.D., Yu. V. Naboikin, L.A. Ogurtsova, A.P. Podgornyi, and F.S. Pokrovskaya, "Generation of light in solutions of dyes in the 550-650nm range", Opt. Spectry., 25, 159 (1968).

47. Kotsubanov, V.D., Yu. V. Naboikin, L.A. Ogurtsova, A.P. Podgornyi, and F.S. Pokrouskaya, "Laser action in solutions of organic luminophors in the 400-650nm range", Opt. Spectry., 25, 406 (1969).
48. Lankard, J.R., and R.J. Von Gutfield, "Organic lasers excited by a pulsed N_2 laser", IEEE J. Quant. Elect., QE-5, 625 (1969).
49. Lempicki, A., and H. Samelson, "Organic laser systems", Lasers, (A.K. Levine, Ed.), Vol. 1, M. Dekker, N.Y., 1966, pp 181-252.
50. L.K. Lidholt, "Dye laser pumped by an ultraviolet nitrogen laser", IEEE J. Quant. Elect., QE-6, 162 (1970).
51. Mack, M.E., "Measurement of nanosecond fluorescence decay times", J. Appl. Phys., 39, 2483 (1968).
52. Mack, M., "Superradiant traveling wave dye laser", Appl. Phys. Lett., 15, 166 (1969).
53. McCumber, D.E., "Theory of phonon-terminated optical masers", Phys. Rev. (A), 134, 299 (1964).
54. McFarland, B.B., "Laser second-harmonic-induced stimulated emission of organic dyes", Appl. Phys. Lett., 10, 208 (1967).
55. McIlrath, T.J., "Absorption from excited states in laser-pumped calcium", Appl. Phys. Lett., 15, 41 (1969).
56. Measures, R.M., "Selective excitation spectroscopy and some possible applications", J. Appl. Phys., 39, 5232 (1968).
57. Miyazoe, Y., and M. Maeda, "Stimulated emission from 19 polymethene dyes-laser action over the continuous range 710-1060mμ", Appl. Phys. Lett., 12, 206 (1968).
58. Murakawa, S., G. Yamaguchi, and C. Yamanaka, "Wavelength shift of dye solution laser", Japan J. Appl. Phys., 7, 681 (1968).
59. Myer, J.A., C.J. Johnson, E. Kierstead, R.P. Sharma, and I. Itzkan, "Dye laser stimulation with a pulsed N_2 laser line at 3371Å", Appl. Phys. Lett., 16, 3 (1970).
60. Neumann, G., and M. Hercher, "Characteristics of short-cavity dye lasers", IEEE J. Quant. Elect., QE-5, 17 (1969).
61. Pappalardo, R., H. Samelson, and A. Lempicki, "Long pulse laser emission from rhodamine 6G using cyclooctatetrene", Appl. Phys. Lett., 16, 267 (1970).
62. Peterson, O.G., W.C. McColgin, and J.H. Eberly, "Triplet state effects in dye lasers at threshold", Phys. Lett., A26, 399 (1969).
63. Peterson, O.G., and B.B. Snavely, "Stimulated emission from flashlamp excited organic dyes in polymethyl methacrylate" Appl. Phys. Lett., 12, 238 (1968).

64. Peterson, O.G., and B.B. Snavely, "Multiple-dye solution lasers", Bull. Am. Phys. Soc., 13, 397 (1968).
65. Peterson, O.G., S.A. Tuccio, and B.B. Snavely, "CW operation of an organic dye solution laser", Appl. Phys. Lett., 17, 245 (1970).
66. Rautian, S.G., and I.I. Sobelmann, "Remarks on negative absorption", Opt. Spectry., 10, 65 (1961).
67. Samelson, H., "Liquid lasers: promising solutions", Electronics, Nov. 11, 1968, pp 142-147.
68. Schaefer, F.P., "Organic dye lasers", invited paper presented at the 1968 Quantum Electronics Conference, Miami, Fla., May 1968.
69. Schaefer, F.P., W. Schmidt, and K. Marth, "New dye lasers covering the visible spectrum", Phys. Lett., 24A, 280 (1967).
70. Schaefer, F.P., W. Schmidt, and J. Volze, "Organic dye solution laser", Appl. Phys. Lett., 9, 306 (1966).
71. Schappert, G.T., K.W. Billman, and D.C. Burnham, "Temperature tuning of an organic dye laser", Appl. Phys. Lett., 13, 124 (1968).
72. Schmeltekoff, A., National Bureau of Standards, Boulder, Colorado, private communication, 1969.
73. Schmidt, W., and F.P. Schaefer, "Self-mode locking of dye lasers with saturable absorbers", Phys. Lett., 26A, 558 (1968).
74. Schmidt, W., and F.P. Schaefer, "Blitzlampen gepumpte Farbstoff-Laser", Z. Naturforsch., 22A 1563 (1967).
75. Shank, C.V., A. Dienes, A.M. Trozzolo, and J.A. Myer, "Near UV to yellow tunable laser emission from an organic dye", Appl. Phys. Lett., 16, 405 (1970).
76. Snavely, B.B., "Flashlamp-excited organic dye lasers", Proc. IEEE, 57, 1374 (1969).
77. Snavely, B.B., and O.G. Peterson, "Experimental measurement of the critical population inversion for the dye solution laser", IEEE J. Quantum Electronics, QE-4, 540 (1968).
78. Snavely, B.B., O.G. Peterson, and R.F. Reithel, "Blue laser emission from a flashlamp-excited organic dye solution", Appl. Phys. Letters, 11, 275 (1967).
79. Snavely, B.B., and F.P. Schaefer, "Feasibility of CW operation of dye lasers", Phys. Lett. 28A, 728 (1969).
80. Soffer, B.H., and V. Evtuhov, "A quasi-continuous dye laser", IEEE J. Quant. Elect., QE-4, 386 (1969).
81. Soffer, B.H., and J.W. Linn, "Continuously tunable picosecond-pulse organic-dye-laser", J. Appl. Phys., 39, 5859 (1968).

82. Soffer, B.H., and B.B. McFarland, "Continuously tunable narrow-band organic dye lasers", Appl. Phys. Letters, 10, 266 (1967).

83. Sorokin, P.P., "Organic lasers", Sci. Am., Feb., 1969, pp 30-40.

84. Sorokin, P.P., W.H. Culver, E.C. Hammond, and J.R. Lankard, "End-pumped stimulated emission from a thiacarbocyanine dye", IBM J. Res. Develop., 10, 401 (1966).

85. Sorokin, P.P., and L.R. Lankard, "Stimulated emission observed from an organic dye, chloroaluminum phtalocyanine", IBM J. Res. Develop., 10, 162 (1966).

86. Sorokin, P.P., and J.R. Lankard, "Flashlamp excitation of organic dye lasers: a short communication", IBM J. Res. Develop. 11, 148 (1967).

87. Sorokin, P.P., J.R. Lankard, E.C. Hammond, and V.L. Moruzzi, "Laser pumped stimulated emission from organic dyes: experimental studies and analytical comparisons", IBM J. Res. Develop., 11, 130 (1967).

88. Sorokin, P.P., J.R. Lankard, V.L. Moruzzi, and E.C. Hammond, "Flashlamp pumped organic dye lasers", J. Chem. Phys., 48, 4726 (1968).

89. Spaeth, M.L., and D.P. Bortfield, "Stimulated emission from polymethine dyes", Appl. Phys. Lett. 9, 179 (1966).

90. Srinivasan, R., "New materials for flash-pumped organic lasers", IEEE J. Quant. Elect. QE-5, 552 (1969).

91. Stockman, D.L., "Stimulated emission considerations in fluorescent organic molecules", Proc. of the 1964 ONR Conf. on Organic Lasers, available as Doc. No. AD 447468 from the Defense Documentation Center for Scientific and Technical Information, Cameron Station, Alexandria, Va.

92. Stockman, D.L., W.R. Mallory, and K.F. Tittel, "Stimulated emission in aromatic organic compounds", Proc. IEEE, 52, 318 (1964).

93. Turro, N.J., Molecular Photochemistry. New York: Benjamin, 1965, pp 54-55.

94. Varga, P., R.G. Kryukov, V.F. Kuprishov, Yu. V. Senatskii, "Emission of polymethene dye used in neodynium glass lasers", Opt. Spectry.26, 545 (1969).

95. Walther, H., and J.L. Hall, "Tunable dye laser with narrow spectral output", Appl. Phys. Lett., 17, 239 (1970).

96. Weber, M.J., and M. Bass, "Frequency and time dependent gain characteristics of dye lasers", IEEE J. Quant. Elect. QE-5, 175 (1969).

97. Winston, H., and R.A. Gudmundsen, "Refractive index effects in proposed liquid lasers", Appl. Opt., 3, 143 (1964).

98. Yariv, A., and J.P. Gordon, "The laser", Proc. IEEE, 51, 4 (1963).

ADDITIONAL BIBLIOGRAPHY

99. Boiteux, M., and O. de Witte, "A transverse flow repetitive dye laser", Appl. Optics, 9, 514 (1970).

100. Capelle, G., and D. Phillips, "Pumping organic dyes with a nitrogen laser", Appl. Optics, 9, 517 (1970).

101. Clark, J.C., and T.J. Davies, "Stimulated emission from organic dye solutions pumped by a small coaxial N_2 laser", Appl. Optics, 9, 1725 (1970).

102. Marling, J.B., D.W. Gregg, and S.J. Thomas, "Effect of oxygen on flashlamp-pumped organic-dye lasers", IEEE J. Quant. Elect., QE-6, 570 (1970).

103. von Gutfield, R.J., B. Welber, and E.E. Tynan, "Increased laser tunability by acidification of organic dyes", IEEE J. Quant. Elect., QE-6, 532 (1970).

PRODUCTION OF NON-EQUILIBRIUM ATOMIC POPULATIONS BY TUNABLE LASER EXCITATION

T.J. McIlrath

Harvard College Observatory

Cambridge, Massachusetts

This paper describes the use of tunable lasers to produce distributions of atomic or molecular populations far from thermal equilibrium. The technique makes it possible to study either atomic or molecular spectra characterizing such distributions or relaxation rates between levels. The principle is simply to produce a flux of photons sufficient to ensure that the rate of production of atoms in excited levels by absorption of photons is comparable to the rate of decay out of the excited levels. Such a flux equalizes the populations of the ground and excited levels coupled by the light, and the coupling of levels yields a ratio of populations that is independent of the thermal properties of the gas.

We have carried out an optical pumping experiment with calcium atoms.[1] Figure 1 shows some of the energy levels of calcium; the ground configuration, $4s^2$ ^1S, and the lowest excited configuration, $4s4p$ ^3P, are connected by the intercombination transition at 6572 Å. A tunable laser using an organic dye solution was pumped with the second harmonic of a Q-spoiled Nd^{3+}:glass laser and tuned to this wavelength. The tunable laser had an output of >30 mJ in 30 ns, with a line width of 0.5 Å.

Figure 2 is a schematic of the experimental arrangement. Calcium vapor was produced in a furnace containing a helium buffer gas, and the laser was used to pump these atoms. A fast-rise flash lamp having a duration of 4 to 6 μs was fired 2 to 3 μs after the calcium atoms were excited by the laser, and an absorption spectrum of the vapor

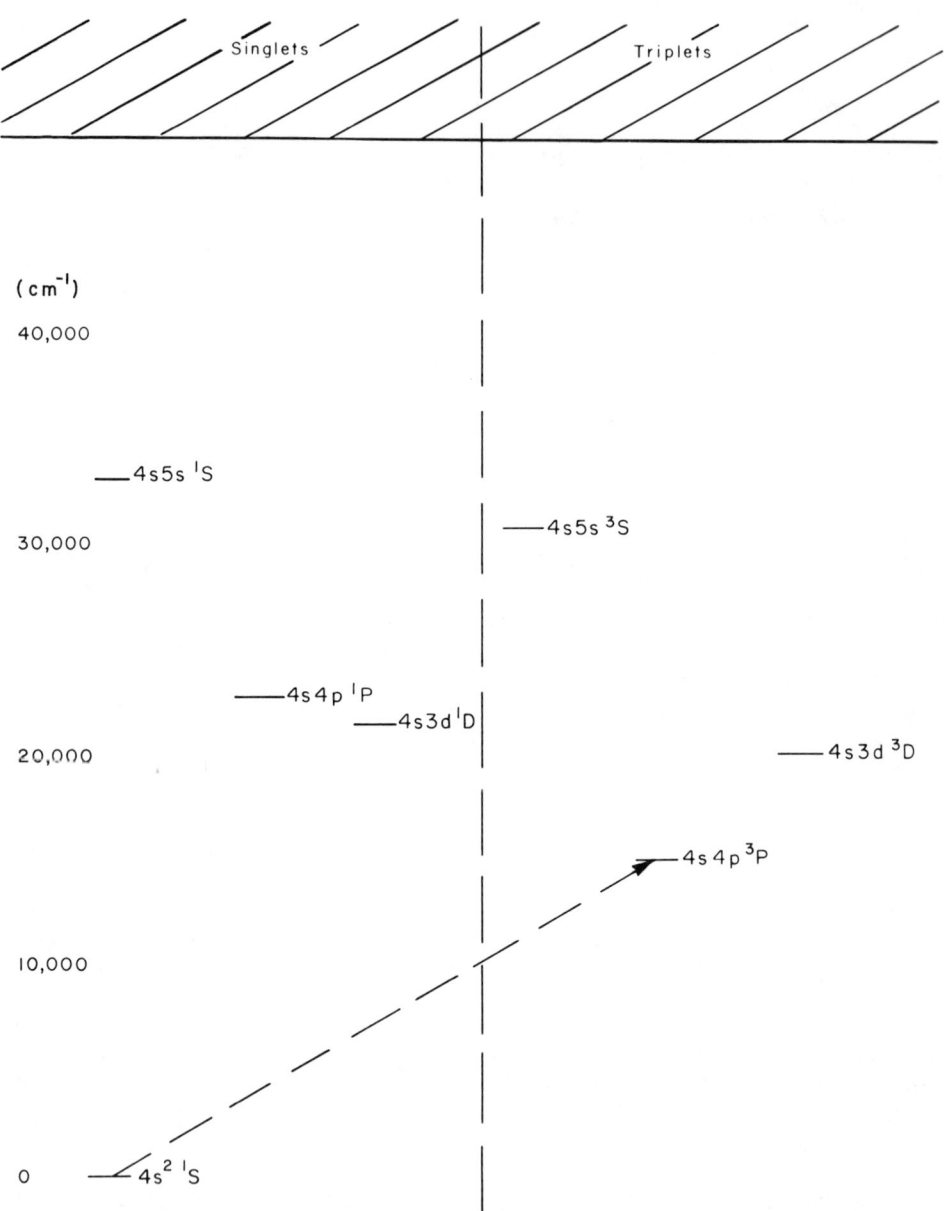

Fig. 1   Partial energy-level diagram of calcium.

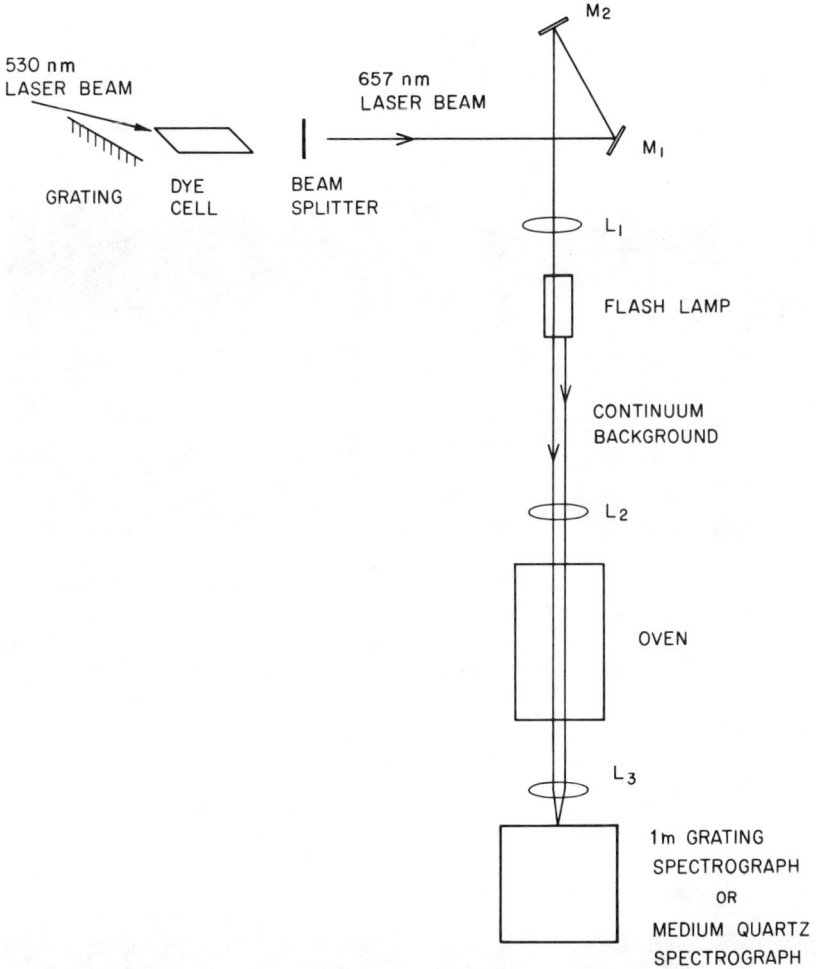

Fig. 2 Arrangement of apparatus for exciting calcium atoms with a tunable laser to obtain excited-state absorption spectra.

was obtained with a one-meter grating spectrograph. This spectrum was compared with one obtained using the same configuration without laser excitation. Figures 3 and 4 show the new lines resulting from the population of the metastable state by laser excitation. We have succeeded in producing between 10^{16} and 10^{17} excited atoms in a cylindrical volume 10 cm long and 0.7 cm across.

It should be emphasized that in obtaining the excited state spectra, the laser was off during and a few micro-

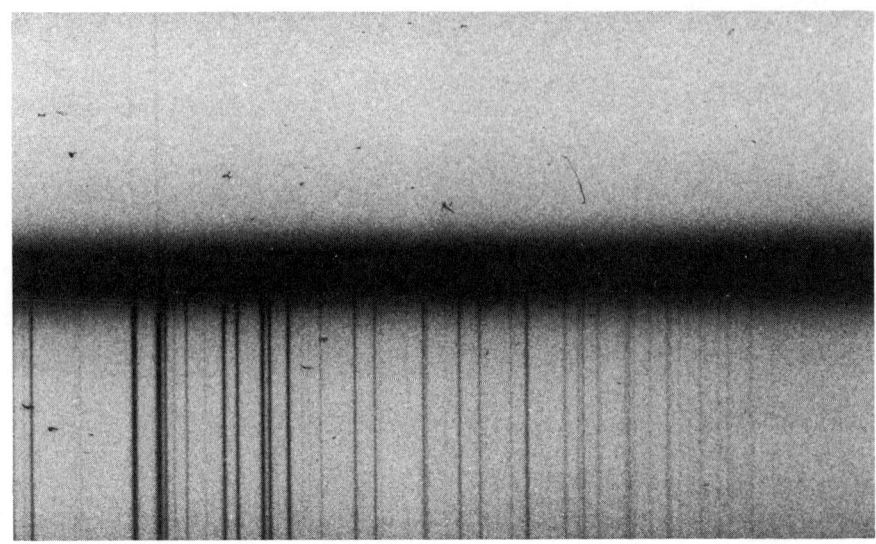

Fig. 3 Absorption spectra of calcium near 3000 Å. (*upper spectrum*) Calcium vapor without laser excitation; (*lower spectrum*) calcium vapor with laser excitation, showing the triplet spectrum out of the 4s4p ^{3}P level.

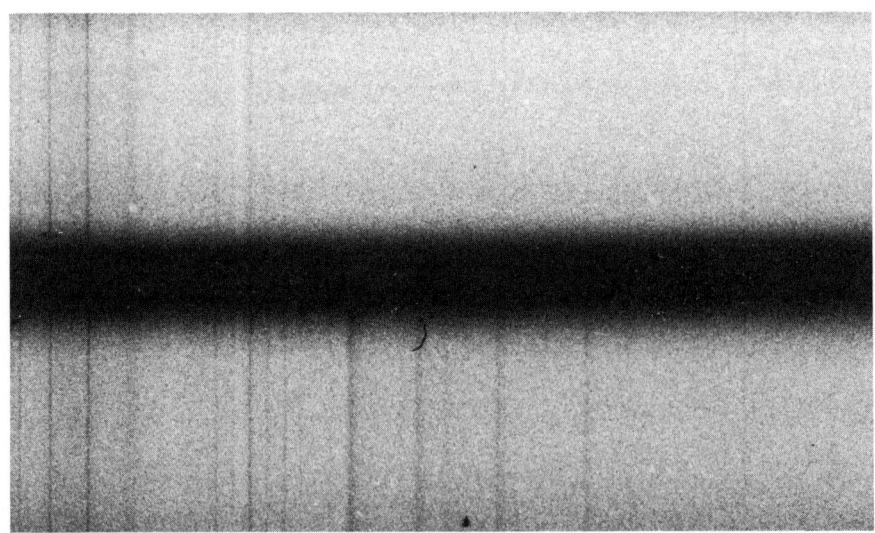

Fig. 4 Absorption spectra of calcium $\lambda\lambda$ 2800-2700. (*upper spectrum*) Calcium vapor without laser excitation; lines are of impurities in furnace and flashlamp. (*lower spectrum*) Calcium vapor with laser excitation, showing transitions from the 4s4p ^{3}P level to the 3d4d ^{3}D and ^{3}S levels. The 3d4d levels are broadened by autoionization.

seconds preceding the firing of the flashlamp used in recording the absorption spectra. With the relaxation rate between fine structure levels $>5 \times 10^9$ s^{-1} at a buffer gas pressure of 500 torr, it is clear that by the time the absorption spectra are taken the method of excitation becomes completely irrelevant. Use of the laser does not affect the absorption spectra in any way except in changing the level populations.

Since only one level of the triplet system, the level with $J = 1$, is populated by the laser, we can measure the fine structure relaxation rate by observing how the singlet absorption spectrum changes into a triplet spectrum with time. We have done this using a 30 ns background source to obtain a lower limit of $\approx 10^8$ s^{-1} for the rate of transitions between fine structure levels with a helium buffer gas pressure of 10 torr. For these measurements we used a broadband dye laser as a background source. Because of uncertainties in the quantitative interpretation of the experiment with a laser background source, we did not pursue the measurement beyond this lower limit.

Let us consider now the effect of the laser and the question of how much light is required to equalize the ground- and excited-state populations. The output of a grating-narrowed dye laser is typically from 0.5 Å to 5 Å wide. It consists of several longitudinal modes with unknown phase and intensity relationships. For an output 1K wide, as in our experiment, there are ≈ 30 modes. One kayser (K) is one cm^{-1} wavenumber and at $\lambda = 6500$ Å corresponds to approximately 0.5 Å. The spacing between modes is comparable to atomic Doppler widths. Because of these features, it is not feasible to precisely calculate the effect of the light on the atoms.

A second complication in estimating the effect of the laser is that the effects of the coherence of the laser radiation must be taken into account. To eliminate cooperative effects between atoms (such as super-radiance) or coherence effects from laser radiation (such as self-induced transparency), we operated at buffer gas pressures of a few hundred torr, where the phase relaxation rate due to collisions is much faster than the pumping rate. In this way, the excited-state atoms could be treated as incoherent with each other and with the laser. By observing the growth of the triplet spectra, we were able to establish a relaxation rate $>10^8$ s^{-1} at 10 torr or $>5 \times 10^9$ s^{-1} at 500 torr, as already mentioned. The widths of the excited-state absorption lines indicate that their relaxation rates are considerably faster. The pumping rate was ≈10^8 s^{-1} with a 30 ns laser pulse. Under these conditions we can get a good approximation to the rate of production of excited states by assuming two-particle atom—photon interactions and by using the photon flux of the laser and the absorption cross section of the atoms to get an excitation rate.

To estimate the pumping rate, let us consider a pulse of I_ν photons cm^{-2} Hz^{-1} in a time τ_L. If the average absorption cross section for an atom in the gas absorbing a photon of frequency ν is σ_ν, then the average number of interactions between an atom and the photon pulse is:

$$P = \int_{-\infty}^{\infty} I_\nu \sigma_\nu d\nu \quad .$$

By an interaction we mean either an absorption or a stimulated emission; therefore P can be greater than 1. P/τ_L is the pumping rate. Assume that I_ν is a constant I_0 over the region for which σ_ν is different from zero. Then,

$$P = I_0 \int_{-\infty}^{\infty} \sigma_\nu d\nu \quad .$$

Because of the laser mode structure, I_ν is clearly not a constant, but at 500 torr the Doppler collision rate is ≈2×10^9 s^{-1}. We can therefore consider the Doppler width as homogeneously broadened on the time scale of the pumping light. This has the same effect as considering the intensity constant over the Doppler width. We can use the relation

$$\int_{-\infty}^{\infty} \sigma_\nu d\nu = \pi r_0 c f$$

to finally obtain

$$P = \pi I_0 r_0 c f$$

where r_0 is the classical radius of the electron, c is the velocity of light, and f is the oscillator strength of the line. To roughly equalize populations, we want $P \geq 1$ with τ_L less than the excited-state decay time. This requires $I_0 \geq (\pi r_0 c f)^{-1}$ photons Hz^{-1}cm^{-2} or, in terms of photons in a one kayser bandwidth, $I_0 \geq 10^{12} f^{-1}$ photons K^{-1}cm^{-2}. For visible light, we need $I_0 \gtrsim 3 \times 10^{-7} f^{-1}$ joules K^{-1}cm^{-2}. A typical laser-pumped dye-laser pulse is 30 ns long, and for a narrowed dye-laser output the line width typically can be ≈0.5 Å wide which for $\lambda = 6500$ Å is ~1K wide. The requirement on the power in the pulse would then be $P_0 \gtrsim 10\ f^{-1}$ watts cm^{-2}, provided the excited-state lifetime τ_A is longer than the laser pulse time τ_L. If τ_L is not less than τ_A, then we get the power by requiring $P\,(\tau_A/\tau_L) \geq 1$. For the calcium intercombination line, τ_A is ≈ 50 μs and $f \approx 10^{-4}$. This leads to a requirement of ≈ 3 mJ cm^{-2} in the 0.5 Å linewidth output of the laser to equalize the ground and excited-state populations. In our experiment, we achieved a maximum output of 30 mJ cm^{-2}. Since we were able to observe a significant decrease in absorption out of the ground state after firing the laser, the indication is that we were indeed moving a significant percentage of the atoms into the excited state.

Over what range can tunable lasers be used? The region from 3500 Å to 10600 Å has been completely covered by tunable lasers;[2] use of second harmonic generation permits the region down to 2500 Å to be covered also. Typical line widths of dye-laser outputs are from 10 Å in the ultraviolet up to 100 Å in the infrared. By using various techniques to narrow the output, one can obtain line widths of less than 0.01 Å.[3] More than one megawatt of power can be achieved from the near ultraviolet to the infrared. Pulse widths can be as short as 10^{-11} s in mode-locked systems[4] or as long as 0.1 ms in flash-lamp pumped systems. Continuous-wave dye lasers have recently been reported.[5] The most efficient and straightforward systems have pulse widths in the range from 2 to 30 ns. The efficiency of dye lasers is as high as 50 percent with laser excitation, and as high as 0.1 percent with flash-lamp excitation. Beam divergences of better than 1 mrad can be achieved.

In summary, probes (tunable lasers) now exist that are strong enough to dominate over natural collisions and radiation processes. These lasers are monochromatic and can operate on a well-defined pair of levels. They allow one to selectively alter the populations in a gas on a very short time scale and in a gross manner. Tunable laser techniques will clearly be of immense value to the study of atomic and molecular physics and the analysis of dynamic systems.

The author gratefully acknowledges the assistance of Dr. W.H. Parkinson through helpful discussions and encouragement. This work was supported by the National Aeronautics and Space Administration under grant NGL 22-007-006.

NOTES

1. T.J. McIlrath, Appl. Phys. Lett., 15, 41 (1969)

2. See, for example, Y. Miyazoe and M. Maeda, Appl. Phys. Lett., 12, 206 (1968); D.W. Gregg, M.R. Querry, J.B. Marling, S.J. Thomas, C.V. Dobler, N.J. Davies, and J.F. Belew, IEEE J. Quantum Electr. QE-6, 270 (1970); H.W. Furumoto and H.L. Ceccon, IEEE J. Quantum Electr. QE-6, 262 (1970); J.A. Myer, I. Itzkan, and E. Kierstead, Nature, 225, 544 (1970).

3. D.J. Bradley, G.M. Gale, M. Moore, and P.D. Smith, Phys. Lett., 26A, 378 (1968).

5. O.G. Peterson, S.A. Tuccio, and B.B. Snavely, Appl. Phys. Lett., 17, 245 (1970).

RECENT STUDIES OF THE CS_2/O_2 COMBUSTION LASER

G.H. Kimbell, T.V. Jacobson, R.D. Suart,

S.J. Arnold and K.D. Foster

Defence Research Establishment Valcartier

P.O. Box 880, Courcelette, P.Q., CANADA

ABSTRACT

The addition of gases such as N_2, N_2O and He were found to have a dramatic effect in changing the performance of both an electrically sparked and a CW laser. Both the output power and spectroscopic output were monitored in the presence of given amounts of diluent gas as a function of fuel/oxygen ratio. While the pumping reaction in the pulsed operation is felt to be $O + CS \rightarrow CO + S$ no unequivocal evidence can be found to rule out the competing reaction $CS + O_2 \rightarrow CS + SO$ in the CW flow laser.

INTRODUCTION

Since the pioneering work of Pimentel (1), chemical laser systems containing halides have received the greatest attention. Of the non-halide reactions known to produce laser action, the CS_2/O_2 system is receiving more and more attention. In 1966, Pollack (2) and in 1968 Gregg and Thomas (3) reported the observation of stimulated emission from vibrationally excited CO produced in flash photolyzed CS_2/O_2 mixtures. Pulsed laser oscillation from CO produced in spark-ignited mixtures was later reported by this laboratory (4). By observing the infrared chemiluminescence of CO in the first overtone region, Hancock and Smith (5) were able to observe a complete population inversion of the vibrational levels of CO when O atoms reacted with CS_2 in a flow system. Shortly thereafter, quasi CW oscillation (6) and CW oscillation (7,8) of CO was realized in flow systems of O atoms and CS_2. Experiments by Foster and Kimbell (9) have since shown that under certain conditions population inversions in free burning CS_2/O_2 flames may be achieved.

The experiments of Hancock and Smith provide evidence that one of the steps responsible for the pumping of the upper vibrational levels of CO, is

$$O + CS = CO + S + 75 \text{ Kcal} \qquad (1)$$

They suggested that as much as 50% of the excess energy of this reaction may appear as vibration in the newly formed C-O bond.

In the case of the flash-initiated reactions, Hancock and Smith have suggested the following mechanism

$$CS_2 + h\nu = CS + S \qquad (2)$$

$$S + O_2 = SO + O \qquad (3)$$

followed by the pumping step (1). Spark excitation, on the other hand, can form atomic O and CS radicals directly by electron impact fragmentation of the parent molecules, (i.e.)

$$CS_2 = CS + S \qquad (4)$$

$$O_2 = 2 O \qquad (5)$$

again followed by reaction (1).

For CW experiments, oxygen has been predissociated in a discharge and added to CS_2 in a flow system. Thus the initial step must be

$$O + CS_2 = CS + SO \qquad (6)$$

again followed by reaction (1)

It has also been suggested that reaction (7)

$$CS + O_2 = CO + SO + 81 \text{ Kcal} \qquad (7)$$

may contribute to the pumping of carbon monoxide (4,7). It should be noted that the mechanisms in all three types of lasers are not completely understood.

TRANSVERSELY SPARKED LASER

Experimental

Experiments on the electrical spark-induced combustion described in Reference 3 have been continued using a greatly improved laser cavity and method of initiation (10). This initiation technique, which entails the use of multiple transverse sparks developed by Beaulieu in his atmospheric pressure CO_2 laser studies (11), allows operation at higher total pressures for a given discharge voltage.

Energy measurements were made with a calibrated pyroelectric detector and pulse shapes determined with an InSb photovoltaic detector cooled to 77°K. Spectroscopic measurements were made with a McPherson Model 2051 monochromator. A schematic of the apparatus used is shown in Figure 1.

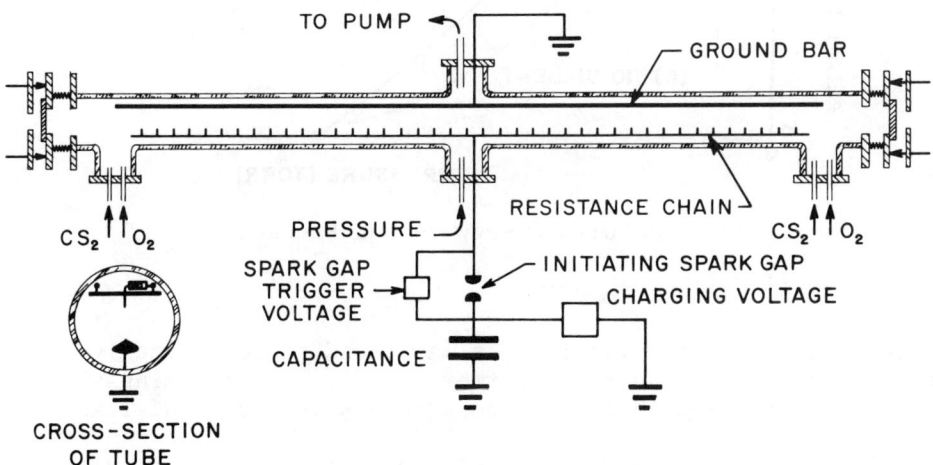

FIGURE 1. Transversely Sparked Laser

Results and Discussion

Figure 2 shows pulse energy per unit concentration as a function of pressure for a mixture of $CS_2/O_2:2/25$. Although maximum pulse energy per unit concentration is reached at a pressure of ~20 torr, laser action is still observed at pressures up to 100 torr. When the combustible mixture at 10 torr is diluted by either nitrogen or helium large increases in pulse energy result. In the case of helium, a maximum pulse energy of 26 millijoules was reached at 30 torr.

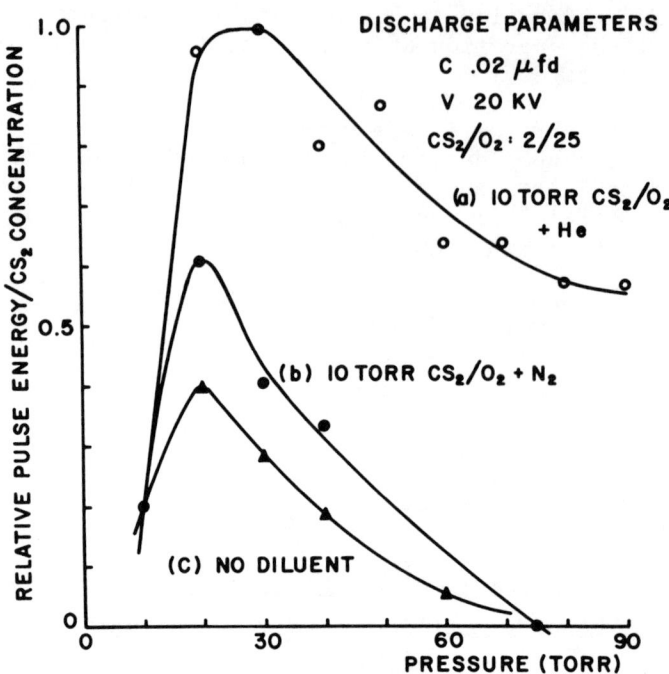

FIGURE 2. Pulse Energy Versus Pressure

The duration of the laser pulse was found to be typically about 40 microseconds under the latter conditions. Assuming a triangular pulse shape this gives peak powers in excess of 1 Kw.

An interesting comparison can be made between these results and those of Osgood et al (12) who measured pulse energies of 3 millijoules and peak powers of 7.7 Kw by Q-switching a CW CO laser. These results were obtained at liquid nitrogen temperatures where condensation of ozone presented a definite explosion hazard in contrast with the simplicity of operation of the present transversely excited device.

Laser oscillation has been found to occur on seventy-nine P branch transitions of CO as shown in Table I. It was also noted that definite cascading occurred similar to that observed by Gregg and Thomas. A typical cascade is shown in Table II.

It is of some interest to speculate on the effectiveness of N_2 and He in increasing the pulse energy. If helium is replaced by argon the pulse energy is reduced. Apparently laser action

TABLE I: Pulsed CS_2/O_2 Laser Transitions

Vibrational Transitions	Rotational Transitions
13-12	P(14) - P(13)
12-11	P(14) - P(10)
11-10	P(15) - P(9)
10-9	P(17), P(15) - P(10)
9-8	P(16) - P(9)
8-7	P(16) - P(12), P(10) - P(8)
7-6	P(15) - P(9)
6-5	P(16) - P(11), P(9) - P(7)
5-4	P(15) - P(9)
4-3	P(15) - P(12), P(10), P(9)
3-2	P(14) - P(7)
2-1	P(12) - P(9)

TABLE II: Cascade Observed for $CS_2/O_2/N_2$

Vibrational-Rotational Transitions	Time of Appearance Relative to Initiation of Lasing (μsec)
11-10 P(9)	0
10-9 P(10)	0.9
9-8 P(11)	1.9
8-7 P(12)	6
7-6 P(13)	8
6-5 P(14)	33
5-4 P(15)	22

is not improved by increasing the average electron energy. Osgood et al (12) have speculated that anharmonic pumping is responsible for the increase in power they observe in the presence of nitrogen. This theory could also apply in this pulsed laser. Helium may contribute to the energy exchange by either lowering the rotational temperature and/or supplying energy to the CO through collisions with metastable atoms.

CW COMBUSTION LASER

Experimental

The continuous wave experiments were conducted in a cavity about a meter in length as shown in Figure 3.

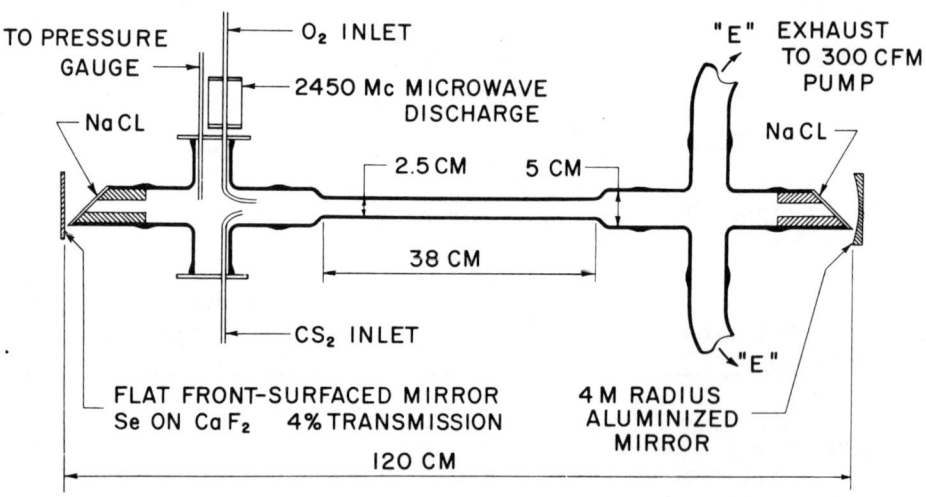

FIGURE 3. Schematic Representation of Chemical Laser Cavity

Oxygen, partially dissociated in a microwave discharge, was introduced through a glass inlet tube. Carbon disulphide, with or without premixed helium, was added through an inlet positioned on the opposite side. The flow velocity varied from 30 to 100 m/s.

CS_2/O_2 COMBUSTION LASERS

Spectroscopic measurements were made as before with a McPherson Model 2051 monochromator, equipped with an InSb photovoltaic detector cooled to 77°K and coupled to a lock-in amplifier.

Results and Discussion

Although laser action could be sustained without helium addition the admixture of this gas greatly stabilized the power output of the laser. The maximum observed output was approximately six milliwatts. Variation in relative power with gas mixture is shown in Figure 4.

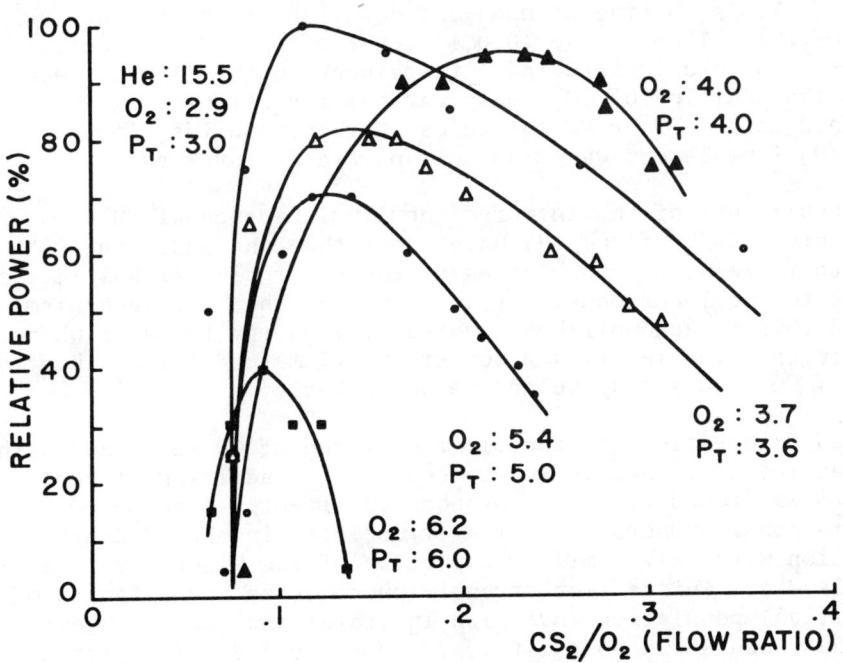

FIGURE 4. Dependence of Relative Laser Power Output upon Feed Ratio and Total Pressure. All Flows are given in 1/min and Pressures in Torr.

A spectroscopic analysis of the coherent radiation from this laser is given in Table III.

TABLE III: Spectral Output - CW CS_2/O_2 Laser

Vibrational Transitions	Rotational Transitions	System
10-9	P(17), P(15) - P(12)	CS_2/O_2:1/5
9-8	P(15) - P(12)	
10-9	P(13) - P(11)	CS_2/O_2/He:1/8/35
9-8	P(12) - P(10)	
9-8	P(12) - P(10)	CS_2/O_2/He:1/29/82
8-7	P(12), P(10), P(8)	

The operation of the DF-CO_2 and HF-CO_2 CW lasers, as developed by Cool (13), is contingent upon the donation of the vibrational energy of the halide to the (0,0,1) level of the CO_2 molecule which is then totally inverted with respect to the (1,0,0) level. Without the addition of CO_2, only partial inversions could be maintained in the DF or HF molecules. Carbon monoxide, formed in the CS_2/O_2 flow laser, was also only partially inverted.

Measurements of the infrared chemiluminescence of CO in a free-burning CS_2/O_2 flame (9) have shown that the addition of N_2 tended to increase the population of the upper vibrational levels relative to the lower ones. It would appear that the mechanism by which this is accomplished involves V-V energy transfer wherein the nitrogen molecule, with a better energy match at low vibrational levels of CO, can act by selective depopulation.

When observations of the first overtone of CO were made using the apparatus described in Reference 14 the addition of cool CO or N_2O was found to have a pronounced effect. With added CO, low vibrational numbers are strongly depleted in the CO overtone IR emission with only a milder depletion of the higher (v ≳ ca. 8) energy levels. This effect strongly promotes an augmentation of the (partial) population inversion in vibrational levels where lasing has been shown to exist i.e. v = 9 → v = 8. This effect is doubtless caused by the close energy match between the fundamental frequencies of CO (or N_2O) with the lower v levels of the chemically pumped CO and the progressively poorer match at higher v levels.

In order to demonstrate this effect, gases with an increasingly better energy match with the fundamental frequency of the CO molecule were added to the flow laser. Thus it was observed that N_2O increased the output power two-fold, COS about six-fold and CO, with the best possible resonance, by ten-fold.

SUMMARY AND CONCLUSIONS

Pulse energies of 26 millijoules and peak powers in excess of one kilowatt at about 5 microns wavelength have been obtained by transversely spark-initiating reactions of CS_2 and oxygen. Both helium and nitrogen have been shown to increase the pulse energy.

In the absence of foreign gases CW laser power of 6 milliwatts has been observed in a flowing CS_2/O_2 combustion laser. Addition of foreign gases such as CO, N_2O and COS have been found to greatly increase the laser performance through an energy transfer mechanism involving selective depopulation of lower vibrational levels of the CO molecule.

REFERENCES

1. Kasper, J.V.V. and Pimentel, G.C., Phys. Rev. Letters, 14, 352 (1965).

2. Pollack, M.A., Appl. Phys. Lett. 8, 237 (1966).

3. Gregg, D.W. and Thomas, S.J., J. Appl. Phys., 39, 4399 (1968).

4. Arnold, S.J. and Kimbell, G.H., Appl. Phys. Letters, 15, 351 (1969).

5. Hancock, G. and Smith, I.W.M., Chem. Phys. Letters, 3, 573 (1969).

6. Wittig, C., Hassler, J.C., Coleman, P.D., Appl. Phys. Letters, 16, 117 (1970).

7. Suart, R.D., Kimbell, G.H. and Arnold, S.J., Chem. Phys. Letters, 5, 519 (1970).

8. Wittig, C., Hassler, J.C., and Coleman, P.D., Nature 226, 845 (1970).

9. Foster, K.D. and Kimbell, G.H., J. Chem. Phys., 53, 2539 (1970).

10. Jacobson, T.V. and Kimbell, G.H., J. Appl. Phys., Dec (1970).

11. Beaulieu, J.A., Appl. Phys. Letters, 16, 504 (1970).

12. Osgood, R.M., Jr., Nichols, E.R., Eppers, W.C., Jr. and Petty, R.D., Appl. Phys. Letters, 15, 69 (1969).

13. See for example Cool, T.A., Falk, T.J. and Stephens, R.R., Appl. Phys. Letters, 15, 318 (1969).

14. Arnold, S.J. and Kimbell, G.H., Chem. Phys. Letters, 3, 469 (1969).

INDEX

Absorption 43,73,74,96,221
Aerodynamics 85
Angle
  Brewster 126,167,210
  convergence 166
Arcs
  free-burning 52
  hollow-cathode 251
  mercury capillary 253
  sources 251
  wall-stabilized 50,55
Argon 10,93,256
Astigmatism 93,96
Atomic
  beam 55
  dipole theory 67
Auto-ionization 53
Avogadro's number 68

Bands
  electronic 107
  Schumann-Runge 42
  vibrational 203
Beamsplitters 90,122,124,129
Beer Law 8
Black Body
  temperature 251
  radiation 254
Boltzmann Law 51,53
Brewster angle 210
Broadening
  collision 74
  Doppler 51,87,98,255
  line 92,253
  Lorentzian 51,101
  natural 71,255
  Stark 78,104,255

Cavity
  resonant 148,187
  reflectors 223

Coefficients
  absorption 8,15,78
  Einstein 8,71
  excitation 73
  extinction 69
  infrared absorption 14
  optical 217
  spectral absorption 5,13
Coherence 119,137,275
Compensation chamber 91
Compressibility 65
Concentration
  species 34,49,51
  dye 223
Constants
  damping 71
  dielectric 67,236
  Hook method 89,97
  Planck's 70
Coulomb
  collisions 190
  forces 178
Cross-sections
  differential 156
  photoionization 26
  Thompson 156,163
Current
  electron 182
  ion 182
  sheet 170

Dale-Gladstone Law 66,68,72
Debye
  radius 182
  length 160
Density
  absolute number 88
  electron 121,124,255
  fluid 66
  ion 253
  number 7,86,190

Dipole
  molecular 67
  moment 67,68
  oscillating 81
Dispersion
  anomalous 49,73,85,87,
        92,113
  curve 107
  function 11
  inverse linear 115
  linear theory 107
  Lorentz theory 68
  negative 70,80,86
  optical 67
Dissociation 26,34
Doppler
  broadening 51,87,98,255
  collision 276
  shift 164
  width 92,101,275,276

Einstein-Boltzmann eq. 63
Electron
  feature 161
  Lorentz theory 68
  thermal speed 164
Electromagnetic 217
Electro-optical 139
Electrostatic 178,217
Emission
  non-equilibrium 34
  stimulated 222,279
Energy
  chemical 197
  dissociation 7
  electrical 216
  electronic 222
  exchange processes 2
  kinetic 178
  levels 203
  photon 222
  selective transfer 203
  storage 216
  thermal 178
  transfer 203,204
  translational 203
  vibrational 198,203

Equilibrium
  Boltzmann 257
  local thermodynamic 35,45,49
  thermal 271
  thermodynamic 29,254
  Saha-Boltzmann 257
Excitation
  electronic 70
  resonant 221

Fabry-Perot Etalon 167,222,255
Field
  electrical 67,178
  magnetic 103,169,178
  optical 138
Filters
  band 124
  dichroic 233
  dielectric band-pass 167
  focal length 95
  focal plane 89,90
  interference, 37,128,167
Flash tube
  quartz 251
  xenon filled 234
Flows
  aerodynamic 85
  hypersonic ionized 125
  laser-produced ionized 178
  non-equilibrium 33
  stratified 217
  supersonic 201
  turbulent 215
Fluorescence
  laser-induced 217,256
  molecular 256
  resonance 222,253
  two-photon 126
Franck-Condon factor 17
Fresnel number 217
Fringe
  bright 90
  central 91
  order 95
  interference 150
  spacing 115
  zero order 93

INDEX

Functions
  autocorrelation 159
  partition 17
  slit 16

Gas
  absorbing 92
  diagnostics 85,86
  diatomic 67
  dynamics 43,178,216,217,221
  shock-heated 88
gf-values 52
Glan-air polarizers 129
Grating
  deflection 93
  diffraction 222,231,236
  Echelle 236
  holographic 96
  spectrograph 273

Hologram 138
Holography 137
Hook method 72,74,85,113
Hönl-London factor 17
Hugoniot relations 190
Hypersonic 125

Image
  converter-camera 128,257
  intensifier 132
  virtual 139
Impurities 61
Incident
  beam 156
  electric vector 158
  wavelength 156
Index
  complex refractive 69
  phase refractive 69
  refractive 66,68,73,78,85
Infrared 133,197,221,222
Intensity
  cosine-squared 75
  emission 8
  modulation 139
  spectral 7

Interactions
  electron-electron 190
  gas dynamic 217
  ion-ion 190
Interferometer
  Jamin 114
  Mach-Zehnder 74,85,89,113,
    122,145
  Michelson 107
  microwave 177,181
Interferometry
  holographic 119,133,137
  Hook 113
  infrared 217
  method 66,80
  Spark 78
  spectral 78
Inversion
  partial 199,209
  population 80
  total 199
Ion
  drift velocity 253
  feature 161
  velocity 164,222

Kerr
  cell shutter 124
  effect 133
Kinetic
  chemical 197
  mechanisms 197
Kramer's continua 5
Kramers-Kronig relations 79

Lasers
  chemical 197,208,214,215
  coherent radiation 238
  combustion 279
  continuous 202
  continuous-wave chemical 197,199
  continuous wave dye 277
  detonation wave chemical 214
  dye 97,107
  electrically excited 216
  emission 199

Lasers (continued)
  fluid dye 222
  fluid mixing 206
  fluid mixing chemical 199,203
  focused 81,119,257
  giant-pulse ruby 128,143
  high power pulsed 165
  liquid dye 221
  mode-locked 127
  molecular 216
  monochromaticity 255
  neodymium glass 126
  oscillation 212
  pulsed 137,157
  pulsed chemical 197,203,212
  pumped 223
  pumped-dye 276
  Q-spoiled cw chemical 212
  Q-switched 122,124,165,177
  Q-switched pulse 178
  Q-switched ruby 120
  Q-switched solid state 223
  radiation 199,275
  semiconductor 221,240
  single mode 129
  sources 137,139,251
  steady state 137
  tunable 271
  xenon 500 dye 276
Lecher-wire elements 177
Level
  energy 70
  pumping 128
  rotational 235
  vibrational 44,204,235,279
Lifetimes
  excited state 277
  radiative 49,55
Light
  monochromatic 115
  source 85,119,251
  spectroscopic sources 244
  tunable source 80
  undispersed white 91,92
  velocity 68
  visible 80

Lines
  absorption 222
  cesium 249
  emission 71,85
  first stokes 248
  intensity 58,71
  resonance 58,87,113,253
  reversal 52
  spectral 71,85
Littrow
  arrangement 89
  lens 89
  spectrograph 96
Lorentzian half-intensity 101
Loschmidt's number 68

Magnetogasdynamic 217
Mean-free-path 5,8,179,180
Metastable 284
Mirrors
  dielectric 126
  transmitting 126
Mixing
  diffusive 198
  injector. 201
  rate 199
Moments
  dipole 67,68
  static 81
  transition 107
Monochromatic radiation 253
Monochromator 167
Mounts
  Czerny-Turner 96
  Elbert 96
  Wadsworth 96

Optical gate 132
Optically thick 5,87,92
Optically thin 51,63
Oscillation
  electromagnetic 217
  laser 282
  pulsed-laser 279

INDEX

Oscillator
  absorption 98
  electron 68,69
  parametric 244,248
  strength 71,87,98

Permittivity 67
Phase
  difference 68
  distribution 139
Photoionization 10,183
Photons
  echoes 82,83
  pulse 276
  scattered 165
Piezo-electric 95
Planck function 5
Plasma
  collisionless 177,178
  dense 182
  electrostatic 178
  fluctuations 159
  frequency 160
  gun 142
  hot 5
  instabilities 177
  laser-produced 121
  luminous 25,120,165
  rarefied flows 177
  theta-pinch 121
  transient 137
Plates
  compensation 92
  interferometric 90
  lithium fluoride 117
  photographic 96
Pockels cell 122,127,172
Polarizability 67,80
Population inversion 197,199,203
                     210,279
Potential
  excitation 51,55
  floating 193
Probability
  absolute transition 49
  de-excitation 254
  transition 7,33
  vibrational 212

Probes
  Langmuir 177,181
  Lecher-wire 181
Pumping
  chemical 203,209
  optical 222,258

Quantum
  defect 6
  mechanics 71
  number 6,82

Radiation
  broadband 222
  diffusion 8
  monochromatic 221
Radiometers 33,37
Radius
  classical 72
  classical electron 156
  Debye 182
Raman
  band energy 248
  emission 148
Ram jets 214
Recombination 26
Refractivity
  change 141
  free electrons 66
  gradients 65
  nonlinear 80
  optical 65
  specific 66
Resolution
  spatial 125
  temporal 128
Resolving power 95,115
Resonance 68,107
Resonant frequency 68
Rotational
  spacings 203
  thermalization 204
Rydberg states 13

Scattering
  laser 156
  Raman 221,222,224
  Rayleigh 221

Scattering (continued)
  resonance 253
  Thomson 119,155,221
  vector 158
Schlieren 65,80,119,149
Self-absorption 11,17,49,51,253
Separation
  energy 203
  Hook 77
  internuclear 11
Shadow technique 65,80
Shock
  heating 79
  incident 37,45
  initiation 215
  reflected 52
Shock waves
  collisionless electrostatic 179
  cylindrical converging 125
  propagation 5
  strong 33
Sodium
  D-line 66
  line-reversal 254
Spark 120,132,216
Species
  atomic 87
  heavy particle 66
  molecular 87
Speckle effect 145
Spectral
  bandwidth 85
  distribution 33
  emissivity 5,11
  microwave regions 213
Spectrograph
  stigmatic 85,86,89
  Czerny-Turner 96
Spectroscopy
  absorption 46,96,222,251
  beam-foil 50
  emission 221
  kinetic 4
  quantitative 86
  shock tube 33
  synthetic emission 15
  vacuum ultraviolet 6

Spectrum
  absorption 7,33,44,92
  atomic 6,9
  continuous 89
  frequency 159
  impurity 5
  molecular 6,7,10
  shock excitation 2,5
  solar 6
  synthetic 1,14
  transition metals 6
  ultraviolet 222
  vacuum ultraviolet 4
  visible 222
States
  excited 70
  internal 67
  metastable 273
  vibrational-rotational 119
Streak
  camera 107
  photograph 120,128,257
Subsonic 178
Supersonic 200,213,214
Super-radiance 275
Susceptibility 67,80,81

Temperature
  brightness 96
  electron 53,188,253,255
  excitation 52
  gas kinetic 52
  high electron 121
  ion 253
  rotational 203
  translational 34,203,255
  ultrasonic 52
  vibrational 203
Temporal resolution 137
Thermal
  conductivity 94
  dissociation 200
  schlieren 231
Thermonuclear 179,257
Theta-pinch 124,169

INDEX

Transitions
  electronic 67,70
  ground state resonance 70
  molecular 203
  rotational 222,283
  Schumann-Runge 256
  vibrational 222,253
  vibration-rotation 203
Tubes
  discharge 79
  expansion 61
  flash 251
  shock 1,2,50,52,62,79
         88,113
  quartz 234
Turbulence 80

Ultrasonic 217
Ultraviolet 33,43,67,72,95
             113,221,253

Vapors
  mercury 79
  potassium 107
  sodium 79,80,115

Vectors
  incident electric 158
  polarization 67
  wave 69
Vibrational
  population-inversion 199
  resonance defect 205
  transitions 222
Voight profile 16
Vorticity 215

Wavelengths
  resonant 67,72,88
  vacuum ultra-violet 50,96
  visible 120
Weight
  molecular 52,68
  statistical 50

Xenon 234,236

Zeeman component 103